# Springback Assessment and Compensation of Tailor Welded Blanks

Focusing on techniques developed to evaluate the forming behaviour of tailor welded blanks (TWBs) in sheet metal manufacturing, this edited collection details compensation methods suited to mitigating the effects of springback. Making use of case studies and in-depth accounts of industry experience, this book gives a comprehensive overview of springback and provides essential solutions necessary to modern-day automotive engineers.

Sheet metal forming is a major process within the automotive industry, with advancement of the technology including utilization of non-uniform sheet metal in order to produce light or strengthened body structures. This is critical in the reduction of vehicle weight in order to match increased consumer demand for better driving performance and improved fuel efficiency. Additionally, increasingly stringent international regulations regarding exhaust emissions require manufacturers to seek to lighten vehicles as much as possible. To aid engineers in optimizing lightweight designs, this comprehensive book covers topics by a variety of industry experts, including compensation by annealing, low-power welding, punch profile radius and tool-integrated springback measuring systems. It ends by looking at the future trends within the industry and the potential for further innovation within the field.

This work will benefit car manufacturers and stamping plants that face springback issues within their production, particularly in the implementation of TWB production into existing facilities. It will also be of interest to students and researchers in automotive and aerospace engineering.

# Springback Assessment and Compensation of Tailor Welded Blanks

Edited by
AB Abdullah and MF Jamaludin

**CRC Press**
Taylor & Francis Group
Boca Raton  London  New York

CRC Press is an imprint of the
Taylor & Francis Group, an **informa** business

MATLAB® is a trademark of The MathWorks, Inc. and is used with permission. The MathWorks does not warrant the accuracy of the text or exercises in this book. This book's use or discussion of MATLAB® software or related products does not constitute endorsement or sponsorship by The MathWorks of a particular pedagogical approach or particular use of the MATLAB® software.

First edition published 2023
by CRC Press
6000 Broken Sound Parkway NW, Suite 300, Boca Raton, FL 33487-2742

and by CRC Press
4 Park Square, Milton Park, Abingdon, Oxon, OX14 4RN

*CRC Press is an imprint of Taylor & Francis Group, LLC*

ISBN: 9780367758349 (hbk)
ISBN: 9780367758417 (pbk)
ISBN: 9781003164241 (ebk)

DOI: 10.1201/9781003164241

Typeset in Times
by codeMantra

# Contents

# Preface

Tailor welded blanks (TWBs) are innovative methods in weight reduction, safety and performance strategies, especially for the automotive industry. Since its introduction in the mid-1980s, the technology has progressed from its initial use in linear-welded steel for two-piece applications to complex non-linear welding of dissimilar materials for multi-section panels. As with other sheet metal forming processes, springback assessment and compensation in the processing of TWBs remain challenging issues to be addressed, particularly as it involves physical and material non-linearities.

In this book, we first present the motivation for the utilization of TWBs as an important solution in the list of strategies for automotive lightweighting. The historical context of fuel economy, environmental regulations and customer demand are discussed to frame the development of current trends and future prospects of various weight reduction technologies. A general review for measurement and compensation of springback is then presented, describing the types of bending in springback problem analysis, types and springback, methods for measurements, factors that affect springback, methods of measurement and compensation strategies. On the theme of TWBs, this book compiles a selection of recent work on the subject of springback for non-uniform blanks which were produced from techniques of forging, friction stir welding (FSW) and low-power laser welding. Non-traditional processing methods on TWBs, such as single-point incremental forming (SPIF) are also discussed, highlighting potential production issues in geometrical accuracy and how to compensate for them by considering optimization of toolpaths and feedback control. Focusing on the development of new types of lightweight TWBs, several of the experimental studies have focused on the evaluation of springback for dissimilar aluminium alloys and aluminium–steel TWBs. Discussion on springback compensation for TWBs includes displacement adjustments and spring-forward methods. Computational approaches are also discussed, which consider finite element analysis and numerical predictions utilizing data-driven modelling approaches. Lastly, this book touches on the issue of sustainability in TWB production and forming processes, which is becoming more relevant in the environmentally responsible industry of the future.

We greatly appreciate the efforts of all the authors who have contributed their excellent chapter articles for this book. Their contributions provide a comprehensive view of the issues related to measurements and mitigations of springback in TWBs, with consideration for new processing methods and material combinations. Each chapter also includes a relevant list of references for further reading. Our intention is for this book to be educational, informative and beneficial as a source of reference to postgraduates, researchers, educators and industrial practitioners, who will further drive the development of innovative TWB technology in the future.

MATLAB® is a registered trademark of The MathWorks, Inc. For product information, please contact The MathWorks, Inc.

3 Apple Hill Drive
Natick, MA 01760-2098 USA
Tel: 508-647-7000
Fax: 508-647-7001
E-mail: info@mathworks.com
Web: www.mathworks.com

# Editors

**AB Abdullah** obtained his PhD from the Universiti Putra Malaysia, Malaysia. Now, he is working at the School of Mechanical Engineering, Universiti Sains Malaysia, with more than 20 years of experience in teaching and research. His research interests include precision sheet metal forming, tool and die design, and wire arc additive manufacturing. He has published more than 100 research papers in various international journals. He has written ten books including as an editor for a research book entitled *Hole-Making Technologies for Composites: Advantages, Limitations and Potential*, published by Woodhead Publishing, Elsevier, UK in 2019. He is actively involved as a reviewer for the *International Journal of Advanced Manufacturing Technology* (Springer, USA), the *Journal of Material Processing Technology and Measurements* (Elsevier), and the *Journal of Testing and Evaluation* (ASTM International). He is a Certified Train the Trainers under HRDF Malaysia and a Professional Engineer (PEng) registered with the Board of Engineers Malaysia.

**MF Jamaludin** is a senior researcher at the Advanced Manufacturing and Material Processing (AMMP Centre), University of Malaya, Kuala Lumpur, Malaysia. He is actively involved in research on precision joining of dissimilar materials using low-power laser welding and friction stir welding processes. With more than 10 years of research experience, he has co-authored 24 publications on various topics related to manufacturing processes and is a co-inventor of 14 patents and industrial designs. His current research engagement is on the formability analysis of dissimilar aluminium TWBs using low-power fibre laser joining.

# Contributors

**MF Adnan**
Metal Forming Research Laboratory
School of Mechanical Engineering
Universiti Sains Malaysia
Penang, Malaysia

**AD Anggono**
Department of Mechanical Engineering
Muhammadiyah University of
 Surakarta
Pabelan, Indonesia

**N Chakraborti**
Faculty of Mechanical Engineering
Czech Technical University in Prague
Prague, Czech Republic

**Z Hussain**
School of Material and Mineral
 Resources Engineering
Universiti Sains Malaysia
Penang, Malaysia

**AI Hussin**
Metal Forming Research Laboratory
School of Mechanical Engineering
Universiti Sains Malaysia
Penang, Malaysia

**Md. Tasbirul Islam**
School of Property, Construction and
 Project Management
RMIT University
Melbourne, Australia

**H Krishnaswamy**
Department of Mechanical Engineering
Indian Institute of Technology Madras
Madras, India

**M Mohamed**
Impression Technologies Ltd
Coventry, United Kingdom
and
Department of Mechanical
 Engineering
Helwan University
Helwan, Egypt

**MN Nashrudin**
Metal Forming Research Laboratory
School of Mechanical Engineering
Universiti Sains Malaysia
Penang, Malaysia

**NM Noor**
School of Mechanical Engineering
Universiti Sains Malaysia
Penang, Malaysia

**B Omar**
Department of Materials Engineering
 and Design
Universiti Tun Hussein Onn Malaysia
Parit Raja, Malaysia

**SS Panicker**
Department of Mechanical Engineering
BITS Pilani-K.K. Birla Goa Campus
Goa, India

**AF Pauzi**
Metal Forming Research Laboratory
School of Mechanical Engineering
Universiti Sains Malaysia
Penang, Malaysia

**KAHA Razak**
Metal Forming Research Laboratory
School of Mechanical Engineering
Universiti Sains Malaysia
Penang, Malaysia

**RI Riza**
Department of Mechanical Engineering
Muhammadiyah University of
    Surakarta
Pabelan, Indonesia

**MZ Rizlan**
Metal Forming Research Laboratory
School of Mechanical Engineering
Universiti Sains Malaysia
Penang, Malaysia

**WA Siswanto**
Department of Mechanical Engineering
Muhammadiyah University of
    Surakarta
Pabelan, Indonesia

**F Yusof**
Faculty of Engineering
Centre of Advanced Manufacturing and
    Material Processing (AMMP Centre)
Universiti Malaya
Kuala Lumpur, Malaysia

# 1 Weight Reduction Strategies in Passenger Car Manufacturing
## Current Trends and Prospects

*Md Tasbirul Islam*
RMIT University

*AB Abdullah and AI Hussin*
Universiti Sains Malaysia

*MF Jamaludin*
Universiti Malaya

## CONTENTS

## 1.1 INTRODUCTION

Stringent government regulations on fuel economy and shifts in customer demands are the two most important driving forces for automakers to manufacture more economic and environmentally friendly vehicles (Mathiyazhagan et al., 2015). In recent years, the average weight of automobiles has increased significantly due to the increasing amount of technical features included in the vehicle (Merklein and Geiger,

2002). However, increased weight of vehicles can result in low fuel economy, which goes against the purpose of regulations. The transportation sector is one of the largest contributors of $CO_2$ emissions, particularly from passenger car fleet (USEPA, 2014a). In the European Union (EU), 12% of the EU's $CO_2$ emission comes from passenger cars (EU, 2016). Thus, weight reduction without compromising structural integrity is the greatest challenge for automakers nowadays. Weight reduction is demanded in both automobile and aeronautic industries for conservation of energy and natural resources (Chen et al., 2011). Han and Clark (1995) have estimated that a 57 kg reduction in the weight of the vehicle is equivalent to an increase in the fuel economy from 0.09 to 0.21 km per litre. Other researchers have projected that a 1% vehicle weight reduction can result in the reduction of fuel consumption by 0.6%–1% (Hayashi, 1996; Pallett and Lark, 2001). In the construction of an automobile, the body-in-white (BiW) accounts for 28% of the vehicle weight, followed by the powertrain and the chassis systems (Lutsey, 2010). Thus, reducing the weight from the body structure directly decreases the weight of the full vehicle (Zhu et al., 2009). Possible weight reduction strategies could be in (1) design optimization by numerical simulation, (2) using light-weight materials and/or material alternatives, and (3) use of advanced manufacturing technologies, such as tailor welded blanks (TWBs) and tailor-rolled blanks (TRBs).

This chapter presents a comprehensive review on reducing automotive weight using three major strategies: (1) design optimization by numerical simulation, (2) material alternatives, and (3) technological innovation. As weight reduction is possible in all major systems of the automobile, this review will extract information on weight reduction (in percentage) that has been achieved, so that future research can be conducted following a set of combination of the strategies. Under the technological innovation, three major technologies are described. This chapter is divided into the following sections:

- Sections 1.2 and 1.3 describe the present weight trends of automobiles, with elaborations on the data obtained for U.S. passenger car fleet. This will also include discussion on the impact of weight on the fuel economy and $CO_2$ emission of passenger cars.
- Section 1.4 describes the possible weight reduction strategies that can be applied to passenger car fleet.
- Section 1.5 proposes future work directions related to weight reduction strategies. One such proposal is on the fabrication of BiW panels using non-uniform thickness sections to reduced weight without compromising crashworthiness, where the production time and costs will be lowered as compared to conventional manufacturing processes.

## 1.2  HISTORICAL PASSENGER CAR WEIGHT AND FUEL ECONOMY TREND IN THE U.S.

Over the years, U.S. passenger car fleet has shifted to lighter-vehicle-weight designs due to regulatory interventions. Figure 1.1 shows the weight trend of light-duty vehicles and the average fuel consumption from 1975 to 2013, obtained from publicly available data sources such as the U.S. Environmental Protection Agency (USEPA)

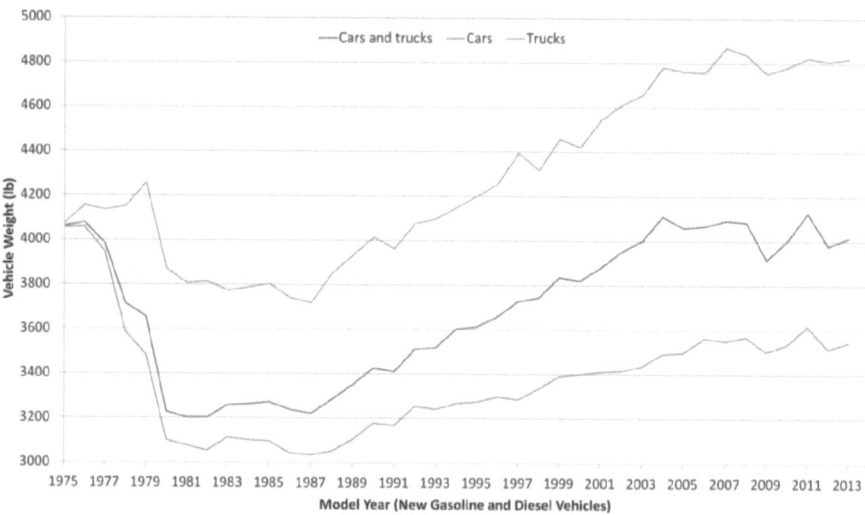

**FIGURE 1.1** U.S. light-duty vehicle trends for weight and adjusted fuel economy for model years 1975–2013 (see USEPA, 2014b).

data on vehicle characteristics (USEPA, 2014b). It can be seen that both increasing and decreasing weight trends of vehicles were evident from this historical shift. A dramatic weight reduction trend is observed during the period from 1975 to 1980, when there was a stringent implementation of federal economy standard as well as the peaking fluctuations of fuel prices. However, a different trend was observed thereafter, where from 1987 to 2013 there is a historical shift toward heavier vehicles with higher fuel economy. The vehicle (for both cars and trucks) weight has increased by 19.77% (14.38% for cars and 22.89% for trucks) over this time period. The highest year-on-year increase of vehicle weight was 3.1%, and the annual average increase in weight was −0.02933% in the time span of 1975–2013. The average weight of the passenger car increased by 1% in model year (MY) 2013.

The increasing weight trend of vehicles occurred for a number of reasons such as consumer shifts toward the vehicle category, size, and differences in vehicle features such as air-conditioning and safety equipment. It is found that added equipment could account for an average of 20%–40% weight increase of U.S. lightweight vehicles (Automotive, 2009). According to the research conducted by Glennan (2007), major vehicle design modifications could result in a weight increase by 177–221 lbs (80–100 kg); even minor facelifts could increase the weight by 44–110 lbs (20–50 kg). Other factors that are responsible for increasing weight trends are larger vehicle sizes and mass characteristics. It is assumed that the increase in size will directly correlate to the increase in weight.

Vehicle efficiency is increased by implementing improvements in vehicles' aerodynamics, engines, and transmission. However, the level of technological innovations that can be implemented is dependent on the budget that varies over time due to regulatory pressures, fluctuating fuel prices, customers' preferences, and the proposition of the automakers (Penna and Geels, 2015).

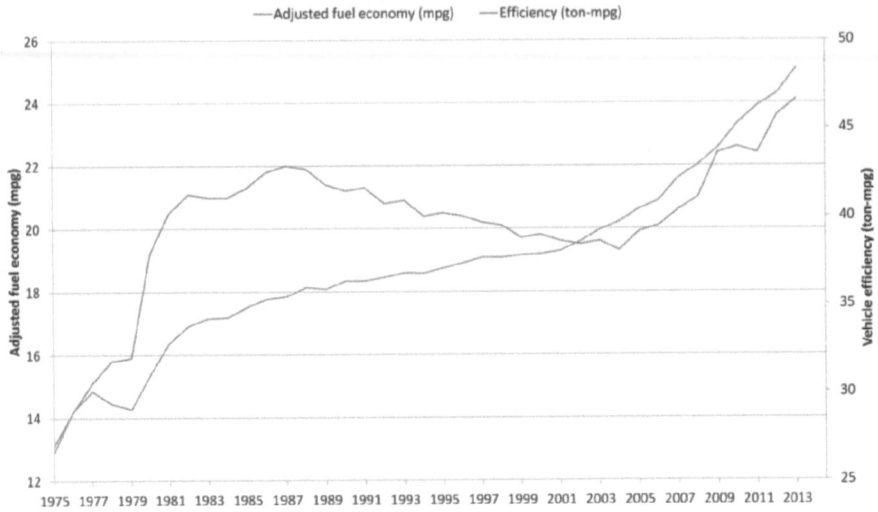

**FIGURE 1.2**   U.S. light-duty vehicle trend for adjusted fuel economy and vehicle efficiency (see USEPA, 2014a).

Figure 1.2 shows the adjusted fuel economy and vehicle efficiency from 1975 to 2013. According to the U.S. Corporate Average Fuel Economy (CAFÉ) standard, fuel economy is measured in miles per gallon (mpg). Vehicle efficiency can also be found by weight-adjusted fuel economy of vehicles, measured in ton-mpg. This simple method is adopted to measure efficiency as obtaining efficient data of individual automobile components such the engines' true efficiency, transmissions, aerodynamics, and vehicle weight can be difficult.

From Figure 1.2, it can be seen that a significant increase in fuel economy was recorded from 1975 to 1987 due to stringent CAFÉ standards, which also restricted the increase in vehicle weights. During this time period, automakers were devoted to fuel economy improvements, which increased from 13.1 to 22 mpg. Vehicles after 1987 started to become heavier but with improved efficiency. A 26.76% increase in efficiency was seen from 1987 to 2013, from 35.431 (ton-mpg) in 1987 to 48.38 (ton-mpg) in 2013. A higher increase in efficiency was from 2001, with a 21.42% increase from 38.0142 (ton-mpg) in 2001 to 48.38075 (ton-mpg) in MY 2013. This increase was due to the steady market growth of light trucks, which has risen from 9.7% in 1979 to 47% in 2001, and further increased to 50% up until 2011 (NHTSA, 2011). On the other hand, a marginal increase in fuel economy was observed over the time span until MY 2004. However, an increase of 20% fuel economy was observed from MY 2004 (19.3 mpg) to MY 2013 (21.1 mpg), reaching 24.1 mpg in MY 2013. This increase in fuel economy can be attributed to the implementation of the "Tier 2 Vehicle and Gasoline Sulfur Program" at the beginning of 2004, developed in collaboration by USEPA and other stakeholders in the U.S. lightweight automobile industry (USEPA, 2005). Overall, from MY 1987 to MY 2013, fuel economy increased by only 8.7%. This indicates that the current vehicle fleet tends to be heavier and faster as compared to having higher fuel economy. Regulatory pressure is one of the key factors that has motivated automakers to produce lightweight vehicles. According to the studies by Knittel (2009), N. Lutsey, and Sperling (2005) and An and DeCicco (2007), it was suggested that if vehicle attributes (i.e. mass, size, and others) were held constant, fuel economy could be improved by new technological developments. However, with the current pace of technological development, new vehicles' technical efficiency is being increased by only 1%–2% per year. Often mass reduction technology would be an important consideration in most technical efficiency improvement techniques over the years (Froes et al., 1998).

## 1.3 IMPACT OF WEIGHT ON FUEL CONSUMPTION AND $CO_2$ EMISSION

Globally, fuel economy and $CO_2$ emission standards are being implemented by many countries, both on mandatory and voluntary basis, such as in the United States (mile/gallon – mandatory), Europe (g $CO_2$/km – voluntary), Canada (L/100 km – voluntary), Japan (km/L – mandatory), China (L/100 km – mandatory), South Korea (km/L – mandatory), Taiwan (km/L – mandatory), and Australia (L/100 km – voluntary) (An and Sauer, 2004). Generally, such regulation concerns with the decrease in the use of conventional fossil fuel (i.e. petroleum such as diesel, gasoline, and petrol) and reduction in $CO_2$ emission (one of the major greenhouse gases) from the transportation sector. As of 2009, 70% of the world's automobile sales are regulated by these regulatory measures (Lutsey, 2010).

Transport sector is one of the major contributors of $CO_2$ emissions, accounting for 25% of the global $CO_2$ emissions. Cars and trucks are the major components in the transport sector, representing around 75% of the transport sector's $CO_2$ emissions (Fan and Lei, 2016). In general, the vehicle mass, fuel type used, and distance travelled by the vehicle are the determining factors for the level emission, as the average vehicle mass is directly related to the energy consumption and the $CO_2$ emission (Zhang et al., 2014). The amount of $CO_2$ emitted from distance travelled is directly proportional to the fuel economy as every litre of gasoline burned releases about 2.4 kg of $CO_2$ (USEPA, 2014a). This amount is obtained from the combustion of hydrocarbons present in conventional fossil fuels. Figure 1.3 shows the relationship between vehicles' weight and $CO_2$ emission rate from 1975 to 2013. During the drastic weight reduction phase from MY 1975 to MY 1980, it was found that for 20% weight reduction, a 31% reduction in $CO_2$ emission was obtainable. However, after 1980, although there is a continuous increase in the weight of the vehicle fleet, the $CO_2$ emission remained at a constant rate. A significant drop in $CO_2$ emissions by 20% was obtained from 2004 to 2013 due to the high fuel economy (increased by 25%) of the fleet (An and Sauer, 2004).

From Figure 1.3, it can be seen that vehicle weight is a significant factor in the level of $CO_2$ emission. Reducing the mass of vehicles can follow two strategies. The first strategy is downsizing, which is reducing the size, and thus the mass, without redesigning the vehicle. The second strategy is to implement mass reduction technologies such as using high-strength materials and mass-optimized structures with redesign work without changing the functionality (or performance) and size of a

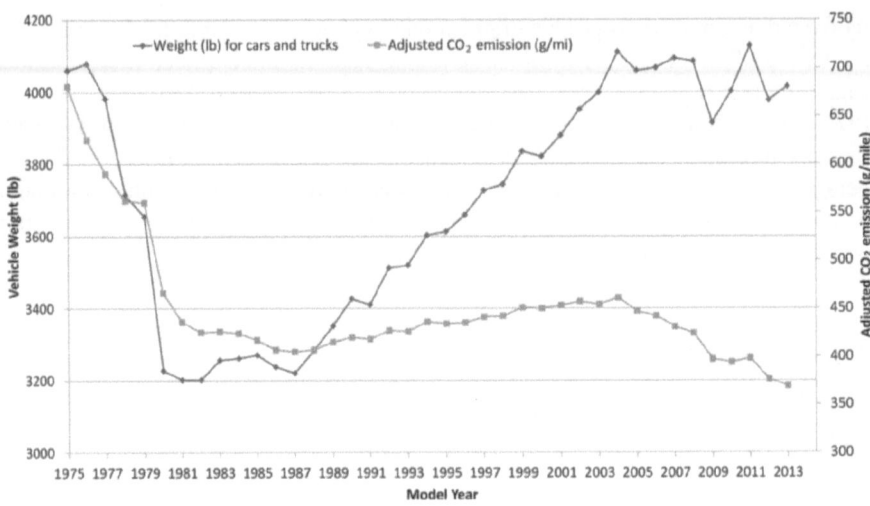

**FIGURE 1.3**   U.S. vehicle weight and adjusted fuel economy (see USEPA, 2014a).

vehicle fleet. If the vehicle performance is kept constant, any vehicle mass reduction will also decrease the fuel consumption rate. Generally 6%–8% reduction in fuel consumption can be obtained with a 10% weight reduction in the vehicle weight (USEPA, 2014a). Decreased fuel consumption also results in the reduction of $CO_2$ emission.

## 1.4   WEIGHT REDUCTION STRATEGIES

From the review of the literature, it can be stated that weight is the most significant factor that affects the fuel economy and $CO_2$ emissions. The weight of a vehicle is contributed by its various material composition and functional components inclusive of the BiW, powertrain, chassis, interior, closure, and other miscellaneous systems such as electrical, lighting, and thermal systems (Stodolsky et al., 1995; Ilyas et al., 2018). Figure 1.4 shows the breakdown of vehicle weight by system and components for a typical vehicle.

Among the systems, BiW is the most important and fundamental component that makes up one-quarter of the weight of a vehicle. Past research and development (R&D) has focused on the optimization of the weight of BiW. In recent years, a greater weight reduction opportunity has been possible due to the use of advanced lightweight materials and new technologies (Jambor and Beyer, 1997). The next two important systems are the powertrain and chassis that carry approximately one-fifth to one-quarter of the weight of a vehicle. In these two systems, significant amounts of materials are being used and new materials developed for these systems can also contribute to the weight reduction of the vehicle (Merklein and Geiger, 2002). For example, a 6% weight reduction can be obtained by the use of advanced materials in the running gear of the powertrain. In general, the use of lightweight materials in the powertrain and chassis can reduce the overall weight of the vehicle by 1%–2% (Jambor and Beyer, 1997).

| Approximate vehicle mass breakdown | System | Major components in system |
|---|---|---|
| | Body-in-white | Passenger compartment frame, cross and side beams, roof structure, front-end structure, underbody floor structure, panels |
| | Powertrain | Engine, transmission, exhaust system, fuel tank |
| | Chassis | Chassis, suspension, tires, wheels, steering, brakes |
| | Interior | Seats, instrument panel, insulation, trim, airbags |
| | Closures | Front and rear doors, hood, lift gate |
| | Miscellaneous | Electrical, lighting, thermal, windows, glazing |

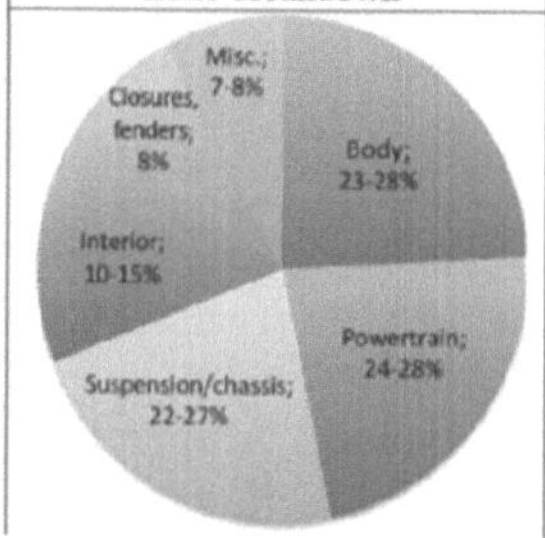

**FIGURE 1.4** Breakdown of vehicle weight by system and components. (Adopted from Lutsey, 2010, and Stodolsky et al., 1995.)

The use of numerical simulations, such as multidisciplinary optimization, generic algorithms, and finite element method, during the development of the vehicle systems can reduce production costs, improve product quality, and assist in identifying weight reduction possibilities. With the use of computer-aided engineering (CAE) optimization and robustness methodologies, weight reduction strategies can be implemented, such as optimized design of additional reinforcements, which leads to an overall weight reduction of the vehicle without compromising structural integrity (Wang et al., 2004). Thus, advanced numerical simulation is one of the paths to estimate weight reduction opportunity virtually, which can be validated by suitable experimentations.

Weight reduction can also be obtained by material substitution by using light material components, redesign of components, and utilizing advanced manufacturing methods. Recent technological development in manufacturing, such as TWBs and TRBs, can also contribute to weight reduction substantially. These strategies were used in the Ultra-Light Steel for Auto Body (ULSAB) project, in which extensive use of TWBs and high-strength steel has resulted in a total weight reduction potential of 26%–32% (Lowe, 1998).

According to Takita and Maruta (2000), Zhanqiu and Runlan (1994), and Bandivadekar (2008), weight reduction from automobiles can be broadly classified into three distinct strategies, namely, (1) design optimization, (2) material alternatives, and (3) technological innovation and/or advanced manufacturing technology. In the following sections, weight reduction strategies will be broadly discussed based on an extensive literature review.

### 1.4.1 Design Optimization by Numerical Simulation

Optimization methods for design applications in the automobile industry are driven by the need to improve product performance, reduce costs, shorten design cycles, and meet the stringent government regulation on fuel economy and safety. Integrating optimization technique at the early stage of the design process can reduce subsequent production costs. Design optimization can be implemented across all components regardless of size, from small parts such as the suspension ring to large parts such as the whole car bodies. Design optimizations have been conducted for gas tank shape

(Chen et al., 1997), tailor blank doors (Song and Park, 2006), structural optimization of b-pillars (Marklund and Nilsson, 2001), and welding locations (Leiva et al., 2001). Table 1.1 summarizes the optimization techniques that have been initiated by researchers for the purpose of weight reduction.

## 1.4.2 Material Alternative

The material composition contributes directly to the weight of a vehicle. New alloys such as advanced high-strength steel (AHSS), aluminium alloys, magnesium alloys, and titanium alloys are potential materials for automotive weight reduction (Ingarao et al., 2011). The choice of material should consider two factors, its strength characteristics and the lightweight performance. In addition, the manufacturing process needs to be considered; thus, the formability and the design of the forming process will also be important considerations. For example, AHSS has high-strength performance, while aluminium and magnesium alloys show better weight reduction

---

**TABLE 1.1**

**Optimization Technique Used in Weight Reduction of Vehicle Components**

| No. | Automobile System | Parts/ Components | Design Optimization Techniques/Methods | Weight Reduction Possibilities (in % or in kg) | References |
|-----|-------------------|-------------------|----------------------------------------|------------------------------------------------|------------|
| 1. | Chassis | Leaf springs | Genetic algorithms (GAs) | 75.6% | Rajendran and Vijayarangan (2001) |
| 2. | Chassis | Leaf springs | Computer algorithm using C-language, finite element results were verified by finite element results using ANSYS software | 85% | Shankar and Vijayarangan (2006) |
| 3. | Closures | The inner panel of a door which is made using TWB | Topology optimization – size optimization, shape optimization by commercial optimization software, GENESIS | 8.72% | Shin et al. (2002) |
| 4. | Closures | The inner panel of the front door which is made using TWB | Topology, size and shape optimizations and design of experiments (DOE) | 11.72% | Lee et al. (2003) |
| 5. | BiW | Car body structure | High-performance computing (HPC) – 254 concurrently operating processors on a Silicon Graphics Inc. (SGI) Origin 2000 computer | 15 kg | Sobieszczanski-Sobieski et al. (2001) |

*(Continued)*

**TABLE 1.1 (*Continued*)**
**Optimization Technique Used in Weight Reduction of Vehicle Components**

| No. | Automobile System | Parts/ Components | Design Optimization Techniques/Methods | Weight Reduction Possibilities (in % or in kg) | References |
|---|---|---|---|---|---|
| 6. | Powertrain | Spur gear train | Optimization algorithms – particle swarm optimization (PSO) and simulated annealing (SA) | 11%, compared to Dong et al. (2005) | Savsani et al. (2010) |
| 7. | BiW | Bus body structure | Computer-aided design (CAD) package UG and finite element (FE) solver ANSYS | 5.7% | Lan et al. (2004) |
| 8. | Closures | Automotive door | The problem was formulated as multiobjective nonlinear mathematical programming, and approximation was done by employing artificial neural networks. | 12.9% | Cui et al. (2008) |
| 9. | BiW | Automotive front-body structure | Robust optimization – four meta-modelling techniques – vector regression, kriging, radial basis functions, and artificial neural networks | 19.45% | Zhu et al. (2009) |
| 10. | Powertrain | Correcting rod | Finite element method | 10% | Shenoy and Fatemi (2005) |
| 11. | BiW | B-pillar | Structural optimization; linear and quadratic response surfaces; finite element programme LS-DYNA | 25% | Marklund and Nilsson (2001) |
| 12. | BiW | Inner panel | Monte Carlo simulation; hybrid genetic/non-linear-programming algorithm | 8.08 kg (in case of minimum weight solution 8.31 kg) | Sandgren and Cameron (2002) |
| 13. | BiW | B-pillar | Metamodel-based optimization – support vector regression | 27.64% | Pan et al. (2010) |
| 14. | BiW | Front side rail | Response surface method | 26.95% | Zhang et al. (2007) |
| 15. | Chassis | Leaf springs | Finite element analysis – ANSYS 10 | 76.4% | Venkatesan and Devaraj (2012) |
| 16. | Closures | Door made with tailor blanks | Multidisciplinary optimization (MDO); response surface method (RSM) | 15% | Song and Park (2006) |

**TABLE 1.2**

**Parameters of Selected Light Weighting Materials (See Ingarao et al., 2011; Kleiner et al., 2006; USDOE, 2014)**

| Sl. no. | Material/Metal | Density | Young Modulus (GPa) | Tensile Strength (MPa) | Specific Strength (= Tensile Strength/ Density) ($10^6$ Nmm/kg) | Mass Reduction Capacity (%) |
|---|---|---|---|---|---|---|
| 1. | Aluminium (Al) | 2.8 | 70 | 150–160 | 52–243 | 30–60 |
| 2. | Magnesium (Mg) | 1.74 | 45 | 100–380 | 57–218 | 30–70 |
| 4. | Steel | 7.83 | 210 | 300–1200 | 38–153 | 15–25 |
| 5. | Titanium (Ti) | 4.5 | 110 | 910–1190 | 202–264 | 40–55 |

capabilities (Ingarao et al., 2011). Table 1.2 shows the material properties and weight reduction capability of several automotive materials. Among the materials, titanium alloy has one of the highest tensile strengths but is denser as compared to aluminium and magnesium alloys. Nevertheless, the mass reduction opportunity using titanium alloy is one of the highest among the materials; however, the per kilogram cost of titanium is very high compared to other automotive alloys (Hartman et al., 1998). Alternatively, aluminium and magnesium alloys can provide the required strength without compromising weight reduction opportunities. For example, aluminium foams are characterized by a low density, high rigidity, good energy absorption, and good recyclability, so the application potential can be seen in components requiring high structural rigidity with a very low weight (Merklein and Geiger, 2002).

The use of aluminium as an alternative to steel can be seen in recent automotive design, primarily for weight reduction purposes (Serrenho et al., 2017). However, to achieve the same structural properties, the average thickness of aluminium has to be increased as compared to steel. It was found that a weight reduction of 50% can be obtained by replacing 0.8 mm of the steel component with 1.2 mm of the aluminium component (Ingarao et al., 2011). Similarly, a 62% weight reduction can be obtained by using magnesium instead of steel (Ingarao et al., 2011). Another work has reported a weight saving as high as 55% by replacing steel with aluminium in BiW applications (Friedman and Kridli, 2000). However, such relationship cannot be concluded as an absolute rule because it will depend on the material properties, strength, and stiffness of part geometry of the whole component. For industrial applications, it is beneficial to determine the actual weight reduction potential of the individual material components. Jambor and Beyer (1997) have found that a weight reduction of 30%–50% is possible if BiW is constructed with all-aluminium instead of steel. Table 1.3 shows a list of component weight reduction potentials for some select production vehicles.

### 1.4.3 TECHNOLOGICAL INNOVATION

#### 1.4.3.1 Tailored Blanks

In modern automobiles, the BiW consists of a significant number of components. The BiW construction should consider three major criteria: weight reduction, structural stiffness, and passive safety. These criteria are major challenges in the field

**TABLE 1.3**

**Component Weight Reduction Potential on Some Selected Production Vehicles**

| Vehicle System | Subcomponent | New Material | Weight Reduction in (lb) | Vehicle Models | References |
|---|---|---|---|---|---|
| Powertrain | Engine, housing, etc. | Alum-Mg-composite | 50.8023 | BMW (R6) | Kulekci (2008) |
| | Cradle system | Aluminium | 9.97903 | GM (Impala) | Taub and Luo (2015) |
| | Intake manifold | Magnesium | 4.53592 | GM (V8); Chrysler | Kulekci (2008) |
| | Camshaft case | Magnesium | 0.907185 | Porsche (911) | Kulekci (2008) |
| | Auxiliaries | Magnesium | 4.98952 | Audi (A8) | Kulekci (2008) |
| | Trans. housing | Aluminium | 3.62874 | BMW (730d); GM (Z06) | Taub and Luo (2015) |
| | Trans. housing | Magnesium | 4.08233–4.53592 | Volvo; Porsche (911); Mercedes; VW (Passat); Audi (A4, A8) | and Kulekci (2008) |
| Body and closures | Frame | Aluminium-intensive body | 90.7185–158.757 | Audi (TT, A2, A8); Jaguar (XJ); Lotus; Honda (NSX, Insight) | Brooke and Evans (2009) |
| | Frame | Aluminium spaceframe | 55.3383 | GM (Z06) | Taub and Luo (2015) |
| | Doors (4) | Aluminium-intensive | 2.26796–22.6796 | Nissan (370z); BMW (7); Jaguar (XJ) | Keith (2010) |
| | Door inner (4) | Magnesium | 10.8862–21.3188 | | and Kulekci (2008) |
| | Hood | Aluminium | 6.80389 | Honda (MDX); Nissan (370z) | Keith (2010) |
| | Lift gate | Magnesium | 2.26796–4.53592 | | Kulekci (2008) |
| Suspension and chassis | Steering wheel | Magnesium | 0.498952 | Ford (Thunderbird, Taurus); Chrysler (Plymouth); Toyota (LS430); BMW (Mini); GM (Z06) | Kulekci (2008) |
| | Steering column | Magnesium | 0.453592–0.907185 | GM (Z06) | Kulekci (2008), |
| | Wheels (4) | Magnesium | 11.7934 | Toyota (Supra); Porsche (911); Alfa Romeo | Kulekci (2008) |
| Interior | Seat frame (4) | Magnesium | 12.7006 | Toyota (LS430); Mercedes (Roadster) | Kulekci (2008) |
| | Instrument panel | Magnesium | 3.17515–5.8967 | Chrysler (Jeep); GM; Ford (Explorer, F150); Audi (A8); Toyota (Century); GM | Kulekci (2008) and Taub and Luo (2015) |

*Source*: Adopted from Lutsey (2010).

of automobile body construction (Salzburger et al., 2001). Weight reduction efforts should not compromise the structural integrity and crashworthiness of the body. For example, the door inner panel is a component of BiW, where criteria such as weight reduction, assembly precision, and component rigidity to avoid sagging during assembly are required (Kusuda et al., 1997). In recent years, tailored blanks and their technology have contributed significantly to weight reduction in BiW (Merklein et al., 2014). The major advantages of using tailored blanks include reduction of weight and production costs, a decreased number of parts that are needed to combine in the assembly processes, and the reduction in material usage (Bocos et al., 2005; Meinders et al., 2000; Merklein and Geiger, 2002). Kusuda et al. (1997) have found that tailored blanks offered the opportunity for rigidity improvement and weight reduction; thus, significant material reduction is possible. It is expected that in the near future, 50% of the automobile body will be manufactured using tailored blanks (Bocos et al., 2005).

Crashworthiness for TWB structures can be determined by test data and simulation. Key indicators for crashworthiness of thin wall structures include total energy absorption (EA), specific energy absorption (SEA), mean crush/bending force efficiency ($F_{mean}$), peak force ($F_{max}$), and crushing force efficiency (CFE) (Zhang et al., 2021; Taghipoor et al., 2020; Xu et al., 2015).

### 1.4.3.1.1  Tailor Welded Blanks

A TWB is made up of combinations of sheet metals that are laser-welded and is characterized by different thicknesses and/or materials within a single work piece. Using TWBs, the final strength, stiffness, and durability of the automobile component can be maintained, but at the same time, reduction of weight can be obtained (Davies et al., 2001). TWB is now widely used in the production of automobile body side frames, door inner panels, and wheelhouse/shock tower panels (Kleiner et al., 2003; Sener et al., 2002). It provides the opportunity for automotive designers to make thickness variations in the body panel that optimizes the material usage. Furthermore, substitution of steel by aluminium alloys coupled with TWB is a logical extension of automobile weight reduction strategies (Davies et al., 2001).

Studies have been conducted on the use of TWB for several automotive components. Sheng (2008) has made small automotive bulb shields from stainless steel TWB using a progressive die and has found that 25% weight reduction was possible. Automobile body panel fabrication using TWB offers the largest opportunity for weight reduction and optimized material use (Chan et al., 2005; Raymond et al., 2004). Hyrcza-Michalska et al. (2010) reported that using TWB in the stamping of draw pieces of different geometries, such as B-pillar, reduces weight as much as 16%. Shin et al. (2002) optimized the inner door panel produced using TWB and found that an 8.72% weight reduction was achieved as compared to the original design inner panel, without compromising on the structural stiffness. In the ULSAB project, a car body manufactured by TWB can be made lighter by 25% with no cost penalty (ULSAB, 2015). Furthermore, new material development coupled with its use in TWB also provides further opportunity for weight reduction. However, TWB technology can be costly, especially if it requires a more complex manufacturing process for certain materials. For example, the use of aluminium as TWB has added processing challenges such as poor weldability, porosity, and hot cracking in the fusion zone or heat-affected zone (HAZ); thus, the final product will have reduced strength and loss of alloying elements (Merklein et al., 2014).

New material joining methods, such as friction stir welding (FSW), can overcome the issues of weldability for the fabrication of TWB. The traditional FSW has been expanded into new processes such as friction stir spot welding, friction stir riveting, and friction stir scribe (Palani et al., 2018). In most cases, FSW is able to produce better blanks in terms of strength and formability as compared to laser welding (Bagheri et al., 2020). FSW allows for long and thicker blanks, although it is currently limited in terms of shape complexity as compared to laser welding (Kashaev et al., 2018; Khan et al., 2021; Mrudula et al., 2021; Raj et al., 2021).

Another important aspect for an automobile component is its crashworthiness. In the case of TWB, the target for reducing weight should not compromise the crashworthiness of the vehicle. It has been found that TWB could also improve the level of crashworthiness (Kusuda et al., 1997). The crashworthiness of TWB blanks is affected by the material properties and grades, the welding direction, and the thickness of corresponding blanks. The research conducted by Xu et al. (2014) showed that the crash behaviour of TWB tubes can be improved by purposely designing the configuration, particularly the welding direction, the combination of material grades, and thicknesses. Suresh et al. (2020) in their investigation on thickness ratio have estimated that 33% weight reduction can be obtained by focusing on this strategy.

One of the issues related to the production of TWS is related to the joining interface. Joining of blanks of different thicknesses creates stress concentration, thus reducing its formability (Gautam and Kumar, 2019; Parente et al., 2016), less drawability (Béres and Danyi, 2017), and discontinuous thickness change in the weld line. Thickness variation can be addressed by using different material grades, thus maintaining the thickness of the blank. However, some researchers have found that welding does not influence the formability of the material, as shown by Singhal (2020), based on the tensile and dome height tests conducted. Similar findings were also obtained by Elshalakany et al. (2017).

### 1.4.3.1.2  Tailor-Rolled Blanks

TRB is another promising technology for weight reduction of automotive components. TRB overcomes the limitations of TWB on the joining of blanks of dissimilar thicknesses as a continuous transition from the thick section to the thin section can be obtained by adjustment of the roll gap. Fabrication of TRBs in a flexible rolling process can be from either (a) longitudinal or (b) transverse thickness directions (Kopp et al., 2005; Hirt and Dávalos-Julca, 2012).

The major advantages of TRB are as follows: (1) no stress peak occurs over the section, thus giving good formability; (2) the thickness variations do not affect production costs; (3) any thickness transition is possible within the process limits; and (4) there are no weld seams and HAZ in the TRB, thus giving good formability characteristics (Duan et al., 2015).

Several studies have been conducted on the evaluation of TRB automotive components. Chatti et al. (2002) have conducted quasi-static component test and drop test on front cross members and crash boxes and whole cross members. It was found that TRB components have reduced the weight of 13.5% and 15.6% less intrusions of structural behaviour as compared to conventional parts. Another study by Kleiner et al. (2006) has shown that a cross member manufactured by TRB was 13% lighter than conventional crash structures having significant structural behaviour in static and dynamic tests. Ingarao et al. (2011) have found that significant component weight reduction is possible with TRB as compared to conventional parts with constant thickness.

## 1.5   CONCLUSION AND FUTURE WORKS

In this chapter, the weight trends of passenger cars are highlighted based on the data for U.S. passenger car fleet. An in-depth review is conducted on the impact of weight on the fuel economy and $CO_2$ emission, and three weight reduction strategies were described.

The greatest challenge for automakers is to combine weight reduction with efficient fuel economy and components of high crashworthiness. For the automotive designers, the thickness of components is among the factors that can be considered for optimization. Past investigations have shown that lightweight materials, such as magnesium and aluminium, having thickness variation in a single blank could enable innovative manufacturing processes that can reduce the weights of components (Ingarao et al., 2011).

Advanced manufacturing technologies such as TRBs and TWBs can substantially reduce the weights of automotive components. However, manufacturing these semi-finished products generally requires considerable costs and efforts. Furthermore, these processes comprise two manufacturing steps. The first is to produce the tailor blanks, and the second is the manufacture of the components using the tailor blanks.

Future research work could be conducted to investigate the manufacturing process of a single blank with thickness variation in a single step. Hot and cold forming operation could be done in such context. Moreover, springback characteristics should be closely observed. To find the structure sufficiently crashworthy, the multidisciplinary design optimization (MDO) method can be employed.

## ACKNOWLEDGEMENT

The authors wishes to acknowledge the support from Universiti Sains Malaysia (USM) through the RUI Grant (1001/PMEKANIK/8014031).

## REFERENCES

An, F., & DeCicco, J. (2007). Trends in technical efficiency trade-offs for the US light vehicle fleet. SAE Technical Paper.

An, F., & Sauer, A. (2004). Comparison of passenger vehicle fuel economy and greenhouse gas emission standards around the world: Pew center on global climate change.

Automotive, C. (2009). Pocket book of steel. CORUS AUTOMOTIVE.

Bagheri, B., Abbasi, M., & Hamzeloo, R. (2020). Comparison of different welding methods on mechanical properties and formability behaviors of tailor welded blanks (TWB) made from AA6061 alloys. *Proceedings of the Institution of Mechanical Engineers, Part C: Journal of Mechanical Engineering Science.* doi: 10.1177/0954406220952504.

Bandivadekar, A. (2008). On the road in 2035: Reducing transportation's petroleum consumption and GHG emissions. Massachusetts Institute of Technology.

Béres, G., & Danyi, J. (2017). Experimental investigation on forming of tailor welded blanks. *Materials Science Forum, 885*, 147–152.

Bocos, J.L., Zubiri, F., Garciandia, F., Pena, J., Cortiella, A., Berrueta, J.M., & Zapirain, F. (2005). Application of the diode laser to welding on tailored blanks. *Welding International, 19*(7), 539–543.

Brooke, L. and Evans, H. (2009). *"Lighten Up!" Automotive Engineering International*, 16–22.

Chan, L.C., Chan, S.M., Cheng, C.H., & Lee, T.C. (2005). Formability and weld zone analysis of tailor-welded blanks for various thickness ratios. *Journal of Engineering Materials and Technology, 127*(2), 179–185. doi: 10.1115/1.1857936.

Chatti, S., Heller, B., Kleiner, M., & Ridane, N. (2002). Forming and further processing of tailor rolled blanks for lightweight structures. *Advanced Technology of Plasticity, 2*, 1387–1392.

Chen, C.J., Maire, S., & Usman, M. (1997). Improved fuel tank design using optimization. *Paper Presented at the ASME McNu97'Design Optimization with Applications in Industry Symposium*, Chicago, IL.

Chen, S., Li, L., Chen, Y., & Huang, J. (2011). Joining mechanism of Ti/Al dissimilar alloys during laser welding-brazing process. *Journal of Alloys and Compounds, 509*(3), 891–898. doi: 10.1016/j.jallcom.2010.09.125.

Cui, X., Wang, S., & Hu, S.J. (2008). A method for optimal design of automotive body assembly using multi-material construction[J]. *Materials and Design*, 29(2), 381–387.

Davies, R.W., Grant, G.J., Khaleel, M.A., Smith, M.T., & Oliver, H.E. (2001). Forming-limit diagrams of aluminum tailor-welded blank weld material. *Metallurgical and Materials Transactions A, 32*(2), 275–283.

Dong, Y., Tang, J., Xu, B., & Wang, D. (2005). An application of swarm optimization to nonlinear programming, *Computers & Mathematics with Applications*, 49(11–12), 1655–1668.

Duan, L., Sun, G., Cui, J., Chen, T., Cheng, A., & Li, G. (2015). Crashworthiness design of vehicle structure with tailor rolled blank. *Structural and Multidisciplinary Optimization, 53*(2), 321–338.

Elshalakany, A.B., Ali, S., Osman, T.A., Megaid, H. & El Mokadem, A. (2017). An experimental investigation of the formability of low carbon steel tailor-welded blanks of different thickness ratios. *The International Journal of Advanced Manufacturing Technology, 88*, 1459–1473.

EU. (2016). Reducing $CO_2$ emissions from passenger cars. Climate Action. Retrieved 11/02/2016, from http://ec.europa.eu/clima/policies/transport/vehicles/cars/index_en.htm.

Fan, F., & Lei, Y. (2016). Decomposition analysis of energy-related carbon emissions from the transportation sector in Beijing. *Transportation Research Part D: Transport and Environment, 42*, 135–145.

Friedman, P.A., & Kridli, G.T. (2000). Microstructural and mechanical investigation of aluminum tailor-welded blanks. *Journal of Materials Engineering and Performance, 9*(5), 541–551.

Froes, F.H., Eliezer, D., & Aghion, E. (1998). The science, technology, and applications of magnesium. *JOM, 50*(9), 30–34.

Gautam, V., & Kumar, A. (2019). Experimental and numerical studies on formability of tailor welded blanks of high strength steel. *Procedia Manufacturing, 29*, 472–480.

Glennan, T.B. (2007). Strategies for managing vehicle mass throughout the development process and vehicle lifecycle. SAE Technical Paper.

Han, H.N., & Clark, J.P. (1995). Lifetime costing of the body-in-white: steel vs. aluminum. *JOM, 47*(5), 22–28.

Hartman, A.D., Gerdemann, S.J., & Hansen, J.S. (1998). Producing lower-cost titanium for automotive applications. *JOM, 50*(9), 16–19.

Hayashi, H. (1996). Forming technology and sheet materials for weight reduction of automobile. University of Miskolc (Hungary), 13–31.

Hirt, G., & Dávalos-Julca, D.H. (2012). Tailored profiles made of tailor rolled strips by roll forming–part 1 of 2. *Steel Research International, 83*(1), 100–105.

Hyrcza-Michalska, M., Rojek, J., & Fruitos, O. (2010). Numerical simulation of car body elements pressing applying tailor welded blanks: Practical verification of results. *Archives of Civil and Mechanical Engineering, 10*(4), 31–44. doi: 10.1016/S1644-9665(12)60029-6.

Ilyas, K., Ismail, D., Fahrettin, O., & Rodney, J.S. (2018). A review of light duty passenger car weight reduction impact on $CO_2$ emission. *International Journal of Global Warming, 15*(3), 333. doi: 10.1504/IJGW.2018.093124.

Ingarao, G., Di Lorenzo, R., & Micari, F. (2011). Sustainability issues in sheet metal forming processes: An overview. *Journal of Cleaner Production, 19*(4), 337–347.

Jambor, A., & Beyer, M. (1997). New cars: New materials. *Materials & Design, 18*(4–6), 203–209. doi: 10.1016/S0261-3069(97)00049-6.

Kashaev, N., Ventzke, V., & Çam, G. (2018). Prospects of laser beam welding and friction stir welding processes for aluminum airframe structural applications. *Journal of Manufacturing Processes, 36*, 571–600.

Keith, D., (2010). HSS, AHSS and aluminum jockey for position in the race to cut auto curb weight. *American Metal Market Monthly*. February 1.

Khan, M.S., Shahabad, S.I., Yavuz, M., Duley, W.W., Biro, E., & Zhou, Y. (2021). Numerical modelling and experimental validation of the effect of laser beam defocusing on process optimization during fiber laser welding of automotive press-hardened steels. *Journal of Manufacturing Processes, 67*, 535–544.

Kleiner, M., Geiger, M., & Klaus, A. (2003). Manufacturing of lightweight components by metal forming. *CIRP Annals-Manufacturing Technology, 52*(2), 521–542.

Kleiner, M., Chatti, S., & Klaus, A. (2006). Metal forming techniques for lightweight construction. *Journal of Materials Processing Technology, 177*(1–3), 2–7. doi: 10.1016/j.jmatprotec.2006.04.085.

Knittel, C.R. (2009). Automobiles on steroids: Product attribute trade-offs and technological progress in the automobile sector. National Bureau of Economic Research.

Kopp, R., Wiedner, C., & Meyer, A. (2005). Flexible rolling for load-adapted blanks. *International Sheet Metal Review, 4*, 20–24.

Kulekci, M.K. (2008). Magnesium and its alloys applications in automotive industry. *International Journal of Advanced Manufacturing Technology, 39*, 851–865.

Kusuda, H., Takasago, T., & Natsumi, F. (1997). Formability of tailored blanks. *Journal of Materials Processing Technology, 71*(1), 134–140.

Lan, F., Chen, J., & Lin, J., (2004). Comparative Analysis for Bus SideStructures and Light Weight Optimization, Journal of Automotive Engineering, 218(10), 1067–1075.

Lee, K.H., Shin, J.K., Song, S.I., Yoo, Y.M., & Park, G.J. (2003). Automotive door design using structural optimization and design of experiments. *Proceedings of The Institution of Mechanical Engineers Part D-journal of Automobile Engineering, 217*(10), 855–865.

Leiva, J.P., Wang, L., Recek, S., & Watson, B.C. (2001). Automobile design using the GENESIS structural optimization program. *Paper Presented at the Nafems Seminar: Advances in Optimization Tecnologies for Product Design*, Chicago, IL.

Lowe, K. (1998). Lighter steel closure panels for cars. *Materials World, 6*(12), 761–762.

Lutsey, N.P. (2010). Review of technical literature and trends related to automobile mass-reduction technology. Institute of Transportation Studies. Retrieved from https://escholarship.org/uc/item/9t04t94w.

Lutsey, N., & Sperling, D. (2005). Energy efficiency, fuel economy, and policy implications. *Transportation Research Record: Journal of the Transportation Research Board, 1941,* 8–17.

Marklund, P., & Nilsson, L. (2001). Optimization of a car body component subjected to side impact. *Structural and Multidisciplinary Optimization, 21*(5), 383–392.

Mathiyazhagan, K., Diabat, A., Al-Refaie, A., & Xu, L. (2015). Application of analytical hierarchy process to evaluate pressures to implement green supply chain management. *Journal of Cleaner Production, 107,* 229–236.

Meinders, T., van den Berg, A., & Huétink, J. (2000). Deep drawing simulations of tailored blanks and experimental verification. *Journal of Materials Processing Technology, 103*(1), 65–73. doi: 10.1016/S0924-0136(00)00420-9.

Merklein, M., & Geiger, M. (2002). New materials and production technologies for innovative lightweight constructions. *Journal of Materials Processing Technology, 125–126,* 532–536. doi: 10.1016/S0924-0136(02)00312-6.

Merklein, M., Johannes, M., Lechner, M., & Kuppert, A. (2014). A review on tailored blanks: Production, applications and evaluation. *Journal of Materials Processing Technology, 214*(2), 151–164.

Mrudula, G., Bhargavi, P., & Krishnaiah, A. (2021). Effect of tool material on mechanical properties of friction stir welded dissimilar aluminium joints. *Materials Today: Proceedings, 38,* 3306–3313. doi: 10.1016/j.matpr.2020.10.042.

NHTSA. (2011). Summary of fule economy performance, Washington, DC U.S. Department of Transportation.

Palani, K., Elanchezhian, C., Ramnath, B.V., Bhaskar, G.B., & Naveen, E. (2018). Effect of pin profile and rotational speed on microstructure and tensile strength of dissimilar AA8011 and AA6061-T6 friction stir welded aluminum alloys. *Materials Today: Proceedings, 5,* 24515–24524.

Pallett, R.J., & Lark, R.J. (2001). The use of tailored blanks in the manufacture of construction components. *Journal of Materials Processing Technology, 117*(1), 249–254.

Pan, F., Zhu, P., & Zhang, Y. (2010). Metamodel-based lightweight design of b-pillar with twb structure via support vector regression, *Composite Structure,* 88(1–2), 36–44.

Parente, M., Safdarian, R., Santos, A.D. et al. (2016). A study on the formability of aluminum tailor welded blanks produced by friction stir welding. *The International Journal of Advanced Manufacturing Technology, 83,* 2129–2141.

Penna, C.C.R., & Geels, F.W. (2015). Climate change and the slow reorientation of the American car industry (1979–2012): An application and extension of the Dialectic Issue LifeCycle (DILC) model. *Research Policy, 44*(5), 1029–1048.

Raj, A., Kumar, J.P., Rego, A.M., & Rout, I.S. (2021). Optimization of friction stir welding parameters during joining of AA3103 and AA7075 aluminium alloys using Taguchi method. *Materials Today : Proceedings, 46.* doi: 10.1016/j.matpr.2021.02.246.

Rajendran, I. & Vijayarangan, S. (2001). Optimal design of a composite leaf spring using genetic algorithms, *Computers & Structures,* 79(11), 1121–1129.

Raymond, S.D., Wild, P.M., & Bayley, C.J. (2004). On modeling of the weld line in finite element analyses of tailor-welded blank forming operations. *Journal of Materials Processing Technology, 147*(1), 28–37. doi: 10.1016/j.jmatprotec.2003.09.005.

Sandgren, E., & Cameron, T.M. (2002). Robust design optimization of structures through consideration of variation, *Computers & Structures,* 80(20–21), 1605–1613.

Salzburger, H.J., Dobmann, G., & Mohrbacher, H. (2001). Quality control of laser welds of tailored blanks using guided waves and EMATs. *IEE Proceedings. Science, Measurement and Technology, 148*(4), 143–148.

Savsani, V., Rao, R.V. and Vakharia, D.P. (2010). Optimal weight design of a gear train using particle swarm optimization and simulated annealing algorithms, *Mechanism and Machine Theory, 45*(3), 531–541.

Sener, J.Y., De Medeiros, C., Lescart, J.C., Marron, G., Antoine, P., & Delfanne, S. (2002). Increasing performances by using multi-thicknesses blanks. SAE Technical Paper.

Serrenho, A.C., Norman, J.B., & Allwood, J.M. (2017). The impact of reducing car weight on global emissions: The future fleet in Great Britain. *Philosophical Transactions of the Royal Society A*, *375*, 20160364. doi: 10.1098/rsta.2016.0364.

Shankar, G.S.S. & Vijayarangan, S. (2006). Mono composite leaf spring for light vehicle-design, end joint analysis and testing. *Materials Science*, 12, 220–225.

Sheng, Z.Q. (2008). Formability of tailor-welded strips and progressive forming test. *Journal of Materials Processing Technology,* *205*(1–3), 81–88. doi: 10.1016/j.jmatprotec.2007.11.108.

Shenoy, P. S., & Fatemi, A. (2005). Connecting rod optimization for weight and cost reduction. *SAE Transactions*, *114*, 523–530. http://www.jstor.org/stable/44718931

Shin, J.-K., Lee, K.-H., Song, S.-I., & Park, G.-J. (2002). Automotive door design with the ULSAB concept using structural optimization. *Structural and Multidisciplinary Optimization, 23*(4), 320–327.

Singhal, H. (2020). Formability Evaluation of Tailor Welded Blanks (TWBs), MSc Thesis, The Ohio State University.

Sobieszczanski-Sobieski, J., Kodiyalam, S. & Yang, R. (2001). Optimization of car body under constraints of noise, vibration, and harshness (NVH), and crash. *Structural and Multidisciplinary Optimization,* 22, 295–306. https://doi.org/10.1007/s00158-001-0150-6

Song, S.I., & Park, G.J. (2006). Multidisciplinary optimization of an automotive door with a tailored blank. *Proceedings of the Institution of Mechanical Engineers, Part D: Journal of Automobile Engineering, 220*(2), 151–163.

Stodolsky, F., Gaines, L., Cuenca, R., & Vyas, A. (1995). *Life-Cycle Energy Savings Potential from Aluminum-Intensive Vehicles* (Vol. 951837). SAE, Warrendale, PA.

Suresh, V.V.N., Suresh, A., Regalla, S.P., Ramana, P.V., & Vamshikrishna, O. (2020). Sustainability aspects in the warm forming of tailor welded blanks. *E3S Web of Conferences 184*, 01042.

Taghipoor, H., Eyvazian, A., Ghiaskar, A., Praveen Kumar, A., Hamouda, A.M., & Gobbi, M. (2020). Experimental investigation of the thin-walled energy absorbers with different sections including surface imperfections under low-speed impact test. *Materials Today: Proceedings*, *27*(2), 1498–1504. doi: 10.1016/j.matpr.2020.03.006.

Takita, M. & Maruta, A. (2000). Trend toward weight reduction of automobile body in Japan. *Paper Presented at the Seoul 2000 FISITA World Automotive Congress*, Seoul, Korea.

Taub, A.I., & Luo, A.A. (2015). Advanced lightweight materials and manufacturing processes for automotive applications. *MRS Bulletin*, 40, 1045–1054. https://doi.org/10.1557/mrs.2015.268ULSAB. (2015). ULSAB programme report. Retrieved 09/02/2016, from http://www.worldautosteel.org/.

USDOE. (2014). Vehicle technologies office: Lightweight materials for cars and trucks. Retrieved 22/12/2015, from http://www.energy.gov/eere/vehicles/vehicle-technologies-office-lightweight-materials-cars-and-trucks.

USEPA. (2005, 24/10/2015). Vehicle standards and regulations. Retrieved 17/12/2015, from http://www3.epa.gov/otaq/standards.htm.

USEPA. (2014a). Light-duty automotive technology, carbon dioxide emissions, and fuel economy trends: 1975 through 2014. Appendix F, US Environmental Protection Agency, Ann Arbor, Michigan, available at: http://www.epa.gov/oms/fetrends.htm# report.

USEPA. (2014b). U.S. greenhouse gas inventory report: 1990–2013. Retrieved 05/02/2016, from http://www3.epa.gov/climatechange/ghgemissions/usinventoryreport.html.Venkatesan, M. & Devaraj, DH. (2012). Design and analysis of composite leaf spring in light vehicle. *International Journal of Modern Engineering Research (IJMER)*, 2(1), 213–218.

Wang, L., Basu, P.K., & Leiva, J.P. (2004). Automobile body reinforcement by finite element optimization. *Finite Elements in Analysis and Design, 40*(8), 879–893. doi: 10.1016/S0168-874X(03)00118-5.

Xu, F., Sun, G., Li, G., & Li, Q. (2014). Experimental study on crashworthiness of tailor-welded blank (TWB) thin-walled high-strength steel (HSS) tubular structures. *Thin-Walled Structures, 74*, 12–27. doi: 10.1016/j.tws.2013.08.021.

Xu, F., Tian, X., & Li, G. (2015). *Experimental Study on Crashworthiness of Functionally Graded Thickness Thin-Walled Tubular Structures*, pp. 1339–1352. doi: 10.1007/s11340-015-9994-3.

Zhang, Y., Zhu, P., & Chen, G. (2007). Lightweight design of automotive front side rail based on robust optimisation. *Thin-Walled Structures, 45*, 670–676.

Zhang, S., Wu, Y., Liu, H., Huang, R., Yang, L., Li, Z., …. Hao, J. (2014). Real-world fuel consumption and $CO_2$ emissions of urban public buses in Beijing. *Applied Energy, 113*, 1645–1655.

Zhang, X., Fu, X., & Yu, Q. (2021). Energy absorption of arched thin-walled structures under transverse loading. *International Journal of Impact Engineering, 157*(July). doi: 10.1016/j.ijimpeng.2021.103992.

Zhanqiu, Y. & Runlan, H. (1994). On weight reduction of automobile. *Automotive Engineering, 6*, 009.

Zhu, P., Zhang, Y., & Chen, G.L. (2009). Metamodel-based lightweight design of an automotive front-body structure using robust optimization. *Proceedings of the Institution of Mechanical Engineers, Part D: Journal of Automobile Engineering, 223*(9), 1133–1147.

# 2 Measurement and Compensation of Springback – A Comprehensive Review

*AB Abdullah*
Universiti Sains Malaysia

*MF Jamaludin*
Universiti Malaya

## CONTENTS

DOI: 10.1201/9781003164241-2

## 2.1   INTRODUCTION

In sheet metal forming processes, springback can result in geometrical deviations from the intended shape (Abdullah et al., 2011). Furthermore, formed parts with severe springback are considered as defects in manufacturing and are rejected. Although springback is very difficult to be eliminated, it can be controlled and minimized. For these reasons, understanding the mechanism of springback is very important. Springback can be defined as a condition that occurs during the forming process of sheet metals or alloys, where the material being formed tends to partially return to its original profile due to the elastic recovery upon unloading (Ramezani et al., 2010; Marretta et al., 2010; Carden et al., 2002). It is influenced not only by the intrinsic properties of the material such as its tensile and yield strengths but also by the material thickness, bend radius, and bend angle (et al., 2004; Li et al., 2002a; Esat et al., 2002).

Past studies have utilized finite element analysis as the main tool to understand, predict, and minimize component defects by controlling the springback (Chou and Hung, 1999; Narkeeran and Lovel, 1999; Schwarze et al., 2011; Shen et al., 2010; Yu, 2009). In addition, powerful numerical tools have been used for predicting and measuring springback such ANN (Bozdemir and Golcu, 2008), optical methods (D'Acquisto and Fratini, 2001; Abdullah et al., 2011), and optimization using statistical methods such as ANOVA and Taguchi methods (Lee and Kim, 2007; Meinders et al., 2008; Song et al., 2007). Recently, the topic on springback measurement and compensation has captured the attention of many researchers, as new metal alloys and tailored sheet metals are being considered in manufacturing processes. Studies on springback can be grouped into physical and numerical approaches. Physical approaches can be further divided into the evaluation of material properties and process characteristics as well as their sensitiveness. Typically, for material properties parameters, the discussion would evolve on the issues of isotropic and kinematic hardening (Vladimirov et al., 2009), the Bauschinger effect (Gau and Kinzel, 2001), evolution of elastic properties (Sun and Wagoner, 2011), and elastic and plastic anisotropy (Li et al., 2002b). For process characteristics, the discussion would focus mostly on the problem of sheet thickness, friction coefficient, blank holder force, tool geometry, contact pressure, temperature, and unloading procedures (Li et al., 2002b; Grèze et al., 2010; Osman et al., 2010; Zhang et al., 2007a). Springback is found in various types of sheet metal forming. Zhan et al. (2011), in their study of creep age forming (CAF), found that the greatest challenge was in predicting the amount of springback, and in their review, they have recommended three significant experimental methods, that is, the three-point bending method, four-point bending method, and cantilever bending devices.

This chapter will review the most common experimental methods used in measuring springback responses. It will also include discussions on twist springback that can also affect the efficiency in sheet metal bending processes. It starts with the introduction, followed by some explanations on the basic concept in springback measurements and calculations. Then, the experimental approaches will be discussed in detail, inclusive of issues and factors that affect springback and the limitations of each evaluation method. The phenomenon of twist springback will then be reviewed, including the methods and challenges in assessing twist forming and its responses. Then, several methods for compensation of springback will be described. Finally, this chapter ends with the conclusion and recommendations for future works.

## 2.2 TYPES OF BENDING IN SPRINGBACK PROBLEM ANALYSIS

Several experimental geometries have been used in the investigation of springback in sheet metal processes. Among the most prevalent types are air bending processes, which include the U-bending and V-bending tests, and the straight flanging test. In recent studies, the split-ring test was used to determine the amount of springback in deep-drawing processes. The prediction of springback is crucial to assist in the designing of dies to obtain precise part tolerances and geometrical control. The most common experimental approaches for the evaluation of springback will be elaborated in the following section.

### 2.2.1 V-Bending

A V-bending operation is commonly performed by compressing the metal strip specimen between a matching V-shaped punch and die. For most air bending tests, also known as free bending, the strip is usually supported by the two shoulders of a stationary die. The required bending angle can be determined from the die opening and punch displacement, as shown in Figure 2.1. The springback is then quantified based on the deviation of the bend, which can be referred to as the springback ratio.

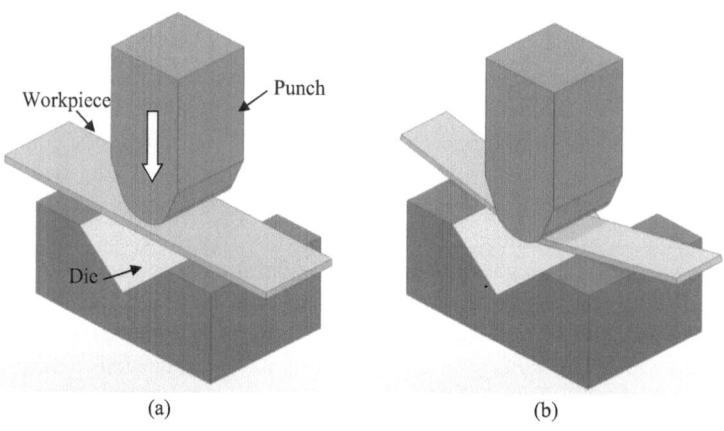

(a)                                    (b)

**FIGURE 2.1**   V-bending process setup: (a) initial stage and (b) final stage.

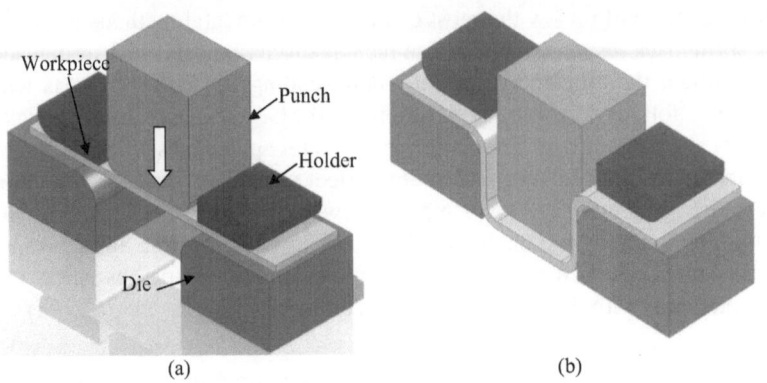

**FIGURE 2.2**   The steps involve in U-bending: (a) initial stage and (b) final stage.

The advantages of the V-bending die are the economical set-up time and relevance to a wide range of part sizes and complex shapes (Thipprakmas, 2010a). Furthermore, as compared to other bending processes, there is no need to change the dies to obtain different bending angles (Gajjar et al., 2007). Thus, the basic advantages of the V-die bending process are (1) a simple tool design, (2) an economical setup time, and (3) an enormous range of sizes and complex shapes that can be fabricated for the part but in contrast result in less accuracy (Thipprakmas, 2010b).

## 2.2.2   U-Bending

A U-bending is characterized by a die with a large depth as compared to its width, which can represent deep-drawing processes. The shape produced by the U-bending is one of the representative parts in sheet metal forming. The configuration of a U-bending test comprises at least a blank, blank holder, punch, and die, as shown in Figure 2.2. The blank is a piece of sheet metal, typically a rectangular shape specimen precut from the stock material, which will be formed into the part. This blank is clamped down by the blank holder over the die, which has a cavity in the external shape of the part. A punch tool moves downward into the blank and draws, or stretches, the material into the die cavity. The movement of the punch is usually hydraulically powered to apply the required force to the blank. Side wall curl is one of the main defects related to the forming of a U-shaped part due to the complicated bending and stretching processes experienced by the side walls (Zhang et al., 2007a). Side wall curve defects can be removed by introducing a large blank holder force (BHF) (Liu et al., 2002; Samuel, 2000). Increasing the BHF increases the flow resistance of the material, turning the stress distribution through the thickness of the side wall to tensile stresses over the whole section. This will make the springback directions of both sides to become consistent, thus decreasing shape distortions (Zhang et al., 2007a). One disadvantage of air bending is its low precision since the sheet does not stay in full contact with the dies. It also requires accurate control of the stroke depth. Furthermore, variations in the thickness of the material and wear on the tools can result in defects in the produced parts.

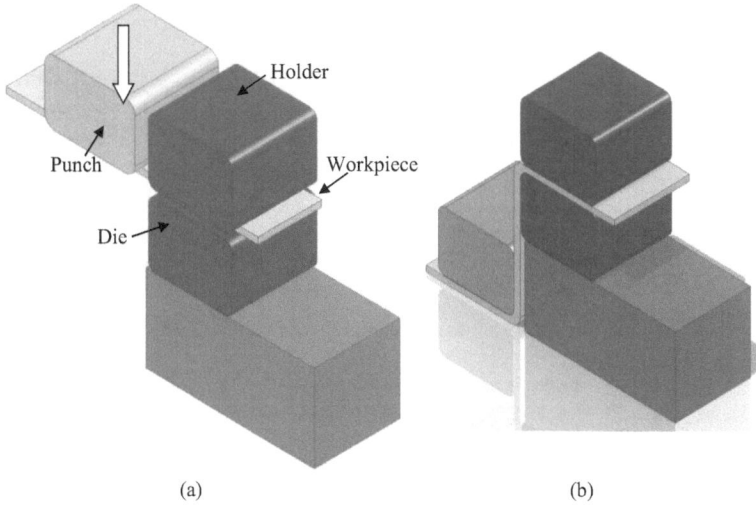

(a)                                                                (b)

**FIGURE 2.3**    Steps involved in draw bending: (a) initial stage and (b) final stage.

### 2.2.3    DRAW BENDING

The draw bend test usually consists of two hydraulic actuators oriented 90° to one another, with a fixed or rolling cylinder at the intersection of their action lines to simulate a tooling radius over which the sheet metal is drawn. The upper actuator (horizontal) is programmed to provide a constant restraining force (or back force), while the lower actuator is set to displace at a constant speed, thus drawing the strip over the cylinder. The process can be simplified as shown in Figure 2.3. The material undergoes tensile loading, bending, and unbending as it is drawn over the tooling. At the end of the test, the material is allowed to spring back by removal of the specimen from the grips of the fixtures, while the parts of the strip outside the formed region would usually remain straight (Hino et al., 2003). Ragai et al. (2005) in their study on the effect of friction on BHF have found that the clamping force may be reduced due to frictional effects. An important advantage of draw bending is that thin-walled tubes can be bent to desired radii smoothly and accurately.

### 2.2.4    STRAIGHT FLANGING

Straight flanging is a forming operation in which one end of the sheet is bent down along a straight line, while the other end is restrained by a pad or blank holder force. The specimen is bent at ~90° using the flanging die set. The test can determine the influences of die corner radius, punch (nose) radius, punch–die clearance, pad force, and sheet material on springback in straight flanging. Visualization and measurements of the process can be captured using photographs to observe phenomena such as sheet thinning, an increase in punch–die clearance, and lifting of the pad. Flange length can be changed easily, and the bending angle can be controlled by the punch stroke, as shown in Figure 2.4.

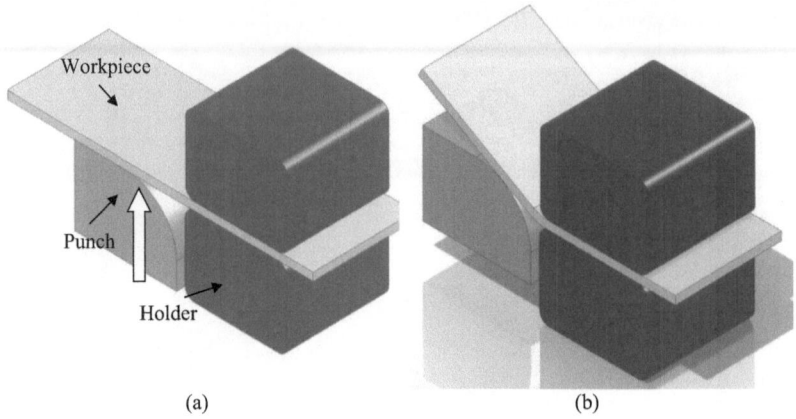

(a)                                                    (b)

**FIGURE 2.4**   The process steps of straight flanging: (a) initial stage and (b) final stage.

Among these variables, a small $R/t$ ratio is the most critical issue (Livatyali and Altan, 2001). Small radius bending is a process which is difficult to model accurately using material models based on simple tension test. The bend test can be applied using a specially designed set-up that generates a bending process without tension. However, the disadvantage of this method is that the proposed bend test is costly, and the empirical parameters are not readily available for most materials. Furthermore, the moment–curvature relation becomes even more complex when tool contact is considered. Thus, an approximate analysis method based on a simple tensile test and in which tool contact is not considered is much more feasible. Similarly in another case, Livatyali et al. (2003) have analysed numerically the effects of die corner radius ($R_d$) on springback successfully and the clearance ratio ($C/t$) with some under-prediction.

### 2.2.5  STRETCH BENDING

Stretch forming is a metal forming process in which a piece of sheet metal is stretched and bent simultaneously over a die to form large, contoured parts (Paulsen and Welo, 2002). Stretch-formed parts are typically large and possess large radius bends, with sizeable variations of possible producible shapes, from a simple curved surface to complex non-uniform cross-sections. It is capable of shaping parts with very high accuracies and smooth surfaces (Yu and Lin, 2007). Typical stretch-formed parts are large, curved panels such as door panels in cars or wing panels on aircraft, window frames, and enclosures.

Stretch forming tests are usually conducted for ductile materials such as aluminium, steel, and titanium. The test is performed on a stretch press, in which the piece of sheet metal is securely gripped along its edges by gripping jaws. Each gripping jaw is attached to a carriage that is pulled by pneumatic or hydraulic force to stretch the sheet, as shown in Figure 2.5. The tooling used in this process is a stretch form block, called a form die, which is a solid contoured piece against which the sheet metal will be pressed (Clausen et al., 2001). The most common stretch presses are oriented

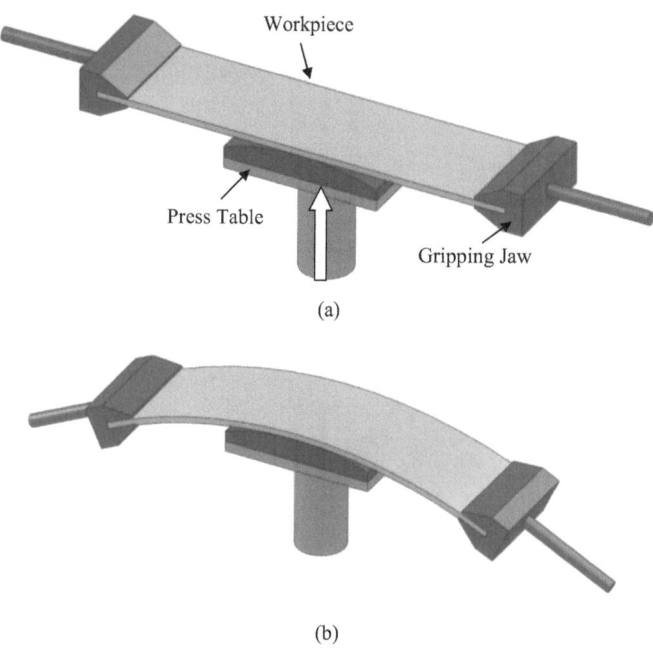

Workpiece

Press Table

Gripping Jaw

(a)

(b)

**FIGURE 2.5**   Illustration of stretch bending (a) setup and labels (b) after complete bending.

vertically, in which the form die rests on a press table that can be raised into the sheet by a hydraulic ram. As the form die is driven into the sheet, which is gripped tightly at its edges, the tensile forces increase, and the sheet plastically deforms into a new shape. Horizontal stretch presses mount the form die sideways on a stationary press table, while the gripping jaws pull the sheet horizontally around the form die.

Studies conducted on the springback of stretched forming have identified several influencing parameters, such as the holding force and the die radius (Ouakdi et al., 2012). In the study by Clausen et al. (2001) on strain hardening properties of the material and the tensile forces, three response parameters were considered, namely, the maximum die force immediately before unloading, the permanent sagging in the mid-section, and the vertical springback displacement in the mid-section during unloading.

## 2.2.6  SPLIT-RING TEST

The split-ring test consists of cutting a ring specimen from the drawn cup and splitting it longitudinally along a radial plane, which is followed by an elastic recovery phase in which springback takes place (Laurent et al., 2009; Thuillier et al., 2002). The variation of the ring's diameter, before and after splitting, directly gives a measure of the springback and the amount of residual stresses present in the formed cup. The test involves four steps: (1) deep drawing of a cylindrical cup from a circular blank with a constant blank holder force, (2) unloading the formed cup from the tooling restraints, (3) cutting a circular ring from the mid-section of the drawn cup, and (4) splitting the ring along a predefined direction to release residual stresses and measure the opening displacement of the ring. The opening of the ring is performed along the rolling direction (RD).

## 2.3  DEFINITION OF SPRINGBACK

Basically, the techniques for measuring springback are based on the fundamental theory of bending. Based on the definition, springback can be described as the deviation of the metal sheet after unloading (Behrouzi et al., 2010).

Springback can be represented as

$$\left[\frac{1}{R_i} - \frac{1}{R_f}\right] = 3\frac{Y}{TE} - 4R_t^2\left(\frac{Y}{TE}\right)^3 \tag{2.1}$$

where $R_i$ and $R_f$ represent the initial and final radii, respectively, while $T$ and $E$ represent the thickness and Young modulus of the workpiece material, respectively. In another case, Osman et al. (2010) have measured the amount of springback-by-springback ratio, $K$, which can be expressed as

$$K = \theta_f / \theta_i \tag{2.2}$$

where $\theta_f$ and $\theta_i$ are the final and initial bend angles, respectively.

In creep age forming (CAF), Jeshvaghani et al. (2011) have measured the amount of springback based on percentage of deflection before and after forming.

$$\text{Springback}(\%) = 100\left(d_{max} / \delta\right) \tag{2.3}$$

where $d_{max}$ is the maximum displacement and $\delta$ is the deflection after unloading. Based on the literature, the springback is shown by the angle deviation in degree or percentage of deviation.

### 2.3.1  SPRINGBACK VERSUS SPRING-GO

Thipprakma and Rojananan (2008) defined spring-go as being opposite to springback, where the angle of bend after unloading is lesser than the angle after loading. Özdin et al. (2014) consider spring-go if the value of the spring angle is negative, while springback has a positive spring angle value. It was also found that the angle tends to become spring-go as the bending angle increases and by using a smaller punch radius. Özdemir et al. (2021) found that spring-go can be observed for non-heat-treated and normalized sheet metals (Cr-Mo steel alloy), while springback is observed in sheet metals that have undergone tempering heat treatment.

## 2.3.2   Types of Springback

Yoshida (2013) classified springback into three modes of flange/wall angel change, sidewall curl, and twist.

### 2.3.2.1   Wall Springback

Nikhare (2016) found that the strength ratio and die and punch radius were the top influencing factors for front edge and weld line springback, while the thickness ratio was the most affecting parameter for back edge springback. Thus, the residual stresses have a small influence on the wall springback (Sedeh and Ghaei, 2021). The lowest value of the wall springback results from the effect of compressive stresses, resulting in an increase in thickness and thus an increase in its stiffness (Kut et al., 2021). It was shown that increasing the die wall angle leads to an increase in springback factor by about 0.003–0.007 experimentally and 0.005–0.0075 numerically due to expansion of the plastic deformation zone through the thickness direction.

### 2.3.2.2   Twist Springback

Twist springback is caused by torsion moments within the cross-section of the part as a consequence of opposing rotations of two cross-sections along their axes (Dezelak et al., 2014). Li et al. (2011) described twist springback in terms of the rotation between two different cross-sections along the axes. From mechanics points of view, twist springback is the result of torsional moments in the cross-section of the workpiece.

### 2.3.2.3   Flange Springback

The flange drawing process is widely used to make flanges from a sheet metal blank using the die, punch, and blank holder. Lee and Kim (2007) described that the occurrence of flange springback was due to the non-parallelity of the formed flanged section with the original blank.

## 2.3.3   Measurement of Springback

Springback can be described as the difference between two angle values measured on the bend/form part. The first angle value is measured when the load is applied to the blank, while the second angle value is measured after the applied load is removed. There are several methods/approaches which can be used using various machines and equipment to measure the angle. Table 2.1 summarizes the springback measurement methods for various types of bending.

## 2.3.4   Factors That Affect the Springback

Springback can be influenced by many factors, including material properties and tool designs. Many past studies were conducted to investigate the influence of these factors on the springback response, which are described in the following sections.

**TABLE 2.1**

**Summary of Springback Measurement Methods for Various Types of Bending**

| Method | Graphical Representation of Springback Measurements (Note: (a) Loading and (b) Unloading Conditions) |
|---|---|
| V-bending | |
| U-bending | |
| Draw bending | |
| Straight flanging | |
| Stretch bending | |
| Split rings | |

### 2.3.4.1   Material Properties

Material properties are one the main factors that influence the formation of springback, which typically include the Young modulus, the yield strength, and hardness values. Karanjule et al. (2018a) have found that increasing the carbon content of the sheet metal increases the strength of the material, which reduces the Young modulus value. This leads to an increase in springback. In terms of hardness, Fukuyasu et al. (1999) have shown that the springback angle decreases with the decrease of hardness, although the changes observed were minimal.

### 2.3.4.2   Thickness

The spring response of the metal will vary as the sheet thickness changes, even for the same type of material. In general, increasing the thickness of the sheet metal decreases the springback value.

### 2.3.4.3   Temperature

Temperature will change the behaviour of the material. Increasing the temperature may lower the stress–strain curve; that is, the yield point will decrease and be similar to UTS. Saito et al. (2017) found that in warm forming, springback was reduced as the temperature increased. Li et al. (2021) have shown that such reduction improves the forming accuracy. Similarly, Cai et al. (2019) discovered that a 90% reduction of springback for aluminium alloy 7075 was obtained as the temperature was increased. Similar behaviours were observed for other materials, such as for titanium alloy (Ti-64) (Ertana and Çertin, 2021).

### 2.3.4.4   Friction

Friction is an inherent feature of the forming process as there are many contact areas where the metal sheet and tool have relative motions to one another (Dametew and Gebresenbet, 2017). Friction is controlled by lubrication, which is affected by material properties, surface finish, temperature, sliding velocity, and contact pressure. The state of lubrication is normally expressed in terms of coefficient of friction. The corners of the die, where the sheet slides and shears continuously, influences this coefficient of friction, which is considered between 0.05 and 0.15. The level of friction and lubrication can influence the part quality and formation of defects, tool wear, and associated manufacturing expenses. Dry lubrication refers to the use of solid lubricants, which are not generally used in metal forming. It is applied when material formability is very high and simple parts can be drawn without lubrication, with minimal effect on part quality (Rigas et al., 2019). On the other hand, film lubrication can be applied when the solid surfaces are close together. In full lubrication, the mating surfaces are fully separated by a thin film of lubrication. Lundh et al. (1998) have defined six contact and friction regions in deep drawing. Poor friction and flow control of sheets could result in wrinkling, while excessive friction leads to crack formation or tearing. Petroleum-based oils and soluble oils are generally used as lubricants in metal forming. Additional factors for the selection of particular lubricants include the method of lubrication, additives, and corrosion control. It is observed that most small-scale sheet metal manufacturers usually select these process parameters by trial and error, based on experience, knowledge, and intuition of the workers. Thus, it may take a long time until the optimum combinations of parameters are found. For economic reasons, they prefer locally available cheaper oils for the lubrication purpose, which may affect the quality of the manufactured parts.

### 2.3.4.5  Blank Holder Force

The blank holder controls the travel of material in the die. It will ensure uniform thickness distribution in the drawn part and thus greatly determine the plastic deformation and product quality. The two opposing failure behaviours of wrinkling and tearing can only be avoided with optimum application of the BHF. If the BHF applied is less than optimum, it allows free flow of the material in the die which leads to wrinkling in the walls. On the other hand, if the BHF is too high, the flow of the sheet metal will be restricted and result in the cracking of the material as the punch forms the profile. The topic of BHF has been the focus of various researchers in the past, with some experimenting on variability of the BHF. Siebel's equation can be used for the calculation of the BHF.

### 2.3.4.6  Die and Punch Geometry

#### 2.3.4.6.1  Punch Nose Radius

In the metal forming process based on mechanical systems, the punch is operated with a flywheel. The punch force determines the deformation required to convert the blank into finished product by flowing it inside the die. The face of the punch and the punch nose radius must be carefully designed based on the blank material and its thickness. A punch nose radius that is too small can pierce or cut the blank instead of forcing the material to bend around the radius (Kakandikar and Nandedkar, 2020). As the punch nose radius increases, it increases the possibility of blank to be stretched rather than be drawn in the blank edge.

The total deformation of the sheet metal in the die depends on the punch nose radius. If the punch nose radius is increased, the punch will not travel up to the end in the die valley region, and this also affects the springback of the sheet metal. Thus, the correct punch nose radius is necessary to reduce the springback effect of the formed blank.

#### 2.3.4.6.2  Die Profile Radius

For flat blank holder, the radius of die profile and its surface are significant in determining the flow of the sheet into the die cavity (Kartik and Rajesh, 2017). If the die profile radius is too small, the sheet being formed may split as the sheet material is being deformed. This is caused by the high restraining forces initiated with bending and unbending of the sheet metal over a tight radius. The tight radius of the die also creates large amounts of heat that may initiate galling, where the sheet metal microscopically welds to the tool. On the contrary, an excessive die radius is also not recommended as it may result in sheet wrinkling between the punch and the die face. Adequate experience and intuition can assist in deciding the suitable range of die profile radii which work properly.

### 2.3.4.6.3    Punch Angle

The punch angle is the angle of deforming tool used to deform the sheet metal in forming operation. Depending on the required shape of the product, the punch may be V-shaped, U-shaped, or curvature-shaped. As the punch angle affects the springback effect of the sheet metal, the correct punch angle should be maintained in the forming operation of the sheet metal.

### 2.3.4.6.4    Grain Direction of the Sheet Metal Material

The grain direction of the sheet metal material plays an important role in the forming operation. A sheet metal material having a uniform grain size and direction gives better results as the deformation is uniform in the required direction.

### 2.3.4.6.5    Die Opening

Forming operations are carried out using the punch and die set configured on a press machine. The sheet metal placed on the die is deformed by the application of force onto the punch. The components of the die can be of V-shape, U-shape, or curvature shape, depending on the desired formed shape. Design of the die, such as the die opening dimensions, has to consider the springback effect of the sheet metal that is dependent on the angle of the die.

### 2.3.4.6.6    Punch Height

Punch height is the distance travelled by the punch to deform the sheet metal. The punch height can be determined by calculating the required force for deformation. The correct punch height is also important as it will also influence the level of springback of the sheet metal formed.

### 2.3.4.6.7    Prebend Conditions of the Sheet Metal

The presence of existing prebend conditions of the sheet metal can also affect the springback response. Existing bends may have already developed residual stresses which affect the springback. Thus, springback of the sheet metal should also consider initial prebend conditions of the sheet metal.

Table 2.2 summarizes most of the recent studies on selected parameters on springback utilizing various tools and methods.

## 2.3.5    MEASUREMENT TOOLS

Table 2.3 summarizes the machines and equipment that have been used to measure the springback in experimental analysis. Most of the studies have used coordinate measuring machine (CMM) and profile projectors for the main machine/equipment. Newer technologies, such as non-contact measurement methods utilizing laser, have good potential as they allow measurements to be made without touching the bent part. Some researchers have fabricated custom-made devices attached to the bending machine, primarily to measure springback, allowing in situ measurements of the springback without the need to dismount the specimen (Kędzierski and Popławski, 2019; Paunoiu et al., 2015). Image processing techniques have also been used for measuring of micro-sized parts (Liu et al., 2019), ultra-size parts (Ha et al., 2021),

**TABLE 2.2**

**Selected Published Work on Optimization of the Various Parameters on Springback**

| References | Material/Tools | Studied Parameters |
|---|---|---|
| Thipprakmas and Phanitwong (2011) | AA1100-O/Taguchi method | Bending angle, material thickness, and punch radius |
| Buang et al. (2015) | DP590/RSM | Blank holder force, clearance, punch travel and rolling direction |
| Jadhav and Bhatt (2016) | Tin-coated sheet/Taguchi method | Die opening, punch and die radius, hole diameter and lubrication |
| Karanjule et al. (2018b) | C-45/Taguchi method, particle swarm optimization (PSO), simulated annealing (SA), and genetic algorithm (GA) | Die semi-angle, land width, and drawing speed |
| Kakandikar and Nandedkar (2018) | D-513/Taguchi, mathematical and GA methods | Lubrication, punch nose radius, BHF, and die profile radius |
| Özdemir et al. (2020) | DP600/Taguchi and RSM methods | Thicknesses and punch radii |
| Adnan et al. (2017) | AA6061/Taguchi method | Bend angle, thickness ratio, and alignment |
| Ling et al. (2016) | AA5052/Taguchi method | Thickness, rolling direction, and cutting type |
| Panda and Pawar (2018) | HSLA 420/Taguchi method | Punch angle, die opening, grain direction, and prebend conditions |

and parts with high complexity due to non-uniform thickness -. Images from the experiment will then be sent for further analysis by using either available CAD software such as AutoCAD (Jadhav and Bhatt, 2016) or custom-made algorithms using MATLAB® (Liu et al., 2019; Nashrudin and Abdullah 2017).

Similarly for finite element (FE) simulation, various pieces of software are available for use, with added flexibility in terms of consideration of increased parameters. Furthermore, simulations can be less costly as compared to experimental studies. Table 2.4 lists recent studies in springback response using various simulation software and parameters.

### 2.3.6 COMPENSATION STRATEGIES

Springback compensation strategies can be described as the steps taken to reduce springback during the forming of a sheet metal part. Weiher et al. (2004) divided the compensation strategies into two: The first is the variation of process parameters, such as BHF and draw beads parametric optimization, and the second is the modification of tool geometries, such as shoulder radius and punch nose radius. However, relegating springback compensation to the shop floor can be time-consuming and create reliability issues due to variations in machine configuration and complexity of the part shape.

**TABLE 2.3**

**Summary of Springback Measurement Methods from Experimental Works**

| References | Equipment/Method Used | Case Study |
|---|---|---|
| Adnan et al. (2017) | Rax Vision Mitutoyo Profile Projector (PC 3000) | V-bending of non-uniform thickness AA6061 |
| Jiang et al. (2018) | OLYMPUS metallographic microscope | Bending of stainless steel orthodontic archwire using a robot |
| Rao and Narayanan (2014) | Profile projector | V-bending of friction stir welding of similar rather than dissimilar aluminium alloy |
| Ha et al. (2020) | Laser | In-line measurement on tube bending |
| Wan Nawang et al. (2015) | Mitutoyo optical measuring machine and Hitachi S-3700 Tungsten Filament Scanning Electron Microscope (W-SEM) | Micro-bending parts using 316L annealed stainless steel strips having thicknesses of 50, 75, and 100 μm |
| Darmawan et al. (2019) | Performed manually by plotting in millimetre papers | Mild steel TWB strips in U-bending |
| Ma et al. (2019) | A 3000iTM series three-coordinate measuring machine (CMM) | V-free bending process on A283-D steel |
| Özdemir (2020) | Coordinate measurement machine | V-bottoming bending processes were carried out by using 4 mm thick X10CrAlSi24 |
| Wang et al. (2008) | Angular displacement transducers (ADT) | Springback control in the air bending process on the press brake machine |
| Kędzierski and Popławski (2019) | Custom-made device | Conducted for the samples made of S235 structural steel |
| Paunoiu et al. (2015) | Custom-made device | Springback evaluation is based on circular bending of a sheet metal |
| Jadhav and Bhatt (2016) | Image processing | AutoCAD®2013 was used to measure angle by drawing two lines on the longer edge of the bend part image |
| Ha et al. (2021) | Image processing | In situ springback monitoring technique for bending of large-size profiles is proposed to overcome the measurement restrictions |
| Liu et al. (2019) | Image processing | Accurate measurement of the springback of micro-bent parts |
| Nashrudin and Abdullah. (2017) | Image processing | Twist bending of non-uniform thickness aluminium plate |
| Shen et al. (2020) | Image processing | The deformation process discretization analysis is carried out based on the gridding method on the deformed specimen |

**TABLE 2.4**

**Summary of Springback Measurement Methods from Simulation Works**

| References | Software | Studied Parameters |
|---|---|---|
| Spathopoulos and Stavroulakis (2020) | Ls-Dyna | Tool radius, BHF, and thickness |
| Paak et al. (2018) | Ls-Dyna | Thickness and various material properties including strength, Young modulus, yield stress, transverse anisotropy, and hardening exponent |
| Fan et al. (2020) | Abaqus | Radius of curvature, die corner radius, and thickness ratio |
| Trzepiecinki and Lemu (2017) | Abaqus | Rolling direction and friction coefficient |
| Phanitwong et al. (2017) | Deform-2D | Bend angle and tool radius |
| Gite et al. (2016) | Ansys | Die valley angle, punch nose radius, and die angle |
| Karabulut and Esen (2021) | Autoform | Wall angles and die radii |
| Sen et al. (2020) | Ansys | Bending angles, punch radius, and thickness |
| Hai et al. (2020) | Abaqus | Punch stroke |
| Aday | Ansys | Material type and thickness |
| Ghimire et al. (2017) | Ansys | Die angle |
| Zhang et al. (2016b) | COPRA | Width of the U-bending, radius of the roll die, lubrication, roll velocity, and thickness |

### 2.3.6.1   Methods and Tools

Springback compensation can be addressed either by experimental methods or by FE simulation. Both approaches have their own advantages and disadvantages. Table 2.5 summarizes the tools and methods adopted by a selection of past studies.

### 2.3.7   CONCLUSIONS AND RECOMMENDATIONS FOR FUTURE STUDIES

The main objective of this chapter is to review the various methods for springback measurements and compensation. Based on the review, parameters that are significant to springback of sheet metals can be categorized into design and process parameters. Several design and process parameters were identified as significant to the springback responses. One such parameter is the thickness of the sheet, which can be challenging to quantify for the springback measurement and improvements of non-uniform thicknesses in tailor welded blanks (TWBs). Furthermore, this complexity increases as the dissimilarity in material is considered, as in new types of tailor blanks made up of different types of materials. These factors of non-uniformity in geometry and material can be the focus of future studies in springback of TWBs.

## ACKNOWLEDGEMENTS

The author would like to acknowledge the support from Universiti Sains Malaysia (USM) through the RUI Grant (1001/PMEKANIK/8014031).

**TABLE 2.5**

**Methods and Tools Used in the Springback Compensation from Selected Publications**

| References | Compensation Method/ Tools | Process | Remarks |
|---|---|---|---|
| Fei et al. (2011) | Profile correction algorithm (FEM) | Incremental forming | The method is applied to the design of tool path surfaces for the workpiece with shallow dishing shapes. The largest dimensional deviation is reduced to 0.22 from 3.06 mm (change to %) after three iterations |
| Wang et al. (2016b) | Displacement adjustment method (FEM validated with experiments) | Stamping | 24 of the 32 measuring points can satisfy geometrical tolerances (75%). The proposed method can effectively reduce the number of die try-outs and try-out time in comparison with the traditional methods |
| Chatti et al. (2009) | Stress superposition (experimental) | Incremental tube forming | The superimposed compressive stresses influence the accumulated elastic energy, leading to a springback reduction. It was observed that the decrement of springback by up to 60% |
| Nowosielski et al. (2013) | Pressing Strategies (FEM) | V-bending | - |
| Birkert et al. (2017) | Physical compensation method (FEM) | Stamping | Compared to established software, namely, reverse displacement method for two case studies, top hat profile, and A-pillar |
| Su et al. (2020) | Curvature correction coefficient (FEM validated with experiments) | Plate bending | Forming theory is unlikely to produce very accurate springback results, but it is very close to the desired result, so that the times of stamping can be reduced as much as possible and improve the processing efficiency |
| Liu et al. (2013) | FEM validated with experiments | U-shape stamping | - |
| Tomáš et al. (2017) | Sequential springback compensation (FEM) | Stamping | Achieve up to 0.5 mm springback reduction, in which effective springback compensation is defined at the maximum range of 0.2 mm |
| Wang et al. (2016a) | Reverse engineering (experimental) | Stamping | The inspection results demonstrated that 15 of 18 measuring points can satisfy the geometrical tolerances (83.33%) |
| Siswanto et al. (2014) | Displacement adjustment method and spring forward algorithm (FEM) | Cold stamping | The results are shown for two dimensions and three dimensions; the springback reduced up to 66% and 55%, respectively. In comparison with Autoform, the HM method shows better performance, while the Autoform can reduce 22% and 35% only for two- and three-dimensional models, respectively |
| Ozturk et al. (2010) | Temperature changes (FEM and experimental) | V-bending | The springback compensation at 300°C was at least 10°C better than room temperature |
| Ma et al. (2019) | Iteration compensation algorithm (experimental) | V-bending and stretch bending | The new method does not depend on material properties and mechanical models |
| Troive et al. (2018) | Virtual compensation method | | Optimal die geometries can be produced in short time |

## REFERENCES

Abdullah, A. B., Sapuan, S. M., Samad, Z., Khaleed, H. M. T. and Aziz, N. A. (2011). Quantitative evaluation of geometric and dimensional error of cold forged AUV propeller blade. *Australian Journal of Basic and Applied Sciences*, 5(9): 1756–1764.

Adnan, M. F., Abdullah, A. B. and Samad, Z. (2017). Springback behavior of AA6061 with non-uniform thickness section using Taguchi method. *International Journal of Advanced Manufacturing Technology*, 89(5–8): 2041–2052.

Behrouzi, A., Dariani, B. M. and Shakeri, M. (2010). Die shape design in channel forming process by compensation springback error, *Majlesi Journal of Mechanical Engineering*, 3(2): 17–26.

Birkert, A., Hartmann, B., Nowack, M., Scholle, M. and Straub, M. (2018). Guideline to optimize the convergence behaviour of the geometrical springback compensation, *Journal of Physics: Conference Series*, 1063: 012128.

Bozdemir, M. and Golcu, M. (2008). Artificial neural network analysis of springback in V bending, *Journal of Applied Sciences*, 8: 3038–3043.

Buang, M. S., Abdullah, S. A. and Saedon, J. (2015). Optimization of springback prediction in U-channel process using response surface methodology, *International Journal of Mechanical and Mechatronics Engineering*, 9(7): 1216–1223.

Carden, W. D., Geng, L. M., Matlock, D. K. and Wagoner, R. H. (2002). Measurement of springback, *International Journal of Mechanical Sciences*, 44(1): 79–101.

Chatti, S., Hermes, M., Weinrich, A., Ben-Khalifa, N. and Tekkaya, A. E. (2009). New Incremental methods for springback compensation by stress superposition, *Production Engineering*, 3(2): 137–144.

Chen, C., Ko, M. and Wenner, M. L. (2008). Experimental investigation of springback variation in forming of high strength steels, *Journal of Manufacturing Science and Engineering*, 130: 1–9.

Chou, I. N. and Hung, C. (1999). Finite element analysis and optimization on springback reduction, *International Journal of Machine Tools and Manufacture*, 39: 517–536.

Clausen, A. H., Hopperstad, O. S. and Langseth, M. (2001). Sensitivity of model parameters in stretch bending of aluminum extrusions, *International Journal of Mechanical Sciences*, 43(2): 427–453.

D'Acquisto, L. and Fratini, L. (2001). An optical technique for springback measurement in axis-symmetrical deep drawing operation, *Journal of Manufacturing Processes*, 3(1): 29–37.

Dametew, A. W. and Gebresenbet, T. (2017). Study the effects of spring back on sheet metal bending using mathematical methods. *Journal of Material Sciences and Engineering*, 6: 382.

Darmawan, A., Anggono, A. D. and Nugroho, S. (2019). Springback phenomenon analysis of tailor welded blank of mild steel in U-bending process. *AIP Conference Proceedings*, 2114: 030021. https://doi.org/10.1063/1.5112425.

Dezelak, M., Stepisnik, A. and Pahole, I. (2014). Evaluation of twist springback prediction after an AHSS forming process. *International Journal of Simulation Modelling*, 13: 171–182.

Ertana, R. and Çetin, G. (2021). Effect of deformation temperature on mechanical properties, microstructure, and springback of Ti-6Al-4V sheets, *Revista de Metalurgia*, 57(4): e209.

Esat, V., Darendeliler, H. and Gokler, M. I. (2002). Finite element analysis of springback in bending of aluminum sheets, *Materials and Design*, 23(2): 223–229.

Fan, L., Gou, J., Wang, G. and Gao, Y. (2020). Springback characteristics of cylindrical bending of tailor rolled blanks. *Advances in Materials Science and Engineering*, 3: 1–11. https://doi.org/10.1155/2020/9371808.

Fei, H., Mo, J. H., Pan, G. and Min, L. (2011). Method of closed loop springback compensation for incremental sheet forming process, *Journal of Central South University of Technology*, 18: 1509–1517.

Fukuyasu, Y., Yokoe, H., Tozawa, Y. and Ono, M. (1999). Effects of thickness and hardness on springback of cold rolled stainless steel strip for spring, *Transactions of Japan Society of Spring Engineers*, 44: 25–29.

Gajjar, H.V., Gandhi, A.H. and Raval, H.K. (2007). Finite element analysis of sheet metal air- bending using hyperform LS-DYNA. *World Academy of Science, Engineering and Technology*, 32: 92–97.

Gau, J. T. and Kinzel, G. L. (2001). An experimental investigation of the influence of the Bauschinger effect on springback predictions, *Journal of Materials Processing Technology*, 108(3): 369–375.

Ghimire, S., Emeerith, Y., Ghosh, R. Ghosh, S. and Barman, R. N. (2017). Finite element analysis of an aluminium alloy sheet in a V-die punch mechanism considering springback effect, *International Journal of Theoretical and Applied Mechanics*, 12(2): 331–342.

Gite, R. E., Phad, K. S. and Bajaj, D. S. (2016). Springback effect analysis of bracket using finite element analysis, *International Advanced Research Journal in Science, Engineering and Technology*, 3(1): 246–255.

Grèze, R., Manach, P. Y., Laurent, H., Thuillier, S. and Menezes, L. F. (2010). Influence of the temperature on residual stresses and springback effect in an aluminium alloy, *International Journal of Mechanical Sciences*, 52(9): 1094–1100.

Ha, T., Ma, J., Blindheim, J., Welo, T., Ringen, G. and Wang, J. (2020). In-line springback measurement for tube bending using a laser system. *Procedia Manufacturing*, 47: 766–773.

Ha, T., Ma, J., Blindheim, J., Welo, T., Ringen, G. and Wang, J. (2021). A computer vision-based, in-situ springback monitoring technique for bending of large profiles. *ESAFORM 2021: 24th International Conference on Material Forming*, Liège, Belgique

Hai, V. G., Minh, N. T. H. and Nguyen, D. T. (2020). A study on experiment and simulation to predict the spring-back of SS400 steel sheet in large radius of V-bending process, *Materials Research Express*, 7: 016562.

Hino, R., Goto, Y. and Yoshida, F. (2003). Springback of sheet metal laminates in draw-bending, *Journal of Materials Processing Technology*, 139(1–3): 341–347.

Jadhav, T. K. and Bhatt, K. K. (2016). Optimization of springback effect in air bending process for tin coated perforated sheet by Taguchi approach, *International Journal of Advance Research in Engineering, Science and Technology*, 3(5).

Jeshvaghani, R. A., Emami, M., Shahverdi, H. R. and Hadavi, S. M. M. (2011). Effects of time and temperature on the creep forming of 7075 aluminum alloy: Springback and mechanical properties, *Materials Science and Engineering: A*, 528(29–30): 8795–8799.

Jiang, J. G., Ma, X. F., Zhang, Y. D., Liu, Y. and Biao, H. (2018). Springback mechanism analysis and experimentation of orthodontic arch-wire bending considering slip warping phenomenon, *International Journal of Advanced Robotic Systems*, 15(3): 1–13.

Kakandikar, G. and Nandedkar, V. (2018). Springback optimization in automotive shock absorber cup with genetic algorithm, *Manufacturing Review*, 5, 1.

Kakandikar, G. and Nandedkar, V. (2020). Multi-objective optimisation of thickness and strain distribution for automotive component in forming process, *International Journal of Computational Intelligence Studies*, 9(1/2): 172–184.

Karabulut, S. and Esen, I. (2021). Finite element analysis of springback of high-strength metal SCGA1180DUB while U-channeling, According to Wall Angle and Die Radius. https://www.researchsquare.com/article/rs-517616/v1.

Karanjule, D. B., Bhamare, S. S. and Rao, T. H. (2018a). Effect of Young's modulus on springback for low, medium and high carbon steels during cold drawing of seamless tubes, *IOP Conference Series: Materials Science and Engineering*, 346: 012043.

Karanjule, D. B., Bhamare, S. S. and Rao, T. H. (2018b). Process parameter optimization for minimizing springback in cold drawing process of seamless tubes using advanced optimization algorithms, *Journal of Soft Computing in Civil Engineering,* 2–3: 72–90.

Kartik, T. and Rajesh, R. (2017). Effect of punch radius and sheet thickness on spring-back in V-die bending, *Advances in Natural and Applied Sciences*, 11(8): 178–183.

Kędzierski, P. and Popławski, A. (2019). Development of test device for springback study in sheet metal bending operation, *AIP Conference Proceedings*, 2078: 020016.

Kut, S., Stachowicz, F. and Pasowicz, G. (2021). Springback prediction for pure moment bending of aluminum alloy square tube, *Materials (Basel, Switzerland)*, 14(14): 3814.

Laurent, H., Grèze, R., Manach, P. Y. and Thuillier, S. (2009). Influence of constitutive model in springback prediction using the split-ring test, *International Journal of Mechanical Sciences*, 51: 233–245.

Lee, S. W. and Kim, Y. T. (2007). A study on the springback in the sheet metal flange drawing, *Journal of Materials Processing Technology*, 187–188: 89–93.

Li, K. P., Carden, W. P. and Wagoner, R. H. (2002a). Simulation of springback, *International Journal of Mechanical Sciences*, 44(1): 103–122.

Li, X., Yang, Y., Wang, Y., Bao, J. and Li, S. (2002b). Effect of the material-hardening mode on the springback simulation accuracy of V-free bending, *Journal of Materials Processing Technology,* 123: 209–211.

Li, H., Sun, G., Li, G., Gong, Z., Liu, D. and L, Q. (2011). On twist springback in advanced high-strength steels, *Materials and Design*, 32(6): 3272.

Li, G., He, Z, Ma, J., Yang, H. and Li, H. (2021). Springback analysis for warm bending of titanium tube based on coupled thermal-mechanical simulation, *Materials (Basel)*, 14(17): 5044.

Liu, G., Lin, Z. Q., Xu, W. L and Bao, Y. X. (2002). Variable blankholder force in U-shaped part forming for eliminating springback error, *Journal of Materials Processing Technology*, 120(1–3): 259–264.

Liu, X., Zhang, M., Huo, X., Ge, X. and Liu, B. (2013). Research on the spring-back compensation method of tailor welded blank U-shaped part, *Applied Mechanics and Materials,* 373–375: 1970–1974.

Liu, X., Du, Y., Lu, X. and Zhao, S. (2019). Springback measurement in micro W-bending, *Proceedings Volume 11343, Ninth International Symposium on Precision Mechanical Measurements,* 1134321. Chongqing, China.

Ling, J. S., Abdullah, A. B. and Samad, Z. (2016). Application of Taguchi method for predicting Springback in V-bending of aluminum alloy AA5052 strip. *Journal of Scientific Research and Development*, 3(7): 91–97.

Livatyali, H. and Altan, T. (2001). Prediction and elimination of springback in straight flanging using computer aided design methods: Part 1. Experimental investigations, *Journal of Materials Processing Technology*, 117(1–2): 262–268.

Livatyali, H., Kinzel, G. L. and Altan, T. (2003). Computer aided die design of straight flanging using approximate numerical analysis, *Journal of Materials Processing Technology*, 142(2): 532–543.

Lundh, H., Bustad, P. A., Carlsson, B., Engberg, G., Gustafsson, L. and Lidgren, R. (1998). Sheet steel forming handbook: Size shearing and plastic forming. G'oteborg, Sweden: SSAB Tunnplat AB.

Ma, R., Wang, C., Zhai, R. and Zhao, J. (2019). An iterative compensation algorithm for springback control in plane deformation and its application, *Chinese Journal of Mechanical Engineering*, 32: 28.

Marretta, L., Ingarao, G. and Di Lorenzo, R. (2010). Design of sheet stamping operations to control springback and thinning: A multi-objective stochastic optimization approach, *Journal of Materials Processing Technology*, 52(7): 914–927.

Meinders, T., Burchitz, I. A., Bonte, M. H. A. and Lingbeek, R. A. (2008). Numerical product design: Springback prediction, compensation and optimization, *International Journal of Machine Tools and Manufacture,* 48(5): 499–514.

Narkeeran, N. and Lovell, M. (1999). Predicting springback in sheet metal forming: An explicit to implicit sequential solution procedure, *Finite Elements in Analysis and Design,* 33(1): 29–42.

Nashrudin, M. N. and Abdullah, AB. (2017). Finite Element Simulation of Twist Forming Process to Study Twist Springback Pattern, *MATEC Web of Conferences* 90, 01026.

Nikhare, C. P. (2016). Parametric study on wall springback in tailor welded long channels. *Proceedings of the ASME 2016 11th International Manufacturing Science and Engineering Conference, Volume 1: Processing,* Blacksburg, Virginia, USA.

Nowosielski, M., Żaba, K., Kita, P. and Kwiatkowski, M. (2013). Compensation of springback effect in designing new pressing technologies, *Proceedings of the 22nd International Conference on Metallurgy and Materials,* 15–17 May, Brno, Czech Republic, EU.

Osman, M. A., Shazly, M., El-Mokaddem, A. and Wifi, A. S. (2010). Springback prediction in V-bending: Modeling and experimentation, *Journal of Achievements in Materials and Manufacturing Engineering,* 38(2): 179–186.

Ouakdi, E. H., Louahdi, R., Khirani, D. and Tabourot, L. (2012). Evaluation of springback under the effect of holding force and die radius in a stretch bending test, *Materials & Design,* 35: 106–112.

Özdemir, M., Dilipak, H. and Bostan, B. (2020). Experimental investigation of deformation and spring-back and Spring-Go amounts of 1.5415 (16MO3) sheet material, *Metallography, Microstructure, and Analysis* 9: 796–806.

Özdemir, M., Dilipak, H. and Bostan, B. (2021). Numerically modelling spring back and spring go amounts and bending deformations of Cr-Mo alloyed sheet material, *Materials Testing,* 62(12): 1265–1272.

Ozdemir, M. (2020). Optimization of Spring Back in Air V Bending Processing Using Taguchi and RSM Method. *Mechanika,* 26(1), 73-81.

Özdin, K., Büyük, E., Abdalov, F., Bayram, H. and Çini, A. (2014). Investigation of Springback and Spring-go of AISI 400 S sheet metal in "V" bending dies depending on bending angle and punch radius, *Applied Mechanics and Materials,* 532: 549–553.

Ozturk, F., Ece, R. E., Polat, N. and Koksal, A. (2010). Effect of warm temperature on springback compensation of titanium sheet. *Materials and Manufacturing Processes,* 25: 1021–1024.

Paak, M., Zoghi, H. and Huhn, S. (2018). A parametric study of springback for compensation strategies, *IOP Conference Series: Materials Science and Engineering,* 418: 012103.

Panda, N. and Pawar, R. S. (2018). Optimization of process parameters affecting on spring-back in V-bending process for high strength low alloy steel HSLA 420 using FEA (HyperForm) and Taguchi Technique, *International Journal of Aerospace and Mechanical Engineering,* 12(1): 28–34.

Paulsen, F. and Welo, T. (2002). A design method for prediction of dimensions of rectangular hollow sections formed in stretch bending, *Journal of Materials Processing Technology,* 128(1–3): 48–66.

Phanitwong, W., Komolruji, P. and Thipprakmas, S. (201). Finite element analysis of springback characteristics on asymmetrical Z-shape parts in Wiping Z-bending process. In *Proceedings of the 6th International Conference on Simulation and Modeling Methodologies, Technologies and Applications,* pp. 225–230. Setubal, Portugal.

Ragai, I., Lazim, D. and Nemes, J. A. (2005). Anisotropy and springback in draw-bending of stainless steel 410: Experimental and numerical study, *Journal of Materials Processing Technology,* 166(1): 116–127.

Ramezani, M., Ripin, Z. M. and Ahmad, R. (2010). Modelling of kinetic friction in V-bending of ultra-high-strength steel sheets, *The International Journal of Advanced Manufacturing Technology,* 46(1–4): 101–110.

Rao, D. B. and Narayanan, G. (2014). Springback of friction stir welded sheets made of aluminium grades during V-bending: An experimental study. *ISRN Mechanical Engineering*, 1–15. https://doi.org/10.1155/2014/681910.

Rigas, N., Junker, F., Berendt, E. and Merklein, M. (2019). Investigation of dry lubrication systems for lightweight materials in hot forming processes. In: Wulfsberg, J. P., Hintze, W. and Behrens, B. A. (eds), *Production at the Leading Edge of Technology*, pp. 43–51. Springer Vieweg: Berlin, Heidelberg.

Saito, N., Fukahori, M., Hisano, D., Hamasaki, H. and Yoshida, F. (2017). Effects of temperature, forming speed and stress relaxation on springback in warm forming of high strength steel sheet, *Procedia Engineering*, 207: 2394–2398.

Samuel, M. (2000). Experimental and numerical prediction of springback and side wall curl in U-bendings of anisotropic sheet metals, *Journal of Materials Processing Technology*, 105: 382–393.

Schwarze, M., Vladimirov, I. N. and Reese, S. (2011). Sheet metal forming and springback simulation by means of a new reduced integration solid-shell finite element technology, *Computer Methods in Applied Mechanics and Engineering*, 200(5–8): 454–476.

Sedeh, M. and Ghaei, A. (2021). The effects of machining residual stresses on springback in deformation machining bending mode. *The International Journal of Advanced Manufacturing Technology*, 114. https://doi.org/10.1007/s00170-021-06816-x.

Sen, H., Yilmaz, S. and Yildiz, R. A. (2020). Springback behavior of DP600 steel: An implicit finite element simulation, *Journal of Engineering Research*, 8(2): 252–264.

Shen, H. Q., Li, S. H. and Chen, G. L. (2010). Numerical analysis of panels dent resistance considering the Bauschinger effect, *Materials and Design*, 31(2): 870–876.

Shen, L., Qiu, C., Wu, X. and Liu, Y. (2020). Thick plate bending and image processing algorithms assistant for deformation process discretization analyses. *Mechanics*, 26(5): 383–389.

Siswanto, W. A., Anggono, A. D., Omar, B. and Jusoff, K. (2014). An alternate method to springback compensation for sheet metal forming, *The Scientific World Journal*, 2014: 301271.

Song, J. H., Huh, H. and Kim, S. H. (2007). Stress-based springback reduction of a channel shaped auto-body part with high-strength steel using response surface methodology, *Journal of Engineering Materials and Technology, Transactions of the ASME*, 129(3): 397–406.

Spathopoulos, S. and Stavroulakis, G. (2020). Springback prediction in sheet metal forming, based on finite element analysis and artificial neural network approach. *Applied Mechanics*, 1: 97–110. https://doi.org/10.3390/applmech1020007.

Su, S., Jiang, Y. and Xiong, Y. (2020). Multi-point forming springback compensation control of two-dimensional hull plate, *Advances in Mechanical Engineering*, 12(4): 1–12.

Sun, L. and Wagoner, R. H. (2011). Complex unloading behavior: Nature of the deformation and its consistent constitutive representation, *International Journal of Plasticity*, 27(7): 1126–1144.

Thipprakmas, S. (2010a). Finite element analysis of punch height effect on V-bending angle, *Materials and Design*, 31(3): 1593–1598.

Thipprakmas, S. (2010b). Finite element analysis on v-die bending process, In: Moratal, D. (ed.), *Finite Element Analysis*. IntechOpen, Rijeka, Croatia.

Thipprakma, S. and Rojananan, S. (2008). Investigation of spring-go phenomenon using finite element method, *Materials and Design*, 29: 1526–1532.

Thipprakmas, S. and Phanitwong, W. (2011). Process parameter design of springback and spring-go in V bending process using Taguchi technique, *Materials and Design*, 32: 4430–4436.

Thuillier, S., Manach, P. Y., Menezes, L. F. and Oliveira, M. C. (2002). Experimental and numerical study of reverse re-drawing of anisotropic sheet metals, *Journal of Materials Processing Technology*, 125–126: 764–771.

Tomáš, P., Michal, V. and František, T. (2017). Methodology of the springback compensation. *In Sheet Metal Stamping Processes, Metal 2017,* May 24th -26th, Brno, Czech Republic, EU.

Trzepiecinki, T. and Lemu, H.G. (2017). Effect of computational parameters on Springback prediction by numerical simulation. *Metals,* 7, 380. https://doi.org/10.3390/met7090380

Troive, L., Bałon, P., Świątoniowski, A., Mueller, T. and Kiełbasa, B. (2018). Springback compensation for a vehicle's steel body panel, *International Journal of Computer Integrated Manufacturing,* 31(2): 152–163.

Vladimirov, I. N., Pietryga, M. P. and Reese, S. (2009). Prediction of springback in sheet forming by a new finite strain model with nonlinear kinematic and isotropic hardening, *Journal of Materials Processing Technology,* 209(8): 4062–4075.

Wan-Nawang, W-A., Qin, Y. and Liu, X. (2015). An experimental study on the springback in bending of w-shaped micro sheet-metal parts. *MATEC Web of Conferences,* 21. https://doi.org/10.1051/matecconf/20152109015

Wang, A., Zhong, K., El Fakir, O., Liu J., Sun, C., Wang L., Lin, J. and Dean, T. A. (2107). Springback analysis of AA5754 after hot stamping: experiments and FE modelling. *The International Journal of Advanced Manufacturing Technology,* 89: 1339–1352. https://doi.org/10.1007/s00170-016-9166-3

Wang, H., Zhou, J., Zhao, T. and Tao, Y. (2016a). Pringback compensation of automotive panel based on three-dimensional scanning and reverse engineering, *The International Journal of Advanced Manufacturing Technology,* 85: 1187–1193.

Wang, H., Zhou, J., Zhao, T. S., Liu, L. Z. and Liang, Q. (2016b). Multiple-iteration springback compensation of tailor welded blanks during stamping forming process, *Materials and Design,* 102: 247–254.

Wang, R., Zhai, F., Wu, B. and Ji, Y. (2017). The research of springback compensation design method based on reverse modelling, *AIP Conference Proceedings,* 1864: 020203.

Weiher, J., Rietman, B., Kose, K., Ohnimus, S. and Petzoldt, M. (2004). Controlling springback with compensation strategies, *AIP Conference Proceedings,* 712, 1011.

Yoshida, T., Sato, K., Isogai, E. and Hashimoto, K. (2013). Springback problems in forming of high-strength steel sheets and countermeasures. *Nippon Steel Technical Report,* 4–10.

Yu, H. Y. (2009). Variation of elastic modulus during plastic deformation and its influence on springback, *Materials and Design,* 30(3): 846–850.

Yu, Z. Q. and Lin, Z. Q. (2007). Numerical analysis of dimension precision of U-shaped aluminium profile rotary stretch bending, *Transactions of Nonferrous Metals Society of China,* 17(3): 581–585.

Zhan, L., Lin, J. and Dean, T. A. (2011). A review of the development of creep age forming: Experimentation, modelling and applications, *International Journal of Machine Tools and Manufacture,* 51(1): 1–17.

Zhang, D. J., Cui, Z. S., Ruan, X. Y. and Li, Y. Q. (2007a). An analytical model for predicting springback and side wall curl of sheet after U-bending, *Computational Materials Science,* 38(4): 707–715.

Zhang, D. J., Cui, Z. S. and Ruan, X. Y. (2007b). An analytical model for predicting sheet springback after V-bending, *Journal of Zhejiang University-Science A,* 8(2): 237–244.

Zhang, Y., Nguyen, H. P. and Jung, D. W. (2016a). Optimization of the spring-back in roll forming process with finite element simulation, *International Journal of Mechanical Engineering and Robotics,* 5(4): 272–275.

Zhang, Z., Zhang, H., Yi, S., Moser, N., Ren, H., Kornel, F. and Ehmann, C. J. (2016b). Springback reduction by annealing for incremental sheet forming, *Procedia Manufacturing,* 5: 696–706.

# 3 Springback of Aluminium and Steel Joint by Friction Stir Welding – A Review

*MZ Rizlan, AB Abdullah, and Z Hussain*
Universiti Sains Malaysia

## CONTENTS

## 3.1   INTRODUCTION

Springback is a common issue in sheet metal forming. The phenomenon occurs due to elastic recovery when the load is removed from the workpiece (da Silva et al., 2016). The elastic recovery is the result of the release of internal elastic stresses within the sheets (Rao and Narayanan, 2014). Forming processes involve the stretching and bending of the metal sheet to obtain the desired form. During bending, one

DOI: 10.1201/9781003164241-3

side of the sheet metal is subjected to compressive stresses, while the other side experiences tensile stresses. Once the load is released, elastic recovery occurs due to elastic stresses from the bending process. This elastic recovery is called springback (Katre et al., 2015).

Springback is a problem in sheet forming which could deteriorate the final part quality due to dimensional changes (Rao and Narayanan, 2014). It causes undesirable geometrical and dimensional changes in the components formed (Gautam and Kumar, 2018). Springback typically degrades the quality and appearance of the products, such as increased variability and tolerances in the subsequent forming operations, assembly, and final parts (Wagoner et al., 2013). In pure bending of homogeneous beams with rectangular cross-section under plane strain conditions, the springback is proportional to the yield stress and inversely proportional to Young's modulus and thickness (Darmawan et al., 2019). Figure 3.1 shows the springback in sheet metal forming.

Springback can be evaluated by unconstrained bending test (Park et al., 2008). Typical methods to study springback responses are V-bending or U-bending tests (Katre et al., 2015). Springback can also be measured by using mechanical bending tests such as three-point air bending, where the specimen is subjected to bending by a cylindrical punch (da Silva et al., 2016). In air bending, the specimen is subjected to sheet metal forming until the predetermined angle is achieved. After a certain loading time, the load is removed, and the bend angle is measured. Springback is the difference between the predetermined bending angle and the bend angle after unloading. Figure 3.2 shows the springback angle.

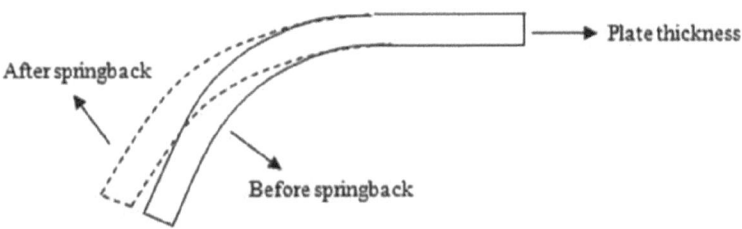

**FIGURE 3.1**    Springback in sheet metal forming.

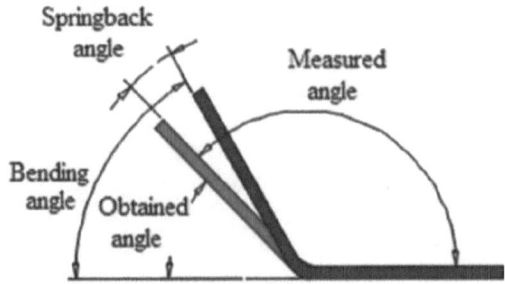

**FIGURE 3.2**    Springback angle.

## 3.2   FACTORS AFFECTING SPRINGBACK

There are various factors that affect springback. Some of these factors are sheet thickness, punch angle, punch radius, punch height, die opening, ratio of die radius to sheet thickness, coining force, prebend conditions of strips, and sheet metal grain direction (Chikalthankar et al., 2014).

Kazan et al. (2009) have studied the effect of the $R/t$ ratio of die radius, $R$, to blank thickness, $t$, on springback angle. Two groups of finite element (FE) simulation models were generated with identical $R/t$ ratios ranging from 1.0 to 5.0. Group G1 has set the blank thickness at a constant of 0.7 mm and the die radius ranging from 0.7 to 5.0 mm. The other group, G2, has set the die radius at a constant of 5.0 mm and blank thicknesses ranging from 1.0 to 5.0 mm. High-strength low-alloy (HSLA) steel was used in the simulation. The springback angles were obtained from the simulation as a response to the predetermined blank thickness and die radius data. The plotted graph of springback angle against $R/t$ ratio is shown in Figure 3.3. It can be observed that after a certain point, the springback angle increases as the $R/t$ ratio increases.

The effect of punch height on the bending angle in V-bending was investigated by Thipprakmas (2010) by using FE analysis. The FE simulation results showed that the effect of punch height on the bending angle was theoretically derived based on material flow analysis and stress distribution. If the punch height is too small, a large gap would be formed between the workpiece and the die as well as in the small reversed bending zone. The effect of stress distribution on the bending allowance zone was not sufficient to make the workpiece open toward the die side. Thus, the obtained angle would be smaller than the required angle.

Similarly, if the punch height is too large, a large, reversed bending zone and no gap between workpiece and the die would form, creating a spring-go effect. This means that the obtained angle would also be smaller than the required angle. However, if the stress distribution on the reversed bending zone is not suppressed, it will cause springback to occur. Thus, the obtained angle is larger than the required angle. Therefore, the balance between compensating the gap between the workpiece and the die as well as stress distribution on bending allowance and the reversed bending zone are crucial to obtain the required bending angle.

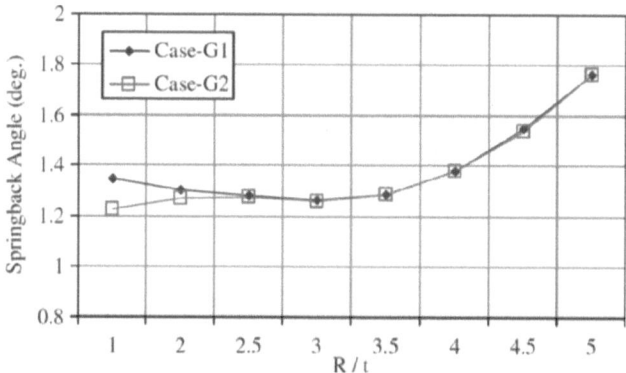

**FIGURE 3.3**   Graph of the springback angle versus $R/t$ ratio. (With permission from Kazan et al., 2009.)

Huang and Leu (1998) have studied the effect of process variables on V-die bending of steel sheets. They concluded that the punch load increases when any one process variable of lubrication, punch radius, or punch speed increases or the strain hardening exponent decreases. When the die radius increases, the punch load initially increases and then decreases. The punch load would slightly decrease as the normal anisotropy increases. Variations in punch widths did not affect the punch load, and the angle of springback was found to be greater for tools with a larger die radius.

Chan et al. (2004) have performed FE analysis of springback in V-bending of the sheet metal forming process. Springback angles of workpieces were studied for different punch angles, punch radii, and die-lip radii. They concluded that springback decreases as the punch angle and punch radius increase. In addition, there is also an optimum punch radius in order to achieve minimum springback.

Thipprakmas and Phanitwong (2011) have used a combination of the finite element method (FEM), the Taguchi technique, and analysis of variance (ANOVA) to determine the most significant parameters between bending angle, material thickness, and punch radius in the V-bending process. It was found that the degree of importance of parameters depends on springback or spring-go. Material thickness has a significant influence in springback, while bending angle is the most influential parameter in spring-go, followed by material thickness.

Gautam et al. (2012) conducted an experimental study to determine the effect of punch corner radii on springback in the V-bending process. In order to compensate for springback, compressive load is applied between punch and the die so that there will be localized compressive stresses on bend curvature. The results of the experiments indicated that the increase in punch corner radius increases the springback. However, the increase in sheet thickness reduces the springback. Springback compensation by localized compressive stress resulted in negligible springback, and this finding was supported with FE analysis simulations. Papeleux and Ponthot (2002) have used FE simulation as well as experiments to study the springback in sheet metal forming. They reported that the springback increases as the sheet thickness increases. Comparison between the FEM simulation and experimental results showed a similar trend.

Chikalthankar et al. (2014) investigated the effect of punch angle, die opening, grain direction of the sheet metal, and prebend strip conditions on springback of deep drawn steel. It was found that punch angle is a significant parameter that affects springback. A lower punch angle creates a lower plastic zone, which results in higher springback. In terms of die opening, it was observed that the supporting flange area increases with increasing die opening, which caused the reduction of springback. Grain direction of sheet metal is associated with yield strength. Springback increases with variation of grain direction from 0° to 90°. Another important factor is the condition of the metal sheet. Springback was found to be less for flat sheets as compared to prebend sheets. This can be due to the impact of residual stresses on springback in prebend sheets as compared to flat sheets.

## 3.3  SPRINGBACK COMPENSATION

Springback can be controlled by increasing sheet tensions to reduce springback and/ or by compensating the tooling shape to achieve final target shape after springback.

Increasing sheet tension reduces the stress gradient through the sheet thickness, which decreases the bending moment and total springback. The variability from material behaviour changes is reduced as well (Wagoner et al., 2013). However, increasing sheet tension tends to cause sheet splitting, especially among newer high-strength materials which typically have lower formability. This problem can be addressed by using die compensation but requires accurate prediction and measurement of springback. However, this technique may not reduce springback scatter caused by typical process and material variations (Wagoner et al., 2013).

## 3.4   FSW OF AL-STEEL

Aluminium and steel joint enables the production of miscellaneous components and structures with hybrid properties. However, the joining of dissimilar materials, particularly aluminium and steel, is problematic due to their huge differences in properties, which are extremely difficult to join using conventional fusion welding methods. However, solid-state welding techniques, such as friction stir welding (FSW), have been shown to successfully join aluminium and steel together.

The main benefit of aluminium alloy and steel joint is the weight reduction, which is a significant factor in energy efficiency, especially for the transportation industry (Ramachandran et al., 2015). To address the demanding criteria of fuel economy standards, the two main trends in the automotive industry are the utilization of high-strength steel for structural body elements such as A- and B-pillars and lightweight aluminium alloys for body components (Tisza and Czinege, 2018). The benefits of aluminium are a lower density, high specific strength, a low crack propagation rate, and high resistance to impact and fatigue loading (Tariq et al., 2019). Aluminium alloy components can be made lighter without strength deterioration.

FSW has been used by Honda Motor Corporation to join cast aluminium and stamped steel for engine parts and C-frames in the production of the Honda Accord 2013 model (Arya et al., 2016). A 25% reduction in body weight was achieved by using FSW components, and a 50% decrease in energy consumption was obtained as compared to utilizing traditional steel subframes. Similarly, Mazda Motor Corporation has also adopted FSW to weld aluminium alloys and steel in the form of direct friction stir spot joining technology to produce the trunk lid of the Mazda MX-5.

The shipbuilding industry can also take advantage of aluminium and steel FSW joint (Corigliano et al., 2018). Current mechanical methods of joining aluminium to steel by bolting and riveting can suffer corrosion due to the presence of crevasses and galvanic potential differences between the two materials.

## 3.5   DIFFICULTY IN AL-STEEL FSW

Joining aluminium alloys using conventional welding methods comes with several problems such as a poor solidification microstructure, high volume fraction of porosity in the fusion zone, and potential segregation of alloying elements. These unwanted issues give rise to significant loss in mechanical properties (Raja et al., 2016). One main drawback of aluminium alloys is their poor weldability, which is caused by lower specific resistivity and lack of formability (Noh et al., 2018). Joining of 5xxx series Al alloy to steel joining is more sensitive to parameters due to the presence of Mg, and formation of thicker IMC is to be expected (Ramachandran et al., 2015).

When using fusion welding, an aluminium oxide layer is formed on the surface of aluminium as soon as it is exposed to air. This oxide layer has to be removed mechanically or by cathodic cleaning to obtain defect-free joints (Çam and İpekoğlu, 2017). If the aluminium oxide layer is not removed, oxide fragments can be trapped in the weld and cause ductility reduction, lack of fusion, and cracks. The oxide layer problem can be minimized during welding by shielding the joint area with non-oxidizing gases such as argon, helium, or hydrogen. Due to high thermal conductivity of aluminium, heat must be supplied at a higher rate to ensure an equal amount of local temperature rise. However, as compared to steel, aluminium has twice the coefficient of linear thermal expansion, which means that to avoid crack formation, heat input must be kept to a minimum. Furthermore, the melting temperature of aluminium alloys is notably lower than that of steel. Hence, the amount of heat required to join aluminium would be much lower compared to steel.

According to Geng et al. (2019), it is difficult to obtain a good weld joint for aluminium alloys using fusion welding owing to the occurrence of solidification cracks, coarse grain porosities, and slag inclusions. In conventional fusion welding of aluminium alloys, dissolution and growth of strengthening precipitates in the weld occur during the thermal cycle. As stated by Salih et al. (2015), problems in aluminium alloy fusion welding include loss of strength, defect formation, and trapped porosity in the cross-section because of shielding gas dissolution or electrode and flux moisture in the molten metal. Lack of fusion in part occurs because of high melting temperatures of up to 2060°C of stable aluminium oxide on the surface. Other than that, stress induced by metal contraction in cooling and differences between the liquidus temperature and eutectic or final solidification temperature generate centre line or solidification cracking. Moreover, heating and cooling cycle variation in HAZ leads to a lower strength joint in heat-treatable alloys.

Similarly, there are several issues in the fusion welding of steel. For example, joint porosity can form when the shielding gas is trapped within the weld, creating tiny holes which weaken joint strength. Excessive heat during fusion welding can warp the parent metal and result in distortion. In addition, distortion can also happen when fabricating long welds since the metal is exposed to a prolonged period of heating. Other defects include the lack of fusion when the parent metal does not fully stick to the weld metal. This is generally caused by poor welding techniques such as excessive heat, incorrect angles, or insufficient arc lengths.

Wang et al. (2017) remarked that when dealing with welding processes involving high heat input such as joining aluminium and steel, severe grain growth results in composition segregation, distortion, and cracks within the weld joint. Although tungsten inert gas (TIG) and metal inert gas (MIG) welding can control the heat input during the welding process, the quality of the welding could still be affected by the large difference in properties between aluminium and steel. The large difference in thermo-mechanical properties and strong affinity to form brittle IMC are the main reasons why it is difficult to joint aluminium alloys and steel by fusion welding (Ramachandran et al., 2015). In the fusion zone, intermetallic compounds are formed, and this is especially difficult in joining dissimilar materials (Ambroziak et al., 2014).

Thus, traditional fusion processes such as arc welding and laser welding are unsuitable for aluminium and steel joining. Solid-state welding processes, such as FSW, can be suitable techniques which can join dissimilar materials that are almost impossible to be join by traditional techniques. FSW can create a strong bond without additional weight as it does not require filler metals or flux.

## 3.6   OPTIMUM PARAMETERS

In FSW of Al-steel, the process parameters must be determined for new combinations of material pairs. FSW parameters should be chosen so that enough heat and stirring are generated to properly plasticize metals and to ensure deep intermixing of materials (Safeen and Spena, 2019).

### 3.6.1   FSW TOOL

Various tool geometries and materials have been used by researchers in FSW of aluminium and steel. For example, tungsten rhenium alloy W25Re tool was used by Campanella et al. (2016) in FSW of AA6016 and DC05 low-carbon steel as it was already proven effective in FSW of titanium as well as to reduce tool pin wear. Zheng et al. (2016) observed that no severe wear occurred after a prolonged use of tungsten carbide tool in FSW of 6061 aluminium alloy and 306 stainless steel.

To minimize tool wear in FSW of AA6061 and DC05 low-carbon steel, Safeen et al. (2020) have chosen FSW tool made of tungsten rhenium with conical pin as the material has proven to be effective in FSW of steel and other wear-resistant materials. Ibrahim et al. (2018) in their study on friction stir diffusion cladding (FSDC) of ASTM A516-70 steel and 5052 aluminium alloy used a tool made of H13 tool steel with concentric circular grooved pin features. The features help to increase the area of surface contact, therefore improving frictional heat generation and material flow.

### 3.6.2   WELDING PARAMETERS

Numerous studies have evaluated the process parameters for FSW. As mentioned by Mishra et al. (2014) in their review paper, tool offset, pin offset, and position of dissimilar alloys affect the maximum temperature and heat distribution. These parameters require careful consideration to ensure appropriate and defect-free joints as well as proper heat generation and distribution. In the FSW study of dissimilar Al 6061-T6 to ultra-low-carbon steel, Boumerzoug and Helal (2017) have identified that the main factors in the FSW process are the tool rotational speed, traverse speed, and vertical pressure on plates. In another study, Ramachandran et al. (2015) investigated the effect of tool axis offset and FSW tool geometry on the mechanical and metallographic characteristics of dissimilar butt welding of aluminium alloys, AA 5052 H32 and HSLA steel, IRSM42-93.

Dehghani et al. (2013) have studied the effect of IMC formation, tunnel formation, and tensile strength in aluminium alloys and mild steel joints using various parameters such as traverse speed, plunge depth, tilt angle, and tool geometry. Meanwhile, Pasha et al. (2014) have evaluated the influence of main variables of the FSW process

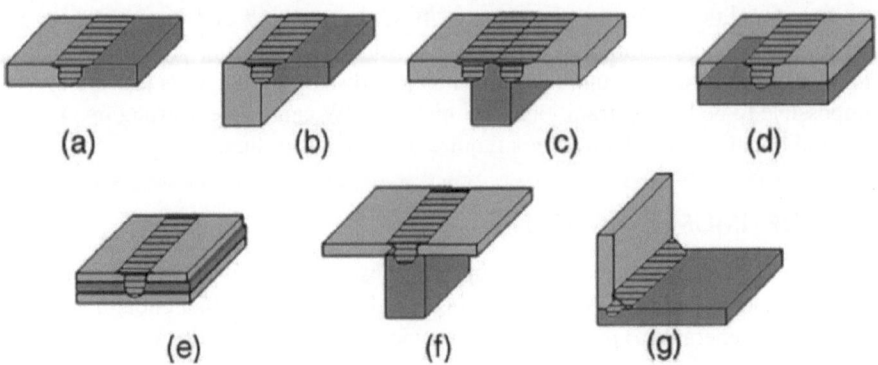

**FIGURE 3.4** Joint configurations for FSW: (a) square butt, (b) edge butt, (c) T butt joint, (d) lap joint, (e) multiple lap joint, (f) T lap joint, and (g) fillet joint.

such as welding (traverse) speed, tool rotational speed, vertical pressure on tool (axial pressure), tool tilt angle, ratio of D/d, and tool design that govern the peak temperature, x-direction force, torque, and process power feature.

Two principal FSW parameters as defined by Kalemba-Rec et al. (2018) are tool rotational speed and welding speed (traverse speed). Other parameters include tool shape and dimensions (pin diameter, pin length, shoulder diameter, shoulder concavity angle), tool tilt angle, plunge depth, vertical force, thickness of welded plates, weld configuration, alloy composition, and its initial temper. Desai and Inamdar (2017) stated that the heat input in the welding process is controlled by rotational speed and welding speed. Mahto et al. (2016) suggested that mechanical properties are influenced by traverse speed and plunge depth.

### 3.6.3 JOINT CONFIGURATION

In FSW, lap and butt joints are the two possible and the most frequently used configurations (Gullino et al., 2019). Possible joint arrangements for FSW are shown in Figure 3.4.

### 3.6.4 MATERIAL PLACEMENT

Material placement in FSW is defined as on the advancing side (AS) and retreating side (RS). As shown in Figure 3.5, the material is in the advancing side when the tool rotation direction is the same as the welding direction. However, for the retreating side, the tool rotation direction is opposite to the welding direction. According to Barbini et al. (2018), plate position is vital for the finishing joint microstructure when the selected mixtures of base materials (BMs) have major differences in mechanical properties. Tool rotational speed and alloy placement were found to greatly influence weld formation, especially in the stir zone (Kalemba-Rec et al., 2018). It was also found that mechanical behaviour and hardness were not affected by weld configuration.

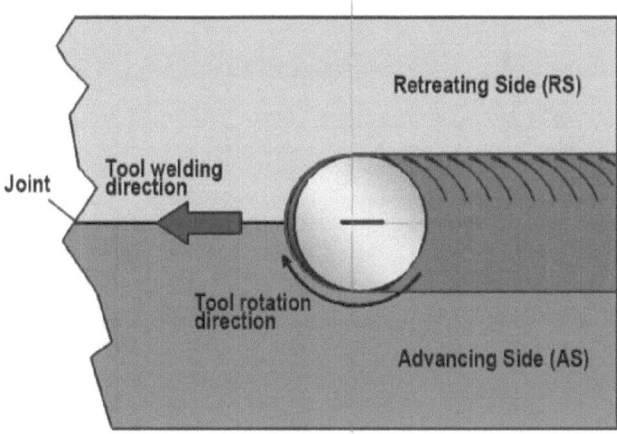

**FIGURE 3.5**   Material placement.

## 3.7   POST-FORMING EVALUATION

Post-forming weld joint evaluations have been done to evaluate the formability and manufacturability of the work piece. Formability is the ability of the metal work piece to undergo plastic deformation without damage (Pavan Kumar et al., 2018). Some of the well-known evaluation methods are bending test, twist forming, and limit dome height (LDH) test.

### 3.7.1   Bending Test

Bending tests such as U-bending and V-bending are used to obtain the springback value. V-shaped dies are frequently used in V-bending because they are simpler and cheaper to manufacture. During the bending process, the bending tool (punch in V-bending) forced the work piece to take the radius $R_b$ and included angle $A_b'$. After the punch is removed, the work piece springs back to radius R and angle A' due to elastic properties of the materials. Parameters involved in V-bending test include bending die angle, holding time, and punch radius. Illustration of V-bending test is shown in Figure 3.6.

### 3.7.2   Twist Forming

In twist forming, the centre of the material strip is fixed, and the twisting moment or torque is then applied at both ends of the material strip. Figure 3.7 shows the twist forming process, in which the material undergoes simultaneous twisting and bending. In this process, twist springback can also manifest upon unloading, in which the twisted material tries to return to its original shape due to elastic recovery.

The value of springback angle is determined from the change in bending angle. Figure 3.8 shows the change in angle before and after springback.

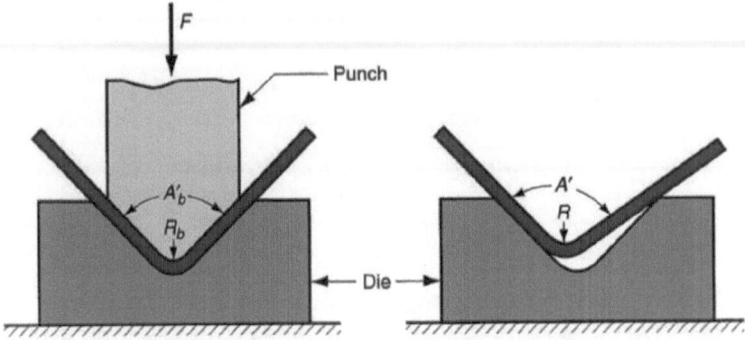

**FIGURE 3.6**   V-bending test.

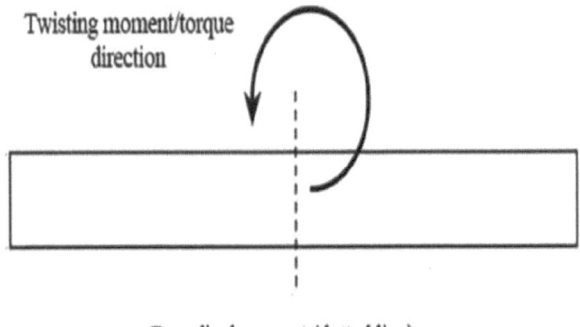

**FIGURE 3.7**   Twist forming process.

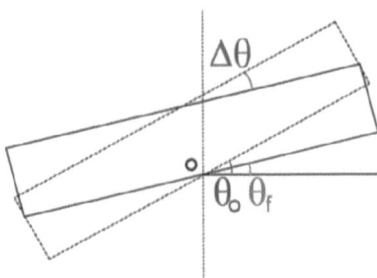

**FIGURE 3.8**   Angle before, $\theta_f$, and after springback, $\theta_o$.

Nashrudin and Abdullah (2016) performed FE simulation of mild steel twist forming using different parameter values of twist angle, hardening constant, and material thickness. The result shows that when thicker mild steel strips were used, the springback angle was reduced due to the increased plastic zone. The springback angle was reduced by increasing the angle of twist. In addition, the torque decreases as the angle of twist increases, resulting in a lower springback angle. Abdullah and Samad (2014) performed twist forming on AA6061-T6 with 1.5 mm thickness by using a torsion test machine. The results show that reducing the strip thickness would decrease the springback angle. Similarly, springback angle reduction was also observed when the workpiece width was decreased.

### 3.7.3 Limiting Dome Height Test

In LDH test, the biaxial stretch forming mode is used to evaluate formability (Noh et al., 2018). LDH simulates common failure strains in the sheet metal forming process due to necking or fracturing (Srinivasu et al., 2017). During LDH test, the material can sustain strains as long as it lies underneath the forming limit curve without failure (Panich and Uthaisangsuk, 2011). As the punch moves into the test specimen, strain concentration would be developed in the weld region (Noh et al., 2018). The failure location and LDH values correspond to the hardness and local properties across the weld zone (Noh et al., 2018).

To measure stretchability, the height of dome at maximum load (near failure) is measured (Srinivasu et al., 2017). LDH test can also be conducted with the aid of ARAMIS for real time recording of deformation by the digital image correlation (DIC) method (Noh et al., 2018). In the evaluation, initial fracture was initiated in weld area bordering the Al6061 sheet. The maximum site of major strain almost coincides with the fracture initiation point. Local properties of the weld region can be used to predict the region showing onset of fracture since fracture initiation in LDH test is at the site of the hardening curve of the weld region with the most degradation. This can be analytically characterized based on local deformation data obtained by the DIC method.

In the LDH test performed by Panich and Uthaisangsuk (2011), principal strains $\phi_1$ and $\phi_2$ are measured for different sample dimensions. The principal strains were then plotted against each other and connected to form a curve. This curve is called the forming limit curve, which is a strain-based failure criterion. The forming limit curve shows the transition from safe material behaviour to material failure. In order to acquire major and minor principal strains, the electrochemical etching process is used to generate circular grid patterns 2.5 mm in diameter on the blank sheet surface. Deformations of circular grids in safe, neck, and crack locations on specimens after forming were then accurately measured using the image analysis programme using a microscope.

LDH test was performed following the ASTM E2218 standard by Noh et al. (2018). The test was carried out on $200 \times 200\,mm^2$ sheets with a punch radius of 50.0 mm and a speed of 0.1 mm/s. The same method was also used by Panich and Uthaisangsuk (2011) since it is the standard test method to determine the forming limit curve. The ASTM E 2218-02 is a test equipped with hemispherical punches 100 mm in diameter. The standard LDH test is shown in Figure 3.9.

The forming limit diagram (FLD), also known as the forming limit curve (FLC), can be used to optimize sheet metal forming (Pavan Kumar et al., 2018). FLD is a strain base criterion which evaluates principal strain at failure and is often used

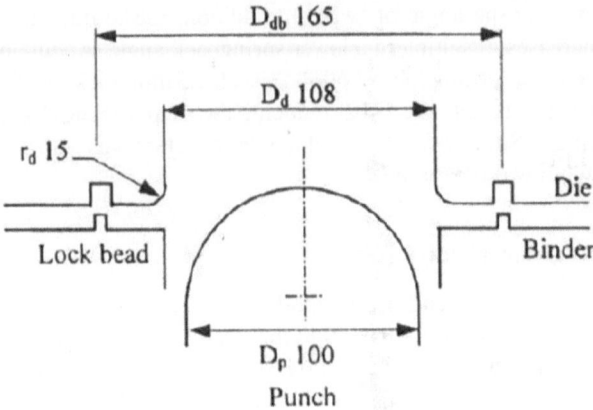

**FIGURE 3.9**   LDH test (see Panich and Uthaisangsuk, 2011).

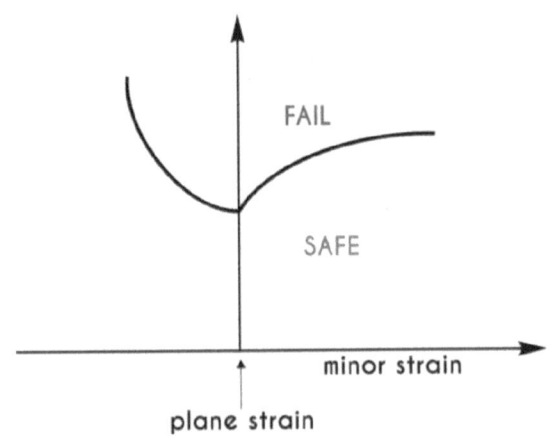

**FIGURE 3.10**   Forming limit diagram.

as a failure criterion for formability prediction (Panich and Uthaisangsuk, 2011). Figure 3.10 shows the FLD to show the safe and failure zones.

In the study by Panich and Uthaisangsuk (2011), the ABAQUS FE programme was used to perform numerical simulations of LDH test in order to determine the forming limit stress diagram, FLSD, by considering both isotropic and anisotropic Hill's yield criteria. In the study by Noh et al. (2018), FE simulation of LDH test was conducted by incorporating the Yld2000-2d anisotropic yield constitutive model for base sheets. To obtain FLC numerically, the Marciniak–Kuczynski (MK) model was used by considering local properties of the weld region. The measured strain value was used to validate the model which showed little difference from the calculated values at the onset of fracture.

## 3.8    SPRINGBACK IN FSW OF AL-STEEL

### 3.8.1    FACTORS AFFECTING SPRINGBACK IN AL-STEEL FSW

In a study of FSW of Al 6061 T6–Al 5052 H32, a decrease in springback was observed for high rotation and welding speeds (Katre et al., 2015). The decreasing trend of springback correlated with modifications in tensile properties such as yield strength to elastic modulus ratio as well as strain hardening exponent of the weld zone with different FSW conditions. Similar observations were also reported by Rao and Narayanan (2014) in their springback study on FSW of dissimilar Al 6061 T6–Al 5052 H32 and similar Al 6061 T6–Al 6061 T6. It was found that springback of friction stir welded sheets reduced with an increase in shoulder diameter, rotation speed, and welding speed and is independent of materials joined. Other factors affecting springback include sheet material, thickness of the sheet, and die width.

Dies with small radii are recommended when dealing with springback; however, a large radius is preferred when considering the mechanical properties of the bent sheet (Rao and Narayanan, 2014). In both experimental and simulation studies, the springback angles were found to increase with increasing bending time (Rao and Narayanan, 2014). In friction stir welded sheets of Al alloys and DP steel, experimental and FE analyses showed that the springback is more influenced by material properties in unconstrained cylindrical bending test as compared to 2D draw bending test (Park et al., 2008).

Various factors affecting the springback in steel also include parameters such as the tool shape and dimension, material properties, contact friction, sheet anisotropy, and sheet thickness. Bakhshi-Jooybari et al. (2009) studied CK67 anisotropic steel and found that springback decreases when the steel sheet thickness increases in V-bending and U-bending. It was also reported that the springback increases when the punch tip radius increases. In addition, a higher bending angle to the rolling direction results in bigger springback. Therefore, 0° orientation to the rolling direction is recommended in V-bending and U-bending.

The effect of bending angle toward springback discussed by Bakhshi-Jooybari et al. (2009) is also supported by Rahmani et al. (2009), which added that an increase in bending orientation results in increased yield strength. Darmawan et al. (2019) also found that apart from bend angle, the sheet blank thickness has a significant effect on the springback in their work on tailor welded blanks (TWBs) of mild steel made of different welding methods. In another study by Chang et al. (2002), springback decreases in TWBs of SCP1 carbon steel with the weld line oriented in the longitudinal direction in U-bending.

According to Gautam and Kumar (2018), springback is greatly affected by weld zone properties and the thicker sheets in TWBs made of sheets with different thicknesses. The ratio of yield stress significantly affects the springback in cylindrical bending and draw bending. It was also observed that springback reduces with an increase in blank holding force. da Silva et al. (2016) focused on the dual-phase DP 600 and DP 780 steels used in automotive industry. It was found that materials with smaller grain diameters have higher mechanical strength due to the Hall–Petch phenomenon, which results in greater springback. Greater springback was also present in smaller steel sheets.

Darmawan et al. (2019) have found that TWBs produced by gas tungsten arc welding (GTAW) and shielded metal arc welding (SMAW) showed lower springback than those produced by gas metal arc welding (GMAW). Higher elasticity of the material was also found to result in higher springback.

Kim et al. (2009) studied on the springback prediction of friction stir welded DP590 steel sheets. Permanent softening behaviour during reverse loading is considered in the combined isotropic-kinematic hardening formula to better predict the springback of friction stir welded DP590 steel sheets. The non-quadratic anisotropic yield function, Yld2000-2d, was used under plane stress conditions. 2-D draw bending tests were compared for verification purposes between the simulation and experimental results for unconstrained cylindrical bending.

### 3.8.2 COMPENSATION

To overcome the problem of springback, the final geometry of the sheet has to be predicted as well as an appropriate tool has to be designed to compensate it (Katre et al., 2015). Among the most commonly used methods to control the springback are overbending, bottoming, and stretch bending. Other methods include heat treatment of materials and changing the working temperature of materials and tools to compensate springback (Katre et al., 2015).

Rao and Narayanan (2014) stated that the springback of friction stir welded sheets of Al 6061-Al 5052 and Al 6061 can be minimized by reducing the punch nose radius. Using a punch with a lower nose radius results in larger plastic deformation, decreasing the relative effect of elastic deformation.

Padmanabhan et al. (2008) performed FE simulation to investigate deep drawing of aluminium-steel TWBs. In the FE model, aluminium alloy 6061-T4 was joined with a range of steels such as mild steel (DC 06) and high-strength steel (AISI 1018, HSLA-340 and DP 600). It was concluded that in the deep drawing process, the optimal blank holder force is essential to increase the draw depth and control the springback.

### 3.8.3 FE SIMULATION

Simulation of springback can be performed by using the FE code in ABAQUS software (Bakhshi-Jooybari et al., 2009). Compared to forming simulations, FE simulation of springback is much more sensitive to numerical tolerances and material models (Wagoner et al., 2013). Simulation of springback in unconstrained bending test was conducted by Park et al. (2008) using the ABAQUS/STANDARD implicit code with user-defined subroutine UMAT. Katre et al. (2014), in their FE simulation study, utilized a commercially available elasto-plastic FE code that uses Langrarian and explicit time integration techniques.

### 3.8.4 MODEL USE IN THE SIMULATION

Critical numerical procedures for springback simulation include the spatial integration scheme, element type, and time integration schemes such as implicit/implicit, explicit/implicit, explicit/explicit, and one-step approaches (Wagoner et al., 2013). Significant material representations in springback simulation include the unloading scheme, strain hardening rule, evolution of plastic properties, plastic anisotropy, Bauschinger effect, and anticlastic curvature (Wagoner et al., 2013).

In the study by Park et al. (2008) on the formability and springback of AA5052-H32 sheets made from the surface friction stir method, the FE prediction of springback from tests such as unconstrained cylindrical bending, 2D draw bending, and draw-bend tests of friction stir welded sheets used the combined isotropic-kinematic hardening law base on the modified Chaboche model and Yld2000-2d yield function. For the simulation, the tools were comprised of a four-node three-dimensional rigid body element (R3D4), while a reduced four-node shell element (S4R) with nine integration points through the thickness was used to simulate the blank (Park et al., 2008). The mesh size used was approximately $1.0 \times 1.0\,mm^2$ (Park et al., 2008). The friction coefficient of no lubrication condition was assumed to be 0.17 and insignificant to the springback simulation (Park et al., 2008).

Katre et al. (2014) have studied the springback of friction stir welded sheets of Al 5052-H32 and Al 6061-T6 using experiments and simulation. FE simulations were performed for the V-bending process using codes that use Langrarian and explicit time integration techniques. Quadrilateral shell elements of Belytschko–Tsay formulation were selected for the mesh, utilizing adaptive meshing to automatically refine the mesh. The average mesh size was 2 mm for the sheet specimen and tools with the tools modelled as rigid bodies. Hollomon's strain hardening law was used to describe stress–strain behaviour of the base material and weld zone, while the plasticity model used Hill's 1990 yield criterion. The 'm' value in the yield criterion was optimized to minimize the error by comparing to experimental results. The sheets were subjected to explicit analysis, and the springback angles were evaluated after bending.

Bakhshi-Jooybari et al. (2009) have evaluated the springback of CK67 steel sheets in V-die and U-die bending processes. Materials properties from tensile tests are used in the simulation. To model the sheet anisotropy, Hill's anisotropy parameters were introduced into the FE code. For anisotropic materials, Hill's yield function is widely used in FE simulation due to its ease of formulation. Punches, dies, and blank holders are assumed to be rigid bodies, while the blank is assumed to be deformable. Quadrangle four-node S4R shell elements were used for sheet modelling. For V-bending simulation, the number of elements in the model is 80, consisting of 5 elements across the width and 16 elements across the length of blank. For U-bending simulation, the number of elements in the model is 40, consisting of 2 elements across the width and 20 elements across the length of blank. For springback measurements, one of the nodes on the blank is selected as SET, and the position history is extracted from loading and unloading steps. Springback is calculated from the difference between blank bend angles during loading and after unloading. Figure 3.11 shows the number of elements in modelling of V-die and U-die bending, while Figure 3.12 shows typical simulation results of springback in V-die and U-die bending.

Gautam and Kumar (2018) investigated the springback in V-bending of interstitial free (IF) steel using experimental and numerical methods. Simulation of the loading step in the bending process uses the ABAQUS-explicit solution procedure as the problems of nonlinear complicated contact can be handled with a reduced

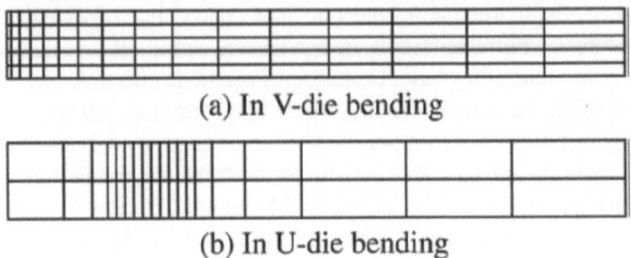

(a) In V-die bending

(b) In U-die bending

**FIGURE 3.11**   The number of elements in modelling of V-die (a) and U-die bending (b). (With permission from Bakhshi-Jooybari et al., 2009.)

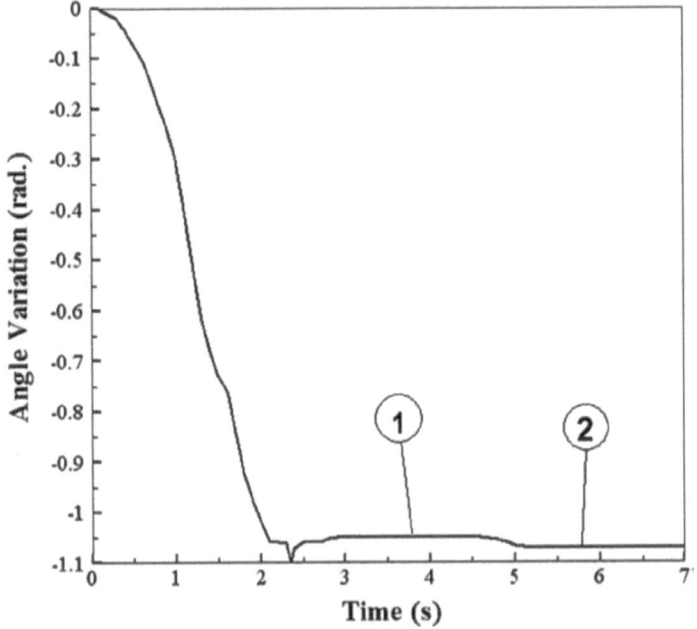

**FIGURE 3.12**   Typical simulation results of springback in V-die and U-die bending; part 1, the SET location before unloading; part 2, location after unloading. (With permission from Bakhshi-Jooybari et al., 2009.)

computational time. Then the completed bending simulation is imported from Explicit to ABAQUS-Implicit for the springback simulation. As the punch and die constraint are removed from the model in unloading, lower nonlinearity is shown by the springback simulation.

The bending process can be treated as a nonlinear problem. The three nonlinearities in bending problems are boundary nonlinearity, material nonlinearity, and geometric nonlinearity. Material nonlinearity is caused by plastic behaviour of anisotropic sheet metals in compliance with the power law of strain hardening in the true stress and true strain area. Boundary nonlinearity occurs when the sheet metal comes in contact with the die shape during the bending operation. When the contact between the sheet and die happens in simulation, there is a large and instantaneous change in the model response, which results in nonlinearity caused by contact boundary conditions. Geometric nonlinearity is a result of geometric changes of the model in analysis. The metal undergoes deflection in the bending process, and the bending load does not maintain perpendicular to the bent sheet. The bending load can be resolved into vertical and horizontal components, causing alteration in model stiffness.

For conventional and TWB specimens, thickness integration by Simpson rule with five-point integration is adopted. Springback prediction relies on integration schemes and integration points through sheet thickness. As springback amount depends on bending moment, which depends on stress distribution in sheet thickness, numerical integration of stress and strain through thickness is required by shell elements to determine bending moment and force.

The punch and die sets are modelled as rigid surfaces since stress variation in the two components is not the focus of the simulation. For dies, the clearances are modelled to be equal with sheet thickness to prevent localized compression. The punch and dies were modelled as an analytical rigid shell, while the blank is modelled as a deformable shell planar with S4R shell elements. The S4R shell element is a four-node thin shell element with reduced integration, hourglass control, and finite member strains. Point mass was assigned to the reference point on the punch and dies to compute the dynamic response.

Hill's plasticity model or Hill's yield potential, an extension of Mises yield function for anisotropic materials, has been used to incorporate sheet anisotropy in FE modelling. Springback in V-bending simulation for the conventional blank and TWB utilizes the Newton–Raphson method by removing constraints such as dies and punches, so that the simulation adopted a static-general procedure (ABAQUS standard) where the nonlinearity is due to geometry only. By using this method, the simulation is divided into a number of load increments and the approximate equilibrium configuration is found at the end of each load increment.

After all constraints were removed from the model, the blank was assigned initial state of bent data file containing history of loading. To ensure that the central node remains stationary, it was assigned as zero velocity. To determine the springback, node coordinates from the load and unloaded frame were captured and plotted using the CAE interface in ABAQUS. The difference of node coordinates will give the change in included angle after unloading.

### 3.8.5 Optimization

As recommended by Katre et al. (2014), anisotropic weld zone assumptions yield slightly better results as compared to isotropic weld zone assumptions. Rao and Narayanan (2014) suggested that for a more accurate springback evaluation, inelastic recovery should be considered as the springback simulations conducted with varied elastic moduli are closer to experimental results.

For an accurate prediction of springback in the bending of TWBs, sheet anisotropy and weld zone properties should be included in the FE simulations (Gautam and Kumar, 2018). It was also found that five to seven integration points are optimal and any further increase in integration points did not affect the springback prediction accuracy. Furthermore, for better results, the mesh size of the weld zone is modelled to be finer than the parent materials, based on maximum von Mises stress in the deformed region where the mesh is refined until there is no significant change (<1.8%) in specific region stress. The explicit method has also built in amplitude curves for smooth loading amplitudes on the punch reference point. This ensures smooth application of bending load as required by quasi-static analysis to enhance accuracy and efficiency.

## 3.9  CONCLUSIONS

This chapter discusses the springback phenomenon in sheet metal forming. The discussion includes the definition, effects of springback toward product quality, and the factors affecting springback. For dissimilar metal joining, such as aluminium to steel, the FSW process is the preferred option due to its numerous advantages. Nevertheless, proper selection of process parameters in FSW is important to create a successful joint which has good mechanical properties. Post-weld tests, such as bending test and LDH test, can be conducted to determine the formability and manufacturability of the friction stir weld joint.

Bending and forming processes of friction stir welded aluminium and steel joints are also subjected to springback. However, appropriate compensation is needed to overcome the dimensional change caused by the springback. FE simulation can be used to predict springback as demonstrated by a number of research studies. By incorporating aspects such as weld zone properties and Hill's yield potential, it is possible to obtain accurate prediction of springback comparable to experimental results.

## ACKNOWLEDGEMENT

The author would like to acknowledge the support from Universiti Sains Malaysia (USM) through RUI Grant (1001/PMEKANIK/8014031).

## REFERENCES

Abdullah, A. B., & Samad, Z. (2014). Measurement of twist springback on AA6061-T6 aluminum alloy strip: A preliminary result. *Applied Mechanics and Materials*, **699**, 44–48. doi: 10.4028/www.scientific.net/amm.699.44.

Ambroziak, A., Korzeniowski, M., Kustroń, P., Winnicki, M., Sokołowski, P., & Harapińska, E. (2014). Friction welding of aluminium and aluminium alloys with steel. *Advances in Materials Science and Engineering*, **2014**. doi:10.1155/2014/981653.

Arya, P. K., Gupta, G., & Rajput, A. K. (2016). A review on friction stir welding for aluminium alloy to steel. *International Journal of Scientific & Engineering Research*, **7**(5), 119–125. doi: 10.4028/www.scientific.net/AMR.922.553.

Bakhshi-Jooybari, M., Rahmani, B., Daeezadeh, V., & Gorji, A. (2009). The study of springback of CK67 steel sheet in V-die and U-die bending processes. *Materials and Design*, **30**(7), 2410–2419. doi:10.1016/j.matdes.2008.10.018.

Barbini, A., Carstensen, J., & Dos Santos, J. F. (2018). Influence of alloys position, rolling and welding directions on properties of AA2024/AA7050 dissimilar butt weld obtained by friction stirwelding. *Metals*, **8**(4). doi:10.3390/met8040202.

Boumerzoug, Z., & Helal, Y. (2017). Friction stir welding of dissimilar materials aluminum AL6061-T6 to ultra low carbon steel. *Metals*, **7**(2). doi:10.3390/met7020042.

Çam, G., & İpekoğlu, G. (2017). Recent developments in joining of aluminum alloys. *International Journal of Advanced Manufacturing Technology*, **91**(5–8), 1851–1866. doi:10.1007/s00170-016-9861-0.

Campanella, D., Spena, P. R., Buffa, G., & Fratini, L. (2016). Dissimilar Al/steel friction stir welding lap joints for automotive applications. *AIP Conference Proceedings*, **1769**. doi:10.1063/1.4963499.

Chan, W. M., Chew, H. I., Lee, H. P., & Cheok, B. T. (2004). Finite element analysis of springback of V-bending sheet metal forming processes. *Journal of Materials Processing Technology*, **148**, 15–24. doi:10.1016/j.jmatprotec.2003.11.038.

Chang, S. H., Shin, J. M., Heo, Y. M., & Seo, D. G. (2002). Springback characteristics of tailor-welded strips in U-bending. *Transactions of Materials Processing*, **12**(5), 440–448. doi:10.5228/kspp.2003.12.5.440.

Chikalthankar, S. B., Belurkar, G. D., & Nandedkar, V. M. (2014). Factors affecting on springback in sheet metal bending : A review. *International Journal of Engineering and Advanced Technology (IJEAT)*, **3**(4), 247–251.

Corigliano, P., Crupi, V., Guglielmino, E., & Mariano Sili, A. (2018). Full-field analysis of AL/FE explosive welded joints for shipbuilding applications. *Marine Structures*, **57**, 207–218. doi:10.1016/j.marstruc.2017.10.004.

da Silva, E. A., Malerba Fernandes, L. F. V., de Jesus Silva, J. W., Ribeiro, R. B., Santos Pereira, M. dos, & Alexis, J. (2016). A comparison between an advanced high-strength steel and a high-strength steel due to the spring back effect. *IOSR Journal of Mechanical and Civil Engineering*, **13**(05), 21–27. doi:10.9790/1684-1305012127.

Darmawan, A. S., Anggono, A. D., & Nugroho, S. (2019). Springback phenomenon analysis of tailor welded blank of mild steel in U-bending process. *AIP Conference Proceedings*, **2114**. doi:10.1063/1.5112425.

Dehghani, M., Amadeh, A., & Akbari Mousavi, S. A. A. (2013). Investigations on the effects of friction stir welding parameters on intermetallic and defect formation in joining aluminum alloy to mild steel. *Materials and Design*, **49**, 433–441. doi:10.1016/j.matdes.2013.01.013.

Desai, N. V., & Inamdar, K. H. (2017). A methodology for friction stir welding of aluminium 6061-T6 and 304L stainless steel. *International Journal of Innovative Research in Science, Engineering and Technology*, **6**(12), 22647–22652. doi:10.15680/IJIRSET.2017.0612074.

Gautam, V., & Kumar, D. R. (2018). Experimental and numerical investigations on springback in V-bending of tailor-welded blanks of interstitial free steel. *Proceedings of the Institution of Mechanical Engineers, Part B: Journal of Engineering Manufacture*, **232**(12), 2178–2191. doi:10.1177/0954405416687146.

Gautam, V., Kumar, P., & Deo, A. S. (2012). Effect of punch profile radius and localised compression on springback in V-bending of high strength steel and its FEA simulation. *International Journal of Mechanical Engineering and Technology (IJMET)*, **3**(3), 517–530.

Geng, H., Sun, L., Li, G., Cui, J., Huang, L. & Xu, Z. (2019). Fatigue fracture properties of magnetic pulse welded dissimilar Al-Fe lap joints, *International Journal of Fatigue*, **121**, 146–154.

Gullino, A., Matteis, P., & Aiuto, F. D. (2019). Review of aluminum-to-steel welding technologies for car-body applications. *Metals*, **9**(3), 1–28. doi:10.3390/met9030315.

Huang, Y.-M., & Leu, D.-K. (1998). Effects of process variables on V-die bending process of steel sheet. *International Journal of Mechanical Sciences*, **40**(7), 631–650. doi: 10.1016/S0020-7403(97)00083-0.

Ibrahim, A. B., Al-Badour, F. A., Adesina, A. Y., & Merah, N. (2018). Effect of process parameters on microstructural and mechanical properties of friction stir diffusion cladded ASTM A516-70 steel using 5052 Al alloy. *Journal of Manufacturing Processes*, **34**, 451–462. doi:10.1016/j.jmapro.2018.06.020.

Kalemba-Rec, I., Kopyściański, M., Miara, D., & Krasnowski, K. (2018). Effect of process parameters on mechanical properties of friction stir welded dissimilar 7075-T651 and 5083-H111 aluminum alloys. *International Journal of Advanced Manufacturing Technology*, **97(5–8)**, 2767–2779. doi:10.1007/s00170-018-2147-y.

Katre, S., Karidi, S., & Narayanan, G. (2014). Springback of friction stir welded sheets: Experimental and prediction. *In 5th International & 26th All India Manufacturing Technology*, pp. 318-1–318-5. Gulwahati, India.

Katre, S., Karidi, S., & Rao, B. D. (2015). Springback and formability studies on friction stir welded sheets. In: Narayanan, R. G. and Dixit, U. S. (Eds.), *Advances in Material Forming and Joining* (pp. 141–165). Springer: Berlin, Germany. doi:10.1007/978-81-322-2355-9.

Kazan, R., Fırat, M., & Tiryaki, A. E. (2009). Prediction of springback in wipe-bending process of sheet metal using neural network. *Materials and Design*, **30**, 418–423. doi:10.1016/j.matdes.2008.05.033.

Kim, J., Lee, W., Chung, K., Park, T., Kim, D., Kim, C., & Kim, D. (2009). Springback prediction of friction stir welded DP590 steel sheet considering permanent softening behavior. *Transactions of Materials Processing*, **18(4)**, 329–335.

Mahto, R. P., Bhoje, R., Pal, S. K., Joshi, H. S., & Das, S. (2016). A study on mechanical properties in friction stir lap welding of AA 6061-T6 and AISI 304. *Materials Science and Engineering A*, **652**, 136–144. doi:10.1016/j.msea.2015.11.064.

Mishra, R., De, P., & Kumar, N. (2014). *Friction Stir Welding and Processing: Science and Engineering*. Springer: Cham. doi:10.1007/978-3-319-07043-8.

Nashrudin, M. N., & Abdullah, A. B. (2016). Finite element simulation of twist forming process to study twist springback pattern. *MATEC Web of Conferences*, **90**. doi:10.1051/matecconf/20179001026.

Noh, W., Song, J. H., Jang, I. J., Gwak, S. H., Kim, C., & Jung, C. Y. (2018). Numerical and experimental investigation for formability of friction stir welded dissimilar aluminum alloys. *IOP Conference Series: Materials Science and Engineering*, **418(1)**, 0–6. doi:10.1088/1757-899X/418/1/012056.

Padmanabhan, R., Oliveira, M. C., & Menezes, L. F. (2008). Deep drawing of aluminium-steel tailor-welded blanks. *Materials and Design*, **29(1)**, 154–160. doi:10.1016/j.matdes.2006.11.007.

Panich, S., Uthaisangsuk, V., Juntarantin, J. & Suranuntchai, S. (2011). Determination of forming limit stress diagram for formability prediction of SPCE 270 steel sheet. *Journal of Materials Science*, **21(1)**, 19–27. Retrieved from https://www.jmmm.material.chula.ac.th/index.php/jmmm/article/view/164/191

Papeleux, L., & Ponthot, J. (2002). Finite element simulation of springback in sheet metal forming. *Journal of Materials Processing Technology*, **126**, 785–791.

Park, S., Lee, C. G., Kim, J., Han, H. N., Kim, S. J., & Chung, K. (2008). Improvement of formability and spring-back of AA5052-H32 sheets based on surface friction stir method. *Journal of Engineering Materials and Technology, Transactions of the ASME*, **130(4)**, 0410071–04100710. doi:10.1115/1.2975233.

Pasha, A., Reddy, R. P., & Ahmad Khan, I. (2014). Influence of process and tool parameters on friction stir welding: Over view. *International Journal of Applied Engineering and Technology*, **4(3)**, 2277–212.

Pavan Kumar, J., Uday Kumar, R., Ramakrishna, B., Ramu, B., & Baba Saheb, K. (2018). Formability of sheet metals: A review. *IOP Conference Series: Materials Science and Engineering*. doi:10.1088/1757-899X/455/1/012081.

Rahmani, B., Alinejad, G., Bakhshi-Jooybari, M., & Gorji, A. (2009). An investigation on springback/negative springback phenomena using finite element method and experimental approach. *Proceedings of the Institution of Mechanical Engineers, Part B: Journal of Engineering Manufacture*, **223**(**7**), 841–850. doi:10.1243/09544054JEM1321.

Raja, S., Hasan, F., & Ansari, A. H. (2016). Effect of friction stir welding on the hardness of Al-6061 T6 aluminium alloy. *International Conference on Advanced Production and Industrial Engineering*, **2016**, 9–13.

Ramachandran, K. K., Murugan, N., & Kumar, S. S. (2015). Effect of tool axis offset and geometry of tool pin profile on the characteristics of friction stir welded dissimilar joints of aluminum alloy AA5052 and HSLA steel. *Materials Science & Engineering A*, **639**, 219–233. doi:10.1016/j.msea.2015.04.089.

Rao, B. D., & Narayanan, R. G. (2014). Springback of friction stir welded sheets made of aluminium grades during V-bending : An experimental study. ISRN Mechanical Engineering.

Safeen, M. W., & Spena, P. R. (2019). Main issues in quality of friction stir welding joints of aluminum alloy and steel sheets. *Metals*, **9**(**5**). doi:10.3390/met9050610.

Safeen, M. W., Russo Spena, P., Buffa, G., Campanella, D., Masnata, A., & Fratini, L. (2020). Effect of position and force tool control in friction stir welding of dissimilar aluminum-steel lap joints for automotive applications. *Advances in Manufacturing*, **8**(**1**), 59–71. doi:10.1007/s40436-019-00290-1.

Salih, O. S., Ou, H., Sun, W., & McCartney, D. G. (2015). A review of friction stir welding of aluminium matrix composites. *Materials and Design*, **86**, 61–71. doi:10.1016/j.matdes.2015.07.071.

Srinivasu, C., Singh, S. K., Jella, G., Jayahari, L., & Kotkunde, N. (2017). Study of limiting dome height and temperature distribution in warm forming of ASS304 using finite element analysis. *Materials Today: Proceedings*, **4**(**2**), 957–965. doi:10.1016/j.matpr.2017.01.107.

Tariq, M., Khan, I., Hussain, G., & Farooq, U. (2019). Microstructure and micro-hardness analysis of friction stir welded bi-layered laminated aluminum sheets. *International Journal of Lightweight Materials and Manufacture*, **2**(**2**), 123–130. doi:10.1016/j.ijlmm.2019.04.010.

Thipprakmas, S. (2010). Finite element analysis of punch height effect on V-bending angle. *Materials and Design*, **31**(**3**), 1593–1598. doi:10.1016/j.matdes.2009.09.019.

Thipprakmas, S., & Phanitwong, W. (2011). Process parameter design of spring-back and spring-go in V-bending process using Taguchi technique. *Materials and Design*, **32**(**8–9**), 4430–4436. doi:10.1016/j.matdes.2011.03.069.

Tisza, M., & Czinege, I. (2018). Comparative study of the application of steels and aluminium in lightweight production of automotive parts. *International Journal of Lightweight Materials and Manufacture*, **1**(**4**), 229–238. doi:10.1016/j.ijlmm.2018.09.001.

Wagoner, R. H., Lim, H., & Lee, M. G. (2013). Advanced issues in springback. *International Journal of Plasticity*, **45**, 3–20. doi:10.1016/j.ijplas.2012.08.006.

Wang, Y., Qi, B., Cong, B., Yang, M., & Liu, F. (2017). Arc characteristics in double-pulsed VP-GTAW for aluminum alloy. *Journal of Materials Processing Technology*, **249**, 89–95. doi:10.1016/j.jmatprotec.2017.05.027.

Zheng, Q., Feng, X., Shen, Y., Huang, G., & Zhao, P. (2016). Dissimilar friction stir welding of 6061 Al to 316 stainless steel using Zn as a filler metal. *Journal of Alloys and Compounds*, **686**, 693–701. doi:10.1016/j.jallcom.2016.06.092.

# 4 Overview and Issues Related to Geometrical Accuracy in the Single-Point Incremental Forming (SPIF) Process of Tailor Welded Blanks

*KAHA Razak, AB Abdullah, and NM Noor*
Universiti Sains Malaysia

## CONTENTS

DOI: 10.1201/9781003164241-4

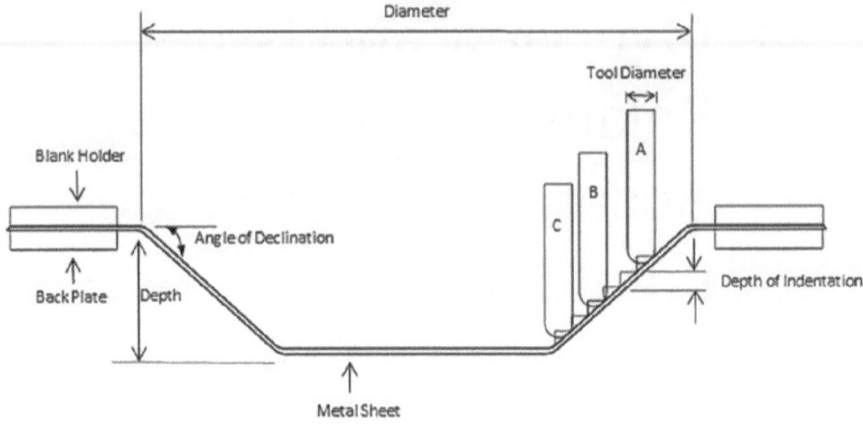

**FIGURE 4.1**    Schematic of the SPIF process. (Reconstructed from Dwivedy and Kalluri, 2010.)

## 4.1  INTRODUCTION

The goal of the manufacturing industry is to prioritize high-volume production of high-quality items having good geometric accuracies and surface finishes. Sheet metal forming processes are usually used in high production rate facilities and tra-ditionally require tooling such as dies, punches, and blank holders. However, this method may not be economical for small batch productions as the tooling costs are very expensive. A cost-effective alternative for sheet metal forming for small batch production and prototype parts is the single-point incremental forming (SPIF). In SPIF, the sheet metal blank is clamped in a custom fixture and a rotating tool follows a predetermined tool path to create shapes at various depths on the workpiece sur-face, schematically shown in Figure 4.1. Automobile sheet metal components such as heat/noise shields, motorcycle seats, vehicle headlights, and gas tanks can be made using this method. Non-automotive parts made using this method include biomedical devices (cranial plate and ankle supports), service panels, and solar cookers (Kumar et al., 2019).

### 4.1.1  SPIF Limitations

However, there are several limitations of SPIF that limit its application and adop-tion in the forming industry. One of the major limitations of the process is its low geometric accuracy, which falls short of compliance to the requirements of industrial applications. For most industrial use, a geometric tolerance of <1.0 mm throughout the entire part surface is expected, with some critical components requiring accuracy levels within ±0.2 mm. Currently, geometric accuracy levels reported in SPIF typi-cally exceed these limits. Allwood et al. (2005) and Reddy et al. (2015) have found the main causes of geometric inaccuracies in SPIF are due to sheet bending effects, persistent local springback around the tool, and global springback that sets in after the tool and clamping are released.

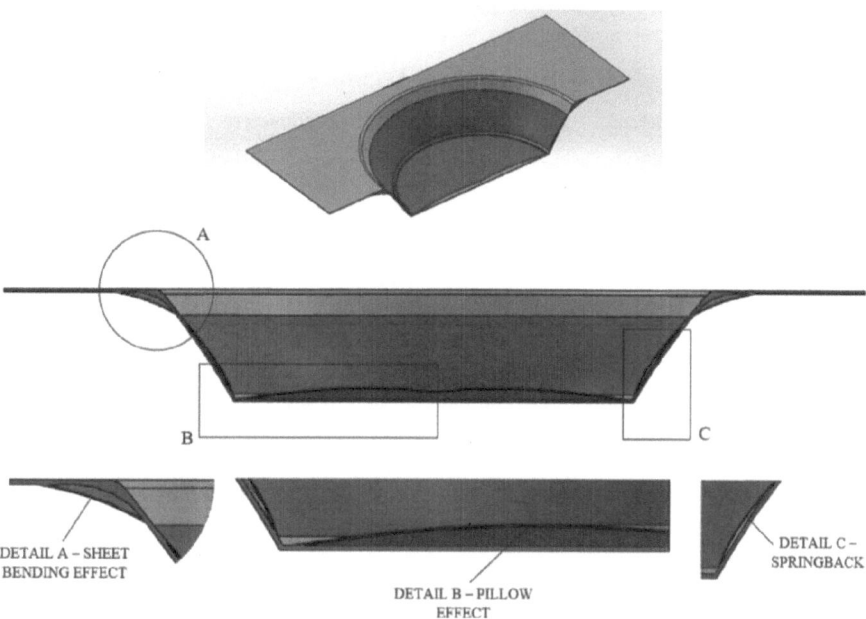

DETAIL A – SHEET BENDING EFFECT

DETAIL B – PILLOW EFFECT

DETAIL C – SPRINGBACK

**FIGURE 4.2** Common geometrical inaccuracies in SPIF.

Allwood et al. (2010) have categorized SPIF part accuracy on three different stages, namely, (1) clamped accuracy, (2) unclamped accuracy and (3) final accuracy. The majority of past studies have focused on improving the clamped accuracy. Micari et al. (2007) used preliminary testing on aluminium alloy 1050-O sheets to show that the measurable differences before and after clamping are minor. They also stated that the part shape, sheet material, and sheet thickness all have substantial influences on the variations in part accuracy caused by unclamping and trimming methods, and changes in part accuracy due to these issues should be investigated on a case-by-case basis. They also defined geometrical errors as the gap between the designed and formed profiles. After releasing the forming tool, the three sorts of geometrical inaccuracies associated with SPIF are present, as illustrated in Figure 4.2. In addition to geometrical accuracy issues, other process limitations for the adoption of SPIF in the forming industry are long processing time, excessive sheet thinning, and rough surface finishes (Lu et al., 2019).

### 4.1.2 Geometrical Accuracy and Dimensional Variation in SPIF

Allwood et al. (2010) proposed an innovative way to minimize geometric accuracy by incorporating partial cut-outs in the sheet blank. However, Essa and Hartley (2011) have discovered that enhanced geometrical accuracy can be obtained by employing a backing plate and kinematic tool and extending the tool path over the base of the sheet. To increase the geometric accuracy of the part formed by the incremental forming process, Wei et al. (2011) suggested a forming error compensation approach.

**TABLE 4.1**

**Parameters of ISF Related to Part Accuracy**

| Types of Parameters | Parameters |
|---|---|
| Process parameters | Tool diameter |
| | Toolpath (step size, toolpath type, etc.) |
| | Feed rate |
| | Spindle speed |
| | Type of lubricant |
| Design process parameters | Sheet thickness |
| | Geometry (forming depth, forming angle) |
| Material parameters | Strain hardening, Young modulus, normal anisotropy |

According to Fiorentino et al. (2011), a two-point incremental forming technique with positive geometry at a low inclination angle tool path can enhance the geometric accuracy of the formed part. Han et al. (2013) developed the PSO-ANN approach to predict springback. To improve the geometry accuracy of the formed part, Fan and Gao (2014) applied electricity to locally increase the temperature of the sheet. The deformation energy and geometrical accuracy of the parts were found to be impacted by sheet thickness and wall angle. However, when the parts were formed with 60° wall angle, the geometric error was shown to be the largest. Subsequently, Najafabady and Ghaei (2016) have enhanced the geometric accuracy of the formed parts by employing alternating electrical current in an electrical heater. Formisano et al. (2017) also compared the geometrical accuracy, formability, and thickness distribution of positive and negative incremental forming processes, determining that positive incremental forming has superior geometrical accuracy, formability, and thickness distribution. Yao et al. (2017) developed a contour compensation method to increase the geometrical accuracy of the formed parts. Li et al. (2017a,b) provided a mathematical model to predict the dimensional errors in the SPIF process. Do et al. (2017) merged positive and negative sheet incremental forming into a single process and used a springback compensation approach to produce repaired complex shape parts. Shrivastava et al. (2018) recently proposed a preheating stage, which can reduce the geometrical accuracy of the formed part in the SPIF process.

### 4.1.3 Parameters Affecting the Geometrical Accuracy

Trzepieciński et al. (2021) have listed several categories of process parameters, design process parameters, and the material parameters of the workpiece that can influence part accuracy in SPIF, as presented in Table 4.1.

## 4.2 EFFECT OF PROCESS PARAMETERS

Many researchers have investigated the impact of SPIF process parameters on the dimensional accuracy. Edwards et al. (2017) investigated the influence of SPIF process factors of heat, feed rate, spindle speed, and vertical step on the springback of a polycarbonate sheet. The result showed that increasing the feed rates and spindle speeds resulted in a decrease in springback. Furthermore, the springback was reduced when heat was applied. Kovine et al. (2017) demonstrated the formation of AA2024-T3 sheets using the SPIF process and acquired product geometry. The process parameters chosen were vertical step, tool path, and lubricant type, while other parameters of rotational spindle speed, feed rate, wall angle tool diameter, and tool coating remained constant. A 3D laser scanning method was utilized to measure the dimensional accuracy, and the results obtained showed that the spiral tool path was successfully created. Dakhli et al. (2017) investigated the effect of two truncated cone shapes on geometry accuracy on mild steel sheet specimens. The process parameters studied were the specimen thickness, vertical step, feed rate, and spindle speed. A 10 mm diameter hemispherical tool was used to generate the spiral tool path. A Next Engine 3D scanner was used to capture the geometry of the generated surface, which can then be compared to the intended geometry of the programmed tool path.

Li et al. (2017a) have demonstrated the SPIF process for three different materials (1060 aluminium, Q235, and DC04 steel) and evaluated their resulting geometrical accuracies. The results showed that the strain strength coefficient and stepping rate have a significant impact on geometrical accuracy. Yao et al. (2017) have investigated the influence of vertical step, thickness, wall angle, and tool diameter on surface roughness, energy of deformation, and geometrical accuracy in the SPIF process. The result showed that the best values were obtained for surface roughness (0.97 μm), deformation energy (1522.4 J), and geometrical error (1.939 mm). The data were optimized for vertical step (0.5 mm), thickness (0.57 mm), wall angle (65°), and tool diameter (16 mm).

### 4.2.1 Tool Diameters

Najm and Paniti (2020) have compared the effect of different sizes and shapes of tools, for both flat-end and hemispherical-end tools, on the geometry and resultant pillow effect for SPIF of AlMn1Mg1 thin sheets. The results show that the flat-end tool improved the geometry accuracy and minimized the pillow effects. However, other studies by Li et al. (2015) and Hussain (2014) concluded that a forming tool with a hemispherical end would result in increased forming accuracy. Jagtap and Kumar (2019) have investigated the influence of compensating effects of tool radius on the accuracy of formed parts and have concluded that the tool offset is a significant parameter, while the tool radius has no effect on the geometric accuracy of the parts. Zhang et al. (2020) have found that the effectiveness of tool diameter on the springback of AZ31B Mg alloys in warm ISF, assisted with oil bath heating, is lower than other parameters.

Bedan and Habeeb (2018) have studied the effect of tool geometry, tool diameter, spindle speed, and wall angle on the dimensional accuracy for SPIF of 0.9 mm thickness aluminium Al050 sheets. Three geometries of forming tools used were hemispherical, ball end, and flat with round corner (see Table 4.2). The geometries of the products were measured using dimensional sensor measuring equipment. It was found that the best dimensional accuracy between the SPIF profile and the original CAD profile was obtained for the parameters of 10 mm ball end tool, 50° wall angle, and 800 rpm tool rotational speed.

**TABLE 4.2**

**Types of Forming Tools**

| Tools | Dimensions (mm) | Types |
|---|---|---|
| 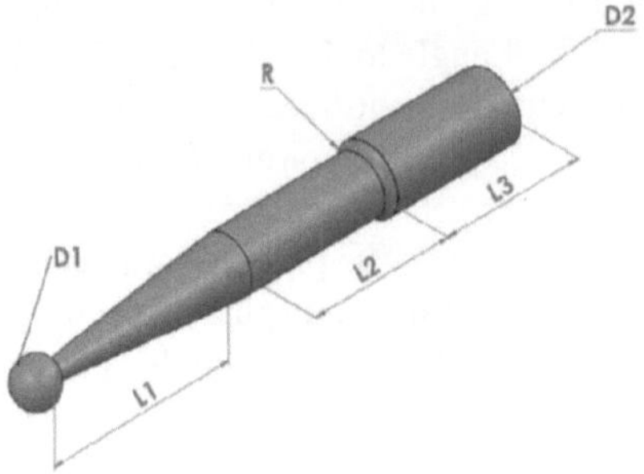 | Type A $L_1=35$ $L_2=28$ $L_3=24$ $D_1=8$ $D_2=12$ $R=2$ | Ball end tool |
| 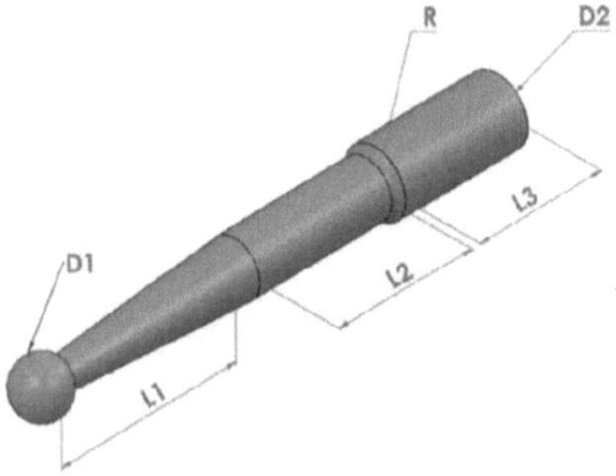 | Type B $L_1=35$ $L_2=28$ $L_3=24$ $D_1=10$ $D_2=12$ $R=2$ | |
| 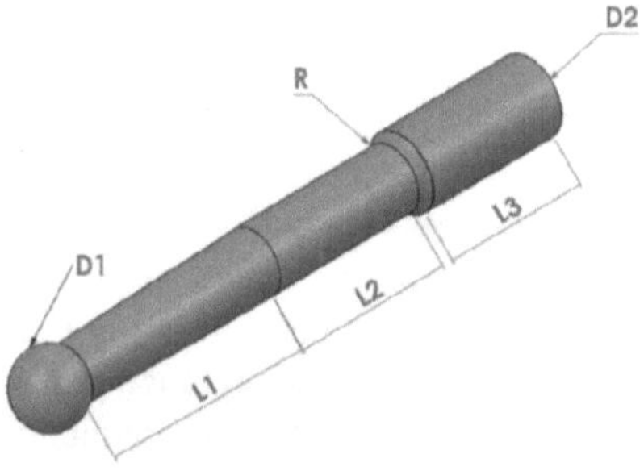 | Type C $L_1=35$ $L_2=28$ $L_3=24$ $D_1=12$ $D_2=12$ $R=2$ | |
| 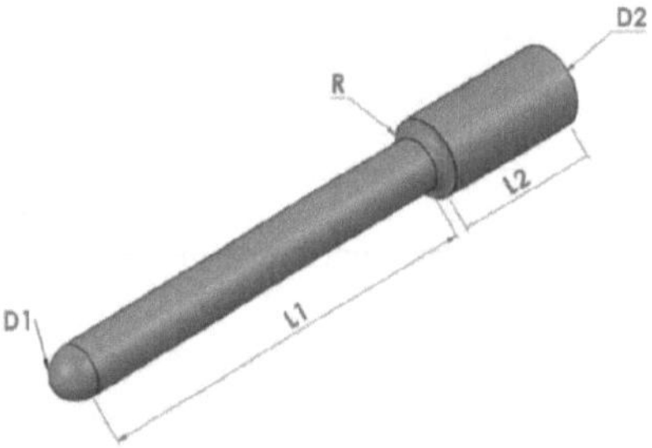 | Type A $L_1=70$ $L_2=24$ $D_1=8$ $D_2=12$ $R=2$ | Hemispherical tool |

(*Continued*)

**TABLE 4.2 (*Continued*)**
**Types of Forming Tools**

| Tools | Dimensions (mm) | Types |
|---|---|---|
| 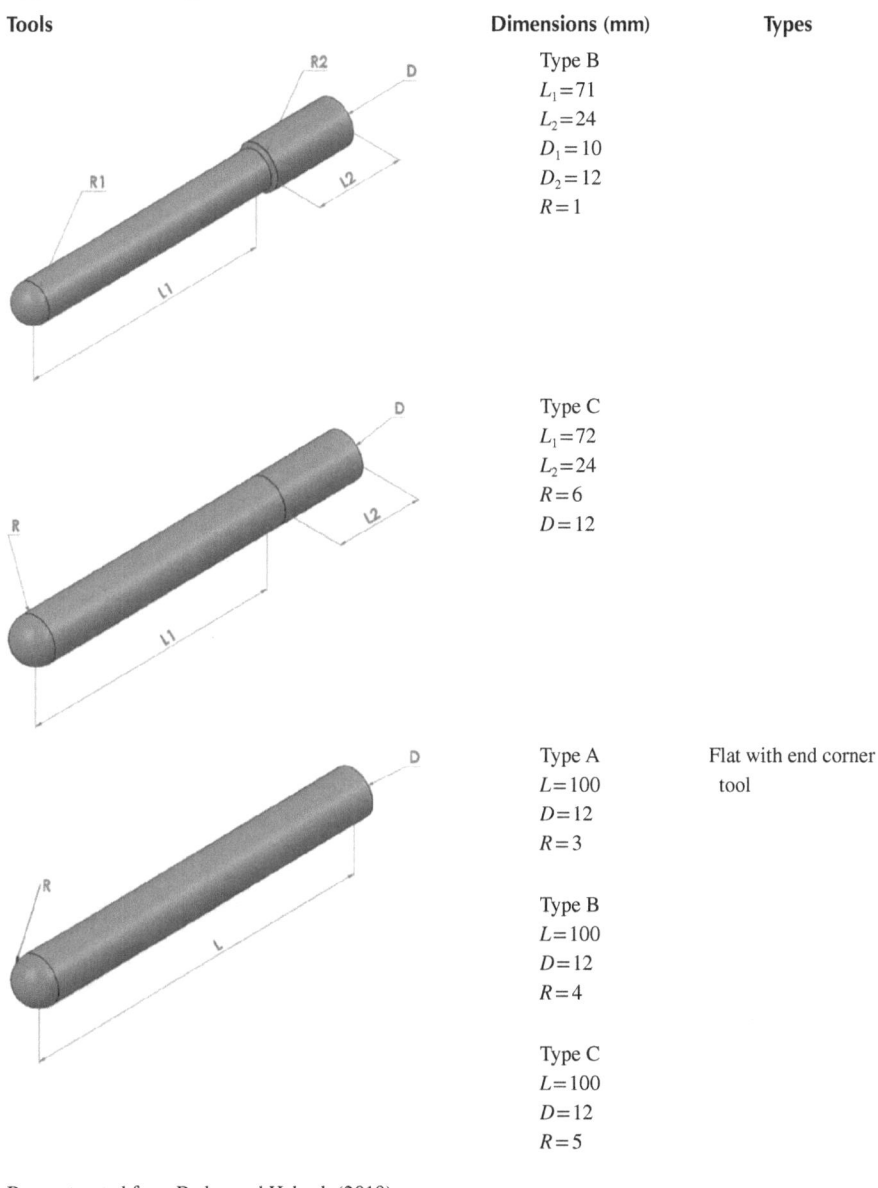 | Type B<br>$L_1 = 71$<br>$L_2 = 24$<br>$D_1 = 10$<br>$D_2 = 12$<br>$R = 1$ | |
| | Type C<br>$L_1 = 72$<br>$L_2 = 24$<br>$R = 6$<br>$D = 12$ | |
| | Type A<br>$L = 100$<br>$D = 12$<br>$R = 3$ | Flat with end corner tool |
| | Type B<br>$L = 100$<br>$D = 12$<br>$R = 4$ | |
| | Type C<br>$L = 100$<br>$D = 12$<br>$R = 5$ | |

Reconstructed from Bedan and Habeeb (2018).

### 4.2.2 TOOLPATH AND TOOLPATH STRATEGIES

In incremental sheet forming, the toolpath is produced usually in CAM software and then sent to a CNC milling machine for the SPIF process. There are various toolpaths available for use such as contour, bidirectional, spiral, and integrated tool-paths. Among these, the contour toolpath is the most widely used for SPIF (Kulin

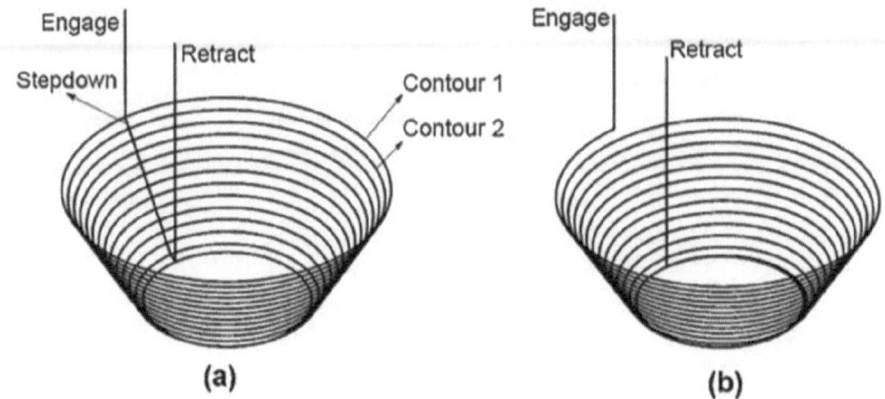

**FIGURE 4.3** Toolpath strategies: (a) z-level contouring toolpath and (b) helical toolpath. (Adapted from Behera et al., 2017.)

and Dodiya, 2019). The development of toolpaths has a direct impact on dimensional accuracy, surface finish, formability, thickness variance, and processing time (Behera et al., 2017).

Recent developments in toolpath generation have seen the use of advanced machine learning and optimization strategies. Akrichi et al. (2019) proposed two types of artificial intelligence (AI) learning approaches, namely, deep learning and shallow learning, to predict geometrical accuracy in the SPIF process. Their aim is to evaluate the performance of deep belief network (DBN), backpropagation neural network (BPNN), and stacked autoencoder (SAE) approaches as prediction methods. Initial sheet thickness, vertical step, toolpath strategy, wall angle, speed rate, and feed rate were selected as input parameters for the SPIF optimization. Each of the parameters has two levels of variations and was tested for z-level contouring and helical toolpaths (see Figure 4.3).

The study concluded that the deep learning technique can be used even with small data and has greater capability in predicting geometrical accuracy in the SPIF process as compared to shallow learning. In another study, Dai et al. (2019) used analysis of variance (ANOVA) to investigate the impact of four parameters of punch diameter, feed rate, vertical step, and spindle speed at four levels to determine the optimal parameters for better geometrical accuracy of cavity parts with a stepped feature area using a single-pass technique. Comparison between a scheme of multi-pass incremental forming (MSIF) with tool compensation and without tool compensation for each pass based on the local geometric deviation showed over 70% reduction in geometric deviation. In the study, the authors proposed three-pass strategies to improve geometrical accuracy (Figure 4.4).

Giraud-Moreau et al. (2018) developed a tool path optimization approach to reduce the differences between the actual and theoretical shapes caused by elastic springback. The optimization of toolpath is based on numerical simulations that account for all stages of the forming process, including tool displacement along the defined path, tool release, and unclamping of the formed part. The proposed optimization

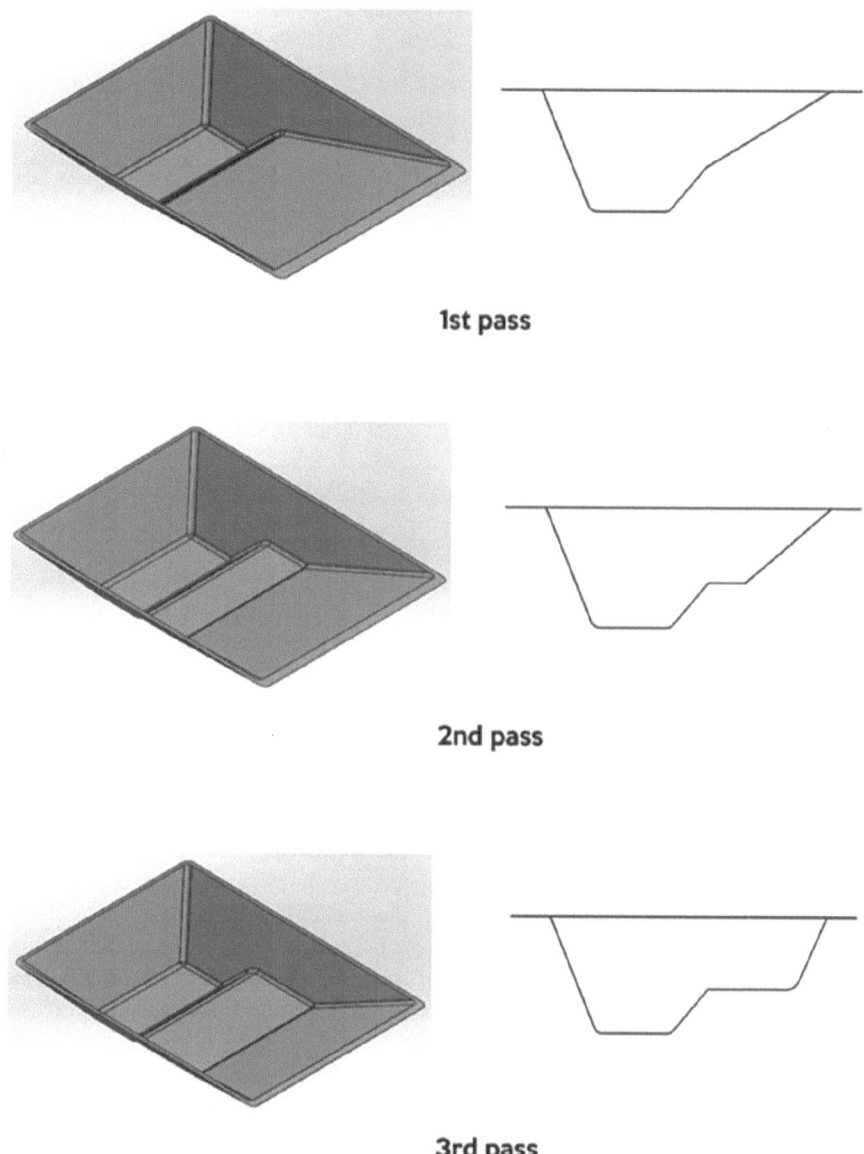

**1st pass**

**2nd pass**

**3rd pass**

**FIGURE 4.4**   Three-pass strategy of incremental forming.

technique consists of specifying the toolpath for each iteration by mirroring the numerical profile acquired in the previous iteration with respect to the theoretical curve. By using this method, the geometry accuracy of the final part formed was improved by 51% compared to the initial tool path.

Zhu et al. (2019) reported the development of toolpath trajectories for large and complex asymmetric parts for five-axis milling machines. The advantages of

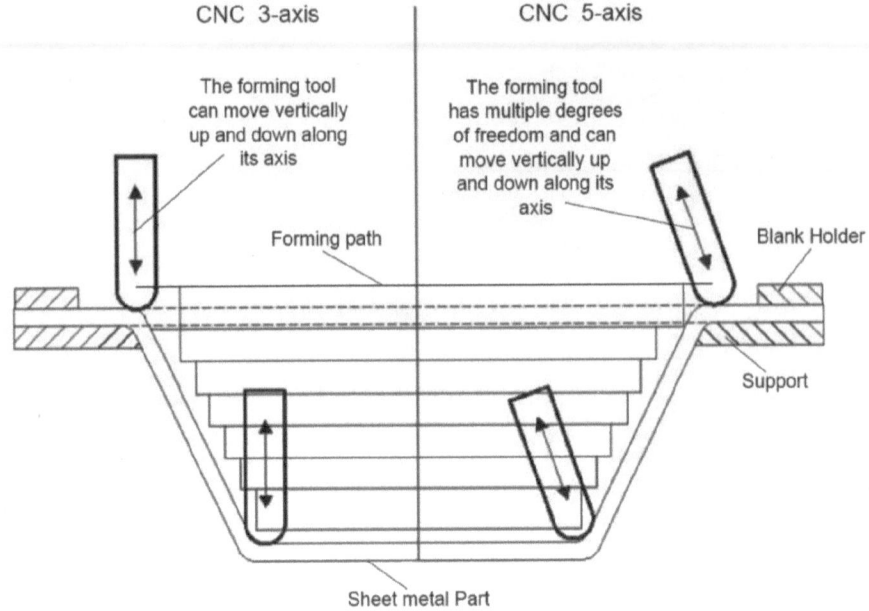

**FIGURE 4.5**   Incremental forming on the three-axis point and five-axis point.

five-axis milling machines over three-axis milling machines include enhanced flexibility in developing intermediate trajectories. As a result, the punch has a high degree of freedom in a five-axis milling machine as compared to a three-axis milling machine, where the punch can only form in the vertical direction (Figure 4.5). In the five-axis milling machine, the punch can press the material perpendicularly as well as in intermediate stages due to its multiple degrees of freedom.

Another variation of incremental forming is the double-sided incremental forming (DSIF), which adds a moving support tool on the opposite side of the sheet. Wang et al. (2018) have utilized DSIF for the production of lightweight parts and proposed a strategy for reducing the springback effect. For an ellipse-shaped part, a bending technique and a squeezing technique can be considered. This reverse bending technique helps to uniformize stress distribution, while the squeezing technique causes the stress values in the area between the two tools to change. Following their theoretical and experimental studies, the authors concluded that both techniques reduce springback, with reverse bending yielding the best result. Rakesh et al. (2016) showed a method to improve the geometric accuracy of formed parts by accounting for tool and sheet deflections caused by forming force. Here, the force equilibrium approach is used to calculate the forming forces needed to predict compensations as well as establish the thickness calculation methodology using the overlap of deformation zones that occur during forming. A variety of geometries (axisymmetric, variable wall angle, free forms, features above and below the initial sheet plane, and numerous features) were used to demonstrate the functions of the proposed methodology. The results showed a significant improvement in part accuracy when adjusted tool paths are used with DSIF, with a maximum error of less than 500 μm between the measured and ideal profiles.

Praveen et al. (2020) have made an analytical study using DSIF to improve the accuracy of small and large parts made from lightweight materials. The study considers both sheet and tool deflections, which are two of the most important elements in DSIF accuracy. The authors employed a mix of small deflection theory and membrane theory to create the adjusted toolpath, which was valid for both small and large parts. They validated the optimized trajectories for frustum-of-cone shaped parts with wall angles ranging from 25° to 60° and opening sizes of 150, 250, and 610 mm, achieving a sheet deflection reduction of about 10% for the compensated toolpath.

Zhang et al. (2015) proposed a novel mixed double-sided incremental sheet forming (MDSIF) toolpath strategy to achieve high geometric accuracy by investigating the effect of forming parameters of the tool positioning and incremental depth. A comparison was made with formed parts by traditional DSIF and accumulative double-sided incremental forming (ADSIF) methods. In MDSIF, the sheet metal is initially formed with ADSIF then reformed with DSIF without the specimen being removed from the clamping mechanism. As a result, in DSIF, increasing the incremental depth or squeeze factor seemed to have a negative influence on part accuracy. Furthermore, MDSIF methods were found to be capable of achieving higher geometric accuracy than ADSIF's previous best desirable geometry. In MDSIF, reducing the magnitude of the squeeze factor leads to improved geometric accuracy.

### 4.2.2.1 Feed Rate

Several major decision criteria need to be considered in designing an optimal process configuration for SPIF, including maximal payload (carrying capacity, defined by the weight the machine tool or robot can lift), flexibility of toolpath, rigidity, and total cost (Alves De Sousa et al., 2014). Most common devices utilized for incremental forming include adapted milling machines, robotics, and special purpose machines (Marabuto et al., 2011). The stiffness of robots is typically around 0.1–120 kN/mm, as compared to milling machines which are generally stiffer at about 200 kN/mm. This results in improved accuracy for SPIF production with milling machines (Verbert et al., 2009). However, most industrial robots have a broader working range, making them more ideal for large-sized parts (Portman, 2011). Past studies in SPIF have tended to use modified CNC milling machines due to their simplicity, high stiffness, and high production rate, although modified CNC machines are restricted in their degrees of freedom (DoF) of the forming tool. On the other hand, the robotic arm has a large working volume and can be operated rapidly. However, the main disadvantages of robotics for SPIF are the low stiffness and relatively low maximum force, which result in a less accurate tool positioning, particularly under heavy loading conditions (Behera et al., 2017).

### 4.2.2.2 Compensation Methods

The absence of custom dies allows SPIF to be simple and flexible (Duflou et al., 2018). However, its apparent simplicity hides highly intricate deformation mechanisms, which makes the process control challenging. The main primary challenge of SPIF is its poor achievable accuracy and process constraints (Maqbool and Bambach,

2018). Several approaches have been proposed to address these inaccuracy issues, such as using more tools, local heating, or local support (Lu et al., 2019). In SPIF, the accuracy is influenced by process parameters, such as the tool type and size; conditions of lubricant and processing speeds; and material and geometric characteristics such as sheet material, sheet thickness, and wall angle (Hussain et al., 2011).

## 4.3   SPIF DEVELOPMENT VARIATIONS

SPIF and TPIF are the two most frequent varieties of ISF. In both operations, the sheet to be formed is clamped on a blank holder throughout the process. The main difference between TPIF and SPIF is the support tool, where TPIF requires a partial or full die for support, whereas SPIF does not require any support tools. Although this makes SPIF more flexible, it has a negative impact on part accuracy. A comparative study found that the presence of a supporting die improves geometry accuracy (Lu et al., 2019). In addition, other variations have also been introduced and studied to minimize part inaccuracies as well as solving other ISF constraints.

### 4.3.1   DOUBLE-SIDE INCREMENTAL FORMING

DSIF introduces a moving supporting tool on the opposite side of the formed sheet, which moves together with the forming tool. An example of a DSIF setup is shown in Figure 4.6a, using two moving tools placed on two industrial robots (KUKA KR360). As seen in Figure 4.6b, the two tools move in sync with a preset spacing between them, with the supporting tool giving flexible support on the other side of the sheet.

Wang et al. (2009) designed a simple DSIF system (Figure 4.7a) with two moving tools attached to a rigid C-frame. A simple sphere shape was chosen as the test shape profile to test both processes of SPIF and DSIF on the same metal sheet specimen, as shown in Figure 4.7b. The results demonstrated that the DSIF process resulted in

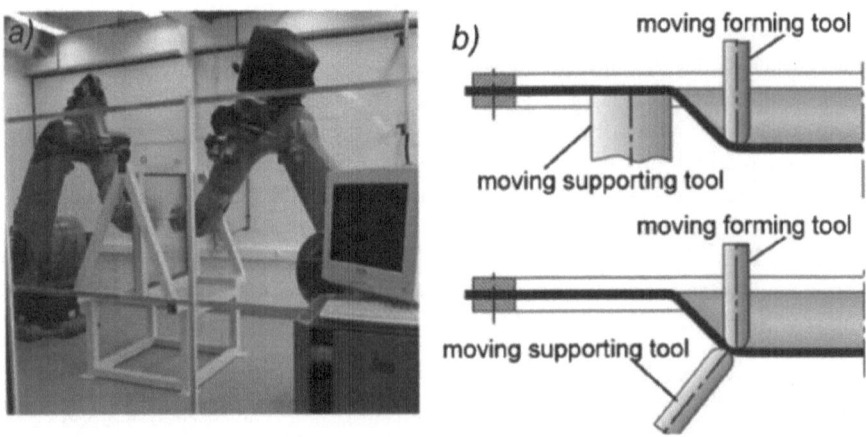

**FIGURE 4.6**   (a) DSIF setup on the KUKA robot and (b) DSIF forming principle. (With permission from Meier et al., 2009.)

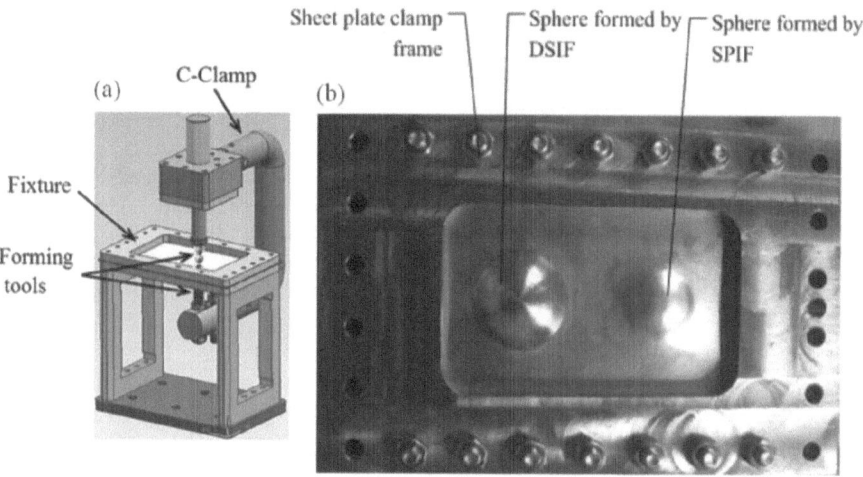

**FIGURE 4.7** (a) DSIF setup using C-frame. (With permission from Meier et al., 2009.) (b) generated spherical surfaces in DSIF and SPIF (see Wang et al., 2009).

superior geometric accuracy than the SPIF procedure, although the general dimensional inaccuracies from the intended profile remained rather large for both processes. However, the limitation of the DSIF technology is that its production has a limited wall angle range due to the restriction of the C-frame structure.

Peng et al. (2019) later introduced an improved DSIF system in which the two tools on either side of the sheet were controlled separately during the forming process. They investigated the effect of tool distance on geometric accuracy in DSIF. The distance was calculated using the squeeze factor(s). More precisely, when $s$ is set to 1.0, the bottom tool barely touches the blank, but $s < 1.0$ means that the blank will be squished between the top and bottom tools. In other experiments, a truncated cone was formed, and material fractures have occurred during the process for both SPIF and DSIF methods (Ai and Long, 2019). Additionally, Malhotra et al. (2011a) introduced numerical simulations to investigate the SPIF and DSIF formation mechanisms as well as the produced shape. Both the actual and simulated results indicated that DSIF resulted in greater geometric precision on the produced part's wall than SPIF. It was concluded that apparent distortion occurred on the component wall in SPIF, which may be reduced in DSIF using the squeezing method due to the localized deformation within a local plastic area around the tool end. To obtain increased geometric accuracy, the gap between the tools should be lower than the thickness of the contact area. Otherwise, the lack of contact between the sheet and the bottom tool may cause the DPIF process to degenerate into the SPIF process.

## 4.3.2 Hybrid SPIF

Enhancements to the SPIF process have been proposed by various researchers to address the constraints of SPIF, particularly the need for improved geometric precision. Several hybrid SPIF variations have been reported, which included a mix of

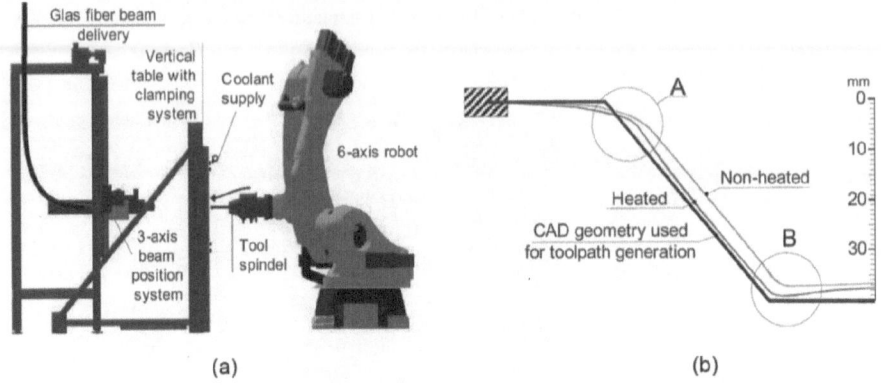

(a)                                                        (b)

**FIGURE 4.8**    (a) Machine structure of laser-assisted SPIF and (b) geometry accuracy comparison in standard (non-heated) SPIF and laser-assisted SPIF. (With permission from Duflou et al., 2007.)

SPIF and stretch forming, laser-assisted SPIF, and electric hot SPIF. Duflou et al. (2007) created a forming machine for laser-assisted SPIF (Figure 4.8) to increase formability and precision. The forming machine was mounted on a six-axis robot, and a laser heat source was positioned on the reverse of the metal sheet as a local heating device. It was found that the laser heat application for local and dynamic heating could lower the forming force in SPIF. Dimensional accuracy obtained via laser-assisted SPIF was also improved as the local heating helped to reduce residual stresses and the springback effect.

## 4.4   MULTI-STAGE STRATEGIES

Multi-stage forming strategies have been introduced in SPIF to improve the geometric accuracy and formability of SPIF, inspired by forming strategies utilized in other sheet forming processes such as sheet metal spinning (Cédric et al., 2020). Normally, a typical SPIF process forms the profile in a single forming stage, following the single-stage forming toolpath. More specifically, the metal sheet is deformed in a single pass, with the tool moving along a toolpath that is identical to the final target shape. In contrast, multi-stage SPIF will first form a preform shape and then subsequently deform into intermediate shapes in multiple stages until the final shape is obtained. The multi-stage approach can be thought of as a separate single-stage forming process in which the toolpath is formed based on the intermediate shapes of each stage (Lu et al., 2019).

Skjoedt et al. (2008) have created a five-stage forming strategy for SPIF to build a cylindrical cup with vertical walls. In terms of forming direction, there are two alternatives in each intermediate stage: forming downward (D) from the top or forming upward (U) from the bottom. To produce the shape, two different approaches were studied: the down-down-down-up (DDDU) approach and the down-up-down-down (DUDD) strategy. As material fracture occurred on the produced component after four forming stages utilizing the DUDD technique. It was concluded that the strain distribution in multi-stage ISF was influenced by both the intermediate forms and the forming direction (D or U). Later, forming limits of multi-stage SPIF were established using experimentally determined fracture forming-limit curves (FFLCs), which were used to successfully form a cylindrical cup with a flat bottom (Skjoedt et al., 2010). Duflou et al. (2008) and Verbert et al. (2008) also investigated a five-stage strategy for forming parts with wall angles >90° and used digital image correlation (DIC) technology to investigate the evolution of the formed profiles, material flow, and thickness distribution at different forming stages. Their findings indicated that the multi-stage technique can produce part geometries that can overcome the constraints of single-step forming.

Bambach et al. (2009) proposed a method for improving the final accuracy of formed and trimmed parts by combining multi-stage forming with a stress-relief annealing process. The proposed multi-stage technique was used to produce a pyramidal benchmark part and a fender section, and the geometric accuracy has improved over single-stage forming. However, due to residual stresses generated during the forming process, the improved geometric accuracy acquired in multi-stage forming is lost after cutting the formed fender section. Then, after multi-stage forming, an extra stress-relief annealing process was used to reduce the final deviations, which resulted in increased part accuracy, although the highest error (4.9 mm) remained significant.

Malhotra et al. (2011b) proposed a multi-stage strategy to reduce final geometric inaccuracy by combining inner centre outward (IO) and outer edge inward (OI) toolpaths for producing each intermediate shape. The researchers provided analytical formulas for predicting and compensating for rigid body motion during the forming process. Simple numerical simulations were used to derive the constants in the analytical model. Based on this, they devised a multi-stage technique and empirically verified its effectiveness in removing the stepped features on the cylinder base, which were first raised and described in Duflou et al. (2008). However, the method for predicting rigid body motion was limited to the stated shape design of a cylindrical cup.

Overall, it was shown that multi-stage forming can be a viable option for increasing geometric accuracy and part formability in SPIF. However, one disadvantage of this method is that the reprocessing of the sheet blank can significantly increase the forming time and may result in reduced final surface quality in some areas of the part. More critically, the current multi-stage techniques are not validated by fundamental models and are only applicable for limited conditions. The majority of the works were based on intuition, experience, or trial and error (Bambach et al., 2009). Deeper knowledge of sheet springback and bending in SPIF is required for the development of more efficient multi-stage forming approaches. However, the multi-stage forming technique is a smart way to improve the material formability and part accuracy of SPIF. Furthermore, the combination of multi-stage methods with other strategies, such as toolpath correction and feedback control in SPIF, would be a potential approach for improving geometric accuracy in SPIF.

## 4.5 TOOLPATH OPTIMIZATION

Toolpath optimization/correction for part accuracy enhancement in SPIF is another strategy that is receiving significant attention (Carette et al., 2019; Li et al., 2019; Dwivedy and Kalluri, 2019). The toolpath has a direct impact on final part geometry

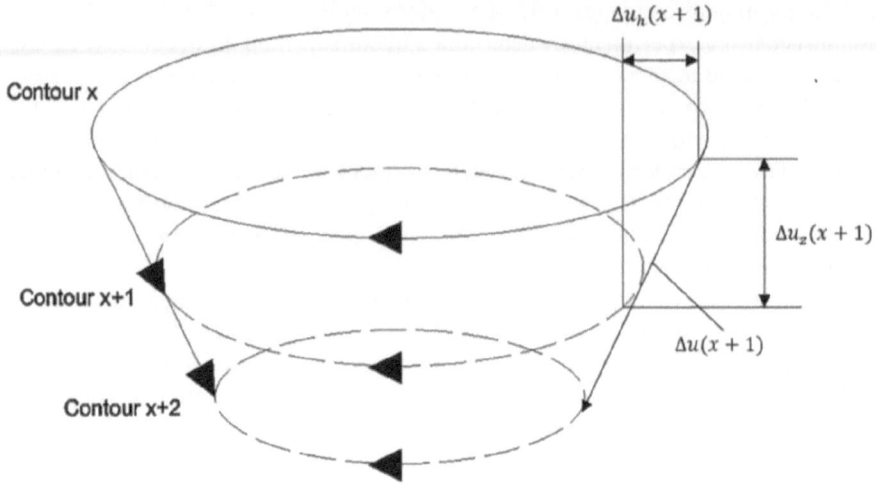

**FIGURE 4.9**   The contour toolpath in SPIF.

as the production of the shape follows the prescribed toolpath. Toolpath optimization/correction can be coupled with the other process strategies to improve the geometrical accuracy even further. A contour toolpath (Figure 4.9) is the most generally used toolpath type, consisting of a series of z-level contours parallel to the horizontal plane created in the milling module of commercial CAM software (Lu et al., 2016; Racz et al., 2019). However, this approach frequently fails to produce pieces with high geometric accuracy. Other toolpath optimization/correction methods have been proposed to overcome this limitation and improve the accuracy of formed parts.

Behera et al. (2013) developed another toolpath compensation technique to improve geometric accuracy in SPIF by predicting the produced shape using multivariate adaptive regression splines (MARS). The formed profile was initially predicted in the designed model based on the detection of the feature category and the interaction between features in the shape design. A corrected single-stage toolpath would be constructed by translating the points in the STL (stereolithography) model, as shown in Figure 4.10, to offset the differences between the predicted and ideal profiles.

A number of case studies showed that adjusted toolpaths were able to minimize the average positive and negative dimensional errors to within 0.4 mm. Finally, it was demonstrated that the MARS toolpath, when integrated with intelligent offsetting of the part features, was capable of improving geometric accuracy even further when compared to using the MARS toolpath alone. Nonetheless, the MARS technique has a significant disadvantage in generating components with features that are close to failure since some zones in the compensated model have larger wall angles than the original design. These angles may surpass the material fracture levels. Further expansion on the work in Behera et al. (2013) was conducted for a case study to produce a human face shape with enhanced geometric accuracy in SPIF using an Al 1050 sheet with a thickness of 1.5 mm (Behera et al., 2014).

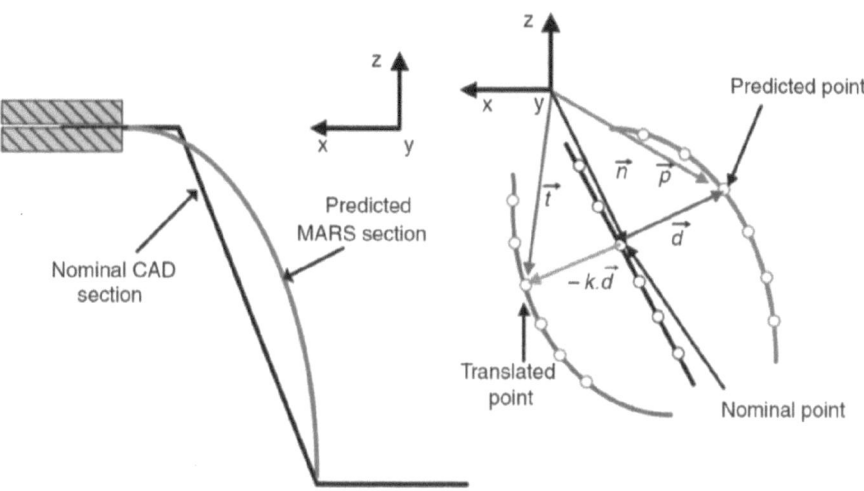

**FIGURE 4.10** Point conversion from the MARS model's estimated shape to the specified CAD shape. (With permission from Behera et al., 2013.)

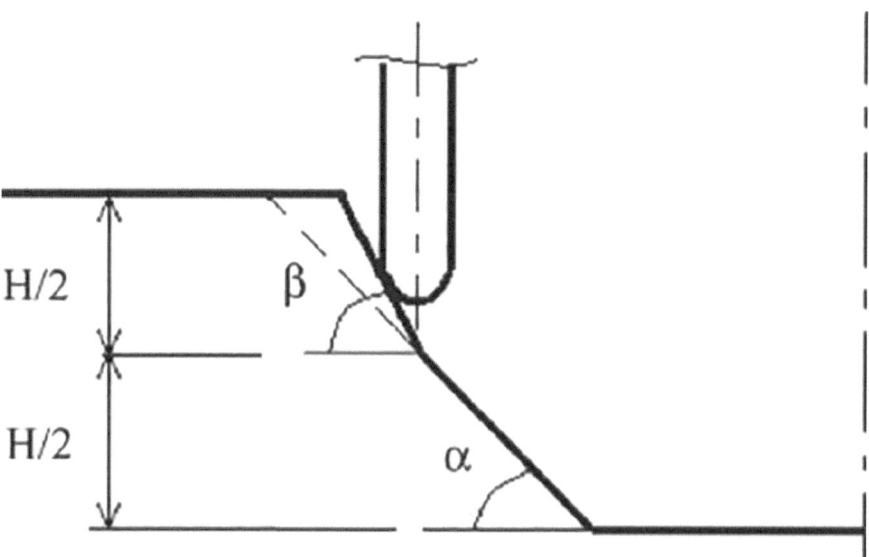

**FIGURE 4.11** The modified toolpath. (With permission from Ambrogio et al., 2004.)

To compensate for the elastic springback, Ambrogio et al. (2004) simply adjusted the toolpath by creating the part of the first half depth on a steeper slope (Figure 4.11). This method was further improved to generate the right shapes utilizing vitiated trajectories; that is, it exploits an inaccurate toolpath to form accurate parts using error compensation (Ambrogio et al., 2005a). Another simple toolpath optimization approach was also proposed for reducing geometric inaccuracy in SPIF (Ambrogio et al., 2005b). In the study, a measurement system comprising of a digital inspector and a CNC open software was utilized to measure the geometry data and alter the tool position. The location of the tool was updated at each loop based on feedback of the error between a specified point on the produced shape profile and the corresponding point on the ideal shape profile. This dynamic tool location update can potentially increase the accuracy of SPIF. The approach is easily implemented as it only considers a single point on the sectional profile to represent the entire shape, and the toolpath modification is based on the offset of this single point.

Malhotra et al. (2010) created a method for automatically generating spiral toolpaths for SPIF. In their method, parallel z-level contours are first produced by adaptive slicing of a CAD model's surface. The spiral path is then constructed via linear interpolation between points on the two subsequent parallel contours, with scallop height and volumetric error limitations. When compared to methods that use commercial CAM toolpaths, the geometric accuracy obtained is comparable or improved, but the processing time is lowered. Lu et al. (2013) proposed yet another feature-based toolpath creation approach. The procedure begins by defining the part's key edges, and parallel contours are designed based on the proper gradients between neighbouring contours and a predetermined scallop height. The spiral toolpath for producing the part would then be developed based on the parallel contours found. The results of one of the test scenarios demonstrated that the proposed feature-based method outperforms the standard z-level-based algorithm in terms of geometric accuracy and surface quality, as well as forming time. However, the achieved geometric inaccuracies are still very considerable (about 1 mm) and should be reduced further.

Lingam et al. (2017) presented a novel tool path generation/optimization technique based on automated feature identification to increase DSIF accuracy. The designed free-form shape was separated into multiple sections with varied features (Figure 4.12a) detected by the automated feature detection prior to toolpath development. The forming sequence of the detected areas is then optimized automatically based on the detected characteristics and the forming mechanics. The outcomes (Figure 4.12b and c) showed that the suggested toolpath generation technique can result in high accuracy in the formation of complex parts with free-form profiles. The maximum inaccuracy between the sectional profiles of the produced and designed parts was 0.3 mm.

To summarize, the toolpath optimization solutions have the potential to improve the geometrical accuracy in ISF. Many strategies for optimizing the toolpath are based on the correction of mistakes measured from previously manufactured components. As a result, many components need to be manufactured prior to obtaining the desired part accuracy.

## 4.6   OTHER STRATEGIES

In addition to the standard strategies discussed previously, there are other alternative strategies that are also being explored for part accuracy improvements, such as parameter investigation/optimization and processing of metal blanks. Ambrogio et al. (2012) have conducted a study to determine the influence of ISF parameters on

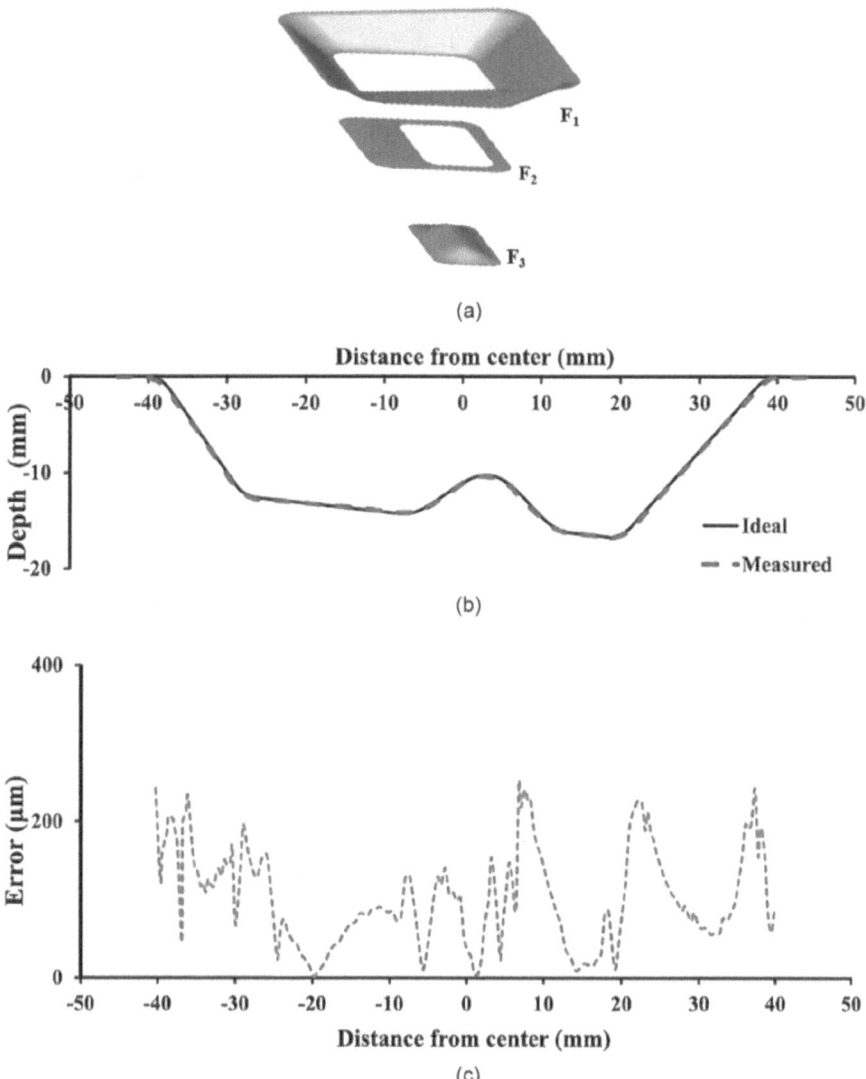

(a)

(b)

(c)

**FIGURE 4.12** (a) Feature recognition, (b) formed cross-sectional profiles compared to the ideal profile, and (c) distribution of errors along the cross-sectional profile. (With permission from Lingam et al., 2017.)

part accuracy. Similarly, Lu et al. (2014) have evaluated the effects of step depth, a significant toolpath parameter set by the user, and found that a smaller step depth value leads to improved geometric precision and part surface quality in the ISF process. However, too small a step depth value should be avoided to avoid material fracture, particularly when making parts with significant wall angles. Similarly, Attanasio et al. (2008) and Radu and Cristea (2013) have conducted studies on the effect of parameter selection on the formed geometric accuracy.

Li et al. (2015) investigated the impact of four important factors (step down, sheet thickness, wall angle, and tool diameter) on the geometric accuracy of ISF using the response surface approach with Box–Behnken design. Gatea et al. (2016) published a complete evaluation of the impact of process factors in incremental sheet formation and have summarized the influence of various processing factors on geometric accuracy and springback. It was determined that springback, a major issue in ISF geometric inaccuracy, could be decreased by reducing the vertical step size and increasing the tool diameter, feed rate, spindle speed, and sheet thickness. Furthermore, increasing the tensile force in the deformed part's wall might minimize ISF springback. Allwood et al. (2010) recommended using partly cut-out blanks with tabs or holes formed at the target part's perimeter. This minimizes geometric deviations by concentrating deformations in the area where the tool moves. However, the results showed that cut-outs did not significantly enhance part accuracy. It was also found that the previous approach of localizing the deformation area surrounding the tool is very important in the production of more accurate parts using ISF.

### 4.6.1 SPIF Process of Tailor Welded Blanks

There are a few issues during SPIF of tailor welded blanks (TWBs) related to the dimensional accuracy. In SPIF or other sheet metal forming processes, dimensional accuracy can be described as deviation of the profile of the formed part as compared to the targeted one. Other similar terminologies are shape accuracy and geometrical error. The most common is the springback, as discussed in the previous sections. In TWB, the control and compensation may be challenging as the material properties and behaviour may be different from those of the common metal sheet. TWB may be combined by dissimilar materials or thicknesses, and these parameters are among the most significant effects to the springback in these specific matters. Many efforts are made in particular to the springback compensation; for example, Oleksik et al. (2021) in their work compared welded blanks to normal blanks and found that formability reduced for welded blanks and the springback increased 3.5 times. Furthermore, as compared to the single-side weld, the double-side weld improved formability of the blank. In another attempt, Ebrahimzadeh et al. (2018) found that springback was affected by the sequence of tool rotary speed and feed rate. Interestingly, they found that SPIF has less formability compared to DPIF techniques for the case of the AA5083 friction stir-welded blank.

### 4.6.2 Conclusions and Recommendation for Future Research

This chapter has presented a comprehensive literature assessment of recent studies in the development of SPIF technology and dimensional accuracy-related issues. This information can be useful for researchers and practitioners interested in the technology, especially on enhancing the accuracy of the process. In this chapter, the various strategies for improving part accuracy in SPIF are reviewed and summarized into four categories: (1) SPIF variation development, (2) multi-stage strategies, (3) toolpath correction/optimization, and (4) other alternative strategies. In particular, the toolpath correction/optimization techniques are gaining interest of researchers as they are proven to be two of the most effective methods for increasing geometric accuracy in SPIF. Newly developed variations and hybrids of SPIF present new directions in the development of the technology, especially in addressing the limitations of geometric accuracy. Multi-stage techniques in SPIF could improve component formability as well as improve geometric accuracy, although currently there are only a limited number of profiles. Further study is needed to build appropriate optimization models for generalized multi-stage techniques to create different sectional profiles. Other researchers have focused on optimizing the process parameters for increased geometric accuracy. Future studies may consider the combinations of the techniques from the five categories, as well as additional unmentioned strategies, for improving geometric accuracy and giving better solutions to the limitations in SPIF processes.

A variety of solutions and recommendations have been presented for improving the dimensional accuracy of tailor blanks in SPIF in general. However, specific springback behaviour is still lacking as the TWB behaviour may not be the same as that of typical blanks. These solutions can reduce dimensional inaccuracies to various degrees. However, dimensional accuracy of SPIF needs to be improved further for it to be suitable for industrial applications. These criteria can be the topic for further investigation in enhancing the dimensional accuracy in SPIF-made components.

First, the scope of investigation could be expanded to fully consider all related SPIF parameters such as the toolpath parameters, sheet thickness, sheet material type, tool diameter, tool feed speed, scallop height, forming forces, clamping, supporting devices, and forming pressure of the clamping, among others. Further studies can also consider unclamped errors and trimmed errors as most earlier studies have only focused on clamped errors. Second, low dimensional accuracy can be overcome by incorporation of various strategies, such as combining the multi-stage strategy with the toolpath optimization technique. The multi-stage strategy is a promising technique to address challenges associated with formality, sheet thinning, and material fracture in SPIF to produce parts that cannot be formed effectively in a single-stage SPIF. Finally, a general database of SPIF parameters for a wide variety of sheet materials, sheet thickness values, and part sizes can be developed for reference to industrial users.

## ACKNOWLEDGEMENT

The author would like to acknowledge the support from Universiti Sains Malaysia (USM) for providing the access to technical papers and journal articles.

## REFERENCES

Ai, S., & Long, H. (2019). A review on material fracture mechanism in incremental sheet forming. *International Journal of Advanced Manufacturing Technology, 104*(1–4), 33–61. doi: 10.1007/s00170-019-03682-6.

Akrichi, S., Abbassi, A., Abid, S., & Ben yahia, N. (2019). Roundness and positioning deviation prediction in single point incremental forming using deep learning approaches. *Advances in Mechanical Engineering, 11*(7), 1–15. doi: 10.1177/1687814019864465.

Allwood, J. M, King, G. P. F., & Duflou, J. (2005). A structured search for applications of the incremental sheet-forming process by product segmentation. *Proceedings of the Institution of Mechanical Engineers, Part B, 219*, 239–244. doi: 10.1243/095440505X8145.

Allwood, J. M, Braun, D., & Music, O. (2010). The effect of partially cut-out blanks on geometric accuracy in incremental sheet forming. *Journal of Materials Processing Technolgy, 210*(11), 1501–1510. doi: 10.1016/j.jmatprotec.2010.04.008.

Alves De Sousa, R. J., Ferreira, J. A. F., Sa De Farias, J. B., Torrão, J. N. D., Afonso, D. G., & Martins, M. A. B. E. (2014). SPIF-A: On the development of a new concept of incremental forming machine. *Structural Engineering and Mechanics, 49*(5), 645–660. doi: 10.12989/sem.2014.49.5.645.

Ambrogio, G., Costantino, I., De Napoli, L., Filice, L., Fratini, L., & Muzzupappa, M. (2004). Influence of some relevant process parameters on the dimensional accuracy in incremental forming: A numerical and experimental investigation. *Journal of Materials Processing Technology, 153–154*(1–3), 501–507. doi: 10.1016/j.jmatprotec.2004.04.139.

Ambrogio, G., De Napoli, L., Filice, L., Gagliardi, F., & Muzzupappa, M. (2005a). Application of incremental forming process for high customised medical product manufacturing. *Journal of Materials Processing Technology, 162–163*(SPEC. ISS.), 156–162. doi: 10.1016/j.jmatprotec.2005.02.148.

Ambrogio, G., Filice, L., De Napoli, L., & Muzzupappa, M. (2005b). A simple approach for reducing profile diverting in a single point incremental forming process. *Proceedings of the Institution of Mechanical Engineers, Part B: Journal of Engineering Manufacture, 219*(11), 823–830. doi: 10.1243/095440505X32797.

Ambrogio, G., Filice, L., & Gagliardi, F. (2012). Improving industrial suitability of incremental sheet forming process. *International Journal of Advanced Manufacturing Technology, 58*(9–12), 941–947. doi: 10.1007/s00170-011-3448-6.

Attanasio, A., Ceretti, E., Giardini, C., & Mazzoni, L. (2008). Asymmetric two points incremental forming: Improving surface quality and geometric accuracy by tool path optimization. *Journal of Materials Processing Technology, 197*(1–3), 59–67. doi: 10.1016/j.jmatprotec.2007.05.053.

Bambach, M., Taleb Araghi, B., & Hirt, G. (2009). Strategies to improve the geometric accuracy in asymmetric single point incremental forming. *Production Engineering, 3*(2), 145–156. doi: 10.1007/s11740-009-0150-8.

Bedan, A. S., & Habeeb, H. A. (2018). Experimental study the effect of tool geometry on dimensional accuracy in single point incremental forming (SPIF) process. *Al-Nahrain Journal for Engineering Sciences, 21*(1), 108. doi: 10.29194/njes21010108.

Behera, A. K., Verbert, J., Lauwers, B., & Duflou, J. R. (2013). Tool path compensation strategies for single point incremental sheet forming using multivariate adaptive regression splines. *CAD Computer Aided Design, 45*(3), 575–590. doi: 10.1016/j.cad.2012.10.045.

Behera, A. K., Lauwers, B., & Duflou, J. R. (2014). Tool path generation framework for accurate manufacture of complex 3D sheet metal parts using single point incremental forming. *Computers in Industry, 65*(4), 563–584. doi: 10.1016/j.compind.2014.01.002.

Behera, A. K., de Sousa, R. A., Ingarao, G., & Oleksik, V. (2017). Single point incremental forming: An assessment of the progress and technology trends from 2005 to 2015. *Journal of Manufacturing Processes, 27*, 37–62. doi: 10.1016/j.jmapro.2017.03.014.

Carette, Y., Vanhove, H., & Duflou, J. (2019). Multi-step incremental forming using local feature based toolpaths. *Procedia Manufacturing, 29*, 28–35. doi: 10.1016/j.promfg.2019.02.102.

Cédric, B., Pierrick, M., & Sébastien, T. (2020). Shape accuracy improvement obtained by μ-SPIF by tool path compensation. *Procedia Manufacturing, 47*, 1399–1402. doi: 10.1016/j.promfg.2020.04.293.

Dai, P., Chang, Z., Li, M., & Chen, J. (2019). Reduction of geometric deviation by multi-pass incremental forming combined with tool path compensation for non-axisymmetric aluminum alloy component with stepped feature. *International Journal of Advanced Manufacturing Technology, 102*(1–4), 809–817. doi: 10.1007/s00170-018-3194-0.

Dakhli, M., Boulila, A., & Tourki, Z. (2017). Effect of generatrix profile on single-point incremental forming parameters. *International Journal of Advanced Manufacturing Technology, 93*(5–8), 2505–2516. doi: 10.1007/s00170-017-0598-1.

Do, V.-C., Lee, B.-H., Yang, S.-H., & Kim, Y.-S. (2017). The forming characteristic in the single-point incremental forming of a complex shape. *Intnational Jornal of Nanomanufacturing, 13*(1), 33–42.

Duflou, J. R., Callebaut, B., Verbert, J., & De Baerdemaeker, H. (2007). Laser assisted incremental forming: Formability and accuracy improvement. *CIRP Annals: Manufacturing Technology, 56*(1), 273–276. doi: 10.1016/j.cirp.2007.05.063.

Duflou, J. R., Verbert, J., Belkassem, B., Gu, J., Sol, H., Henrard, C., & Habraken, A. M. (2008). Process window enhancement for single point incremental forming through multi-step toolpaths. *CIRP Annals: Manufacturing Technology, 57*(1), 253–256. doi: 10.1016/j.cirp.2008.03.030.

Duflou, J. R., Habraken, A. M., Cao, J., Malhotra, R., Bambach, M., Adams, D., … Jeswiet, J. (2018). Single point incremental forming: State-of-the-art and prospects. *International Journal of Material Forming, 11*(6), 743–773. doi: 10.1007/s12289-017-1387-y.

Dwivedy, M., & Kalluri, V. (2019). The effect of process parameters on forming forces in single point incremental forming. *Procedia Manufacturing, 29*, 120–128. doi: 10.1016/j.promfg.2019.02.116.

Ebrahimzadeh, P., Baseri, H., & Mirnia, M. J. (2018). Formability of aluminum 5083 friction stir welded blank in two-point incremental forming process. *Proceedings of the Institution of Mechanical Engineers, Part E: Journal of Process Mechanical Engineering, 232*(3), 267–280.

Edwards, W. L., Grimm, T. J., Ragai, I., & Roth, J. T. (2017). Optimum process parameters for springback reduction of single point incrementally formed polycarbonate. *Procedia Manufacturing, 10*, 329–338. doi: 10.1016/j.promfg.2017.07.002.

Essa, K., & Hartley, P. (2011). An assessment of various process strategies for improving precision in single point incremental forming. *International Journal of Material Forming, 4*(4), 401–412. doi: 10.1007/s12289-010-1004-9.

Fan, G., & Gao, L. (2014). Numerical simulation and experimental investigation to improve the dimensional accuracy in electric hot incremental forming of Ti-6Al-4V titanium sheet. *International Journal of Advanced Manufacturing Technology, 72*(5–8), 1133–1141. doi: 10.1007/s00170-014-5769-8.

Fiorentino, A., Attanasio, A., Marzi, R., Ceretti, E., & Giardini, C. (2011). On forces, formability and geometrical error in metal incremental sheet forming. *International Journal of Materials and Product Technology, 40*(3–4), 277–295. doi: 10.1504/IJMPT.2011.039936.

Formisano, A., Boccarusso, L., Capece Minutolo, F., Carrino, L., Durante, M., & Langella, A. (2017). Negative and positive incremental forming: Comparison by geometrical, experimental, and FEM considerations. *Materials and Manufacturing Processes, 32*(5), 530–536. doi: 10.1080/10426914.2016.1232810.

Gatea, S., Ou, H., & McCartney, G. (2016). Review on the influence of process parameters in incremental sheet forming. *International Journal of Advanced Manufacturing Technology, 87*(1–4), 479–499. doi: 10.1007/s00170-016-8426-6.

Giraud-Moreau, L., Belchior, J., Lafon, P., Lotoing, L., Cherouat, A., Courtielle, E., … Maurine, P. (2018). Springback effects during single point incremental forming: Optimization of the tool path. *AIP Conference Proceedings, 1960*. doi: 10.1063/1.5035035.

Han, F., Mo, J. H., Qi, H. W., Long, R. F., Cui, X. H., & Li, Z. W. (2013). Springback prediction for incremental sheet forming based on FEM-PSONN technology. *Transactions of Nonferrous Metals Society of China (English Edition)*, 23(4), 1061–1071. doi: 10.1016/S1003-6326(13)62567-4.

Hussain, G. (2014). Experimental investigations on the role of tool size in causing and controlling defects in single point incremental forming process. *Proceedings of the Institution of Mechanical Engineers, Part B: Journal of Engineering Manufacture*, 228(2), 266–277. doi: 10.1177/0954405413498864.

Hussain, G., Lin, G., & Hayat, N. (2011). Improving profile accuracy in SPIF process through statistical optimization of forming parameters. *Journal of Mechanical Science and Technology*, 25(1), 177–182. doi: 10.1007/s12206-010-1018-8.

Jagtap, R., & Kumar, S. (2019). Incremental sheet forming: An experimental study on the geometric accuracy of formed parts. In: U. Chandrasekhar, L.-J. Yang, & S. Gowthaman (Eds.), *Innovative Design, Analysis and Development Practices in Aerospace and Automotive Engineering (i-dad 2018): Volume 1*, Lecture Notes in Mechanical Engineering. Springer: Berlin, Germany. doi: 10.1007/978-981-13-2697-4_9.

Kovine, P., Bayram, H., & Köksal, N. S. (2017). Investigation of the geometrical accuracy and thickness distribution using 3D laser scanning of AA2024-T3 sheets formed by SPIF. *Materials and Technology*, 51(1), 111–116. doi: 10.17222/mit.2015.296.

Kulin, S. A., & Dodiya, P. H. R. (2019). A review on development of toolpath strategies in incremental sheet forming. *International Journal of Technical Innovation in Modern Engineering & Science (IJTIMES)*, 4(6), 1–7.

Kumar, A., Gulati, V., & Kumar, P. (2019). Experimental investigation of forming forces in single point incremental forming. *Lecture Notes in Mechanical Engineering*, 10(1), 423–430. doi: 10.1007/978-981-13-6412-9_41.

Li, Y., Lu, H., Daniel, W. J. T., & Meehan, P. A. (2015). Investigation and optimization of deformation energy and geometric accuracy in the incremental sheet forming process using response surface methodology. *International Journal of Advanced Manufacturing Technology*, 79(9–12), 2041–2055. doi: 10.1007/s00170-015-6986-5.

Li, Z., Lu, S., & Chen, P. (2017a). Improvement of dimensional accuracy based on multistage single point incremental forming of a straight wall cylinder part. *International Journal of Precision Engineering and Manufacturing*, 18(9), 1281–1286. doi: 10.1007/s12541-017-0151-z.

Li, Z., Lu, S., Zhang, T., Mao, Z., & Zhang, C. (2017b). Analysis of geometrical accuracy based on multistage single point incremental forming of a straight wall box part. *International Journal of Advanced Manufacturing Technology*, 93(5–8), 2783–2789. doi: 10.1007/s00170-017-0723-1.

Li, C. L., Ni, Y. K., Haw, Y., & Lu, Y. H. (2019). Tool path and sheet thickness distribution of axisymmetric cup in single point incremental forming. *Key Engineering Materials*, 823, 1–7. doi: 10.4028/www.scientific.net/KEM.823.1.

Lingam, R., Prakash, O., Belk, J. H., & Reddy, N. V. (2017). Automatic feature recognition and tool path strategies for enhancing accuracy in double sided incremental forming. *International Journal of Advanced Manufacturing Technology*, 88(5–8), 1639–1655. doi: 10.1007/s00170-016-8880-1.

Lu, B., Chen, J., Ou, H., & Cao, J. (2013). Feature-based tool path generation approach for incremental sheet forming process. *Journal of Materials Processing Technology*, 213(7), 1221–1233. doi: 10.1016/j.jmatprotec.2013.01.023.

Lu, H. B., Li, Y. L., Liu, Z. B., Liu, S., & Meehan, P. A. (2014). Study on step depth for part accuracy improvement in incremental sheet forming process. *Advanced Materials Research*, 939, 274–280. doi: 10.4028/www.scientific.net/AMR.939.274.

Lu, H., Kearney, M., Li, Y., Liu, S., Daniel, W. J. T., & Meehan, P. A. (2016). Model predictive control of incremental sheet forming for geometric accuracy improvement. *International Journal of Advanced Manufacturing Technology*, 82(9–12), 1781–1794. doi: 10.1007/s00170-015-7431-5.

Lu, H., Liu, H., & Wang, C. (2019). Review on strategies for geometric accuracy improvement in incremental sheet forming. *International Journal of Advanced Manufacturing Technology, 102*(9–12), 3381–3417. doi: 10.1007/s00170-019-03348-3.

Malhotra, R., Reddy, N. V., & Cao, J. (2010). Automatic 3D spiral toolpath generation for single point incremental forming. *Journal of Manufacturing Science and Engineering, Transactions of the ASME, 132*(6), 1–10. doi: 10.1115/1.4002544.

Malhotra, R., Bhattacharya, A., Kumar, A., Reddy, N. V., & Cao, J. (2011a). A new methodology for multi-pass single point incremental forming with mixed toolpaths. *CIRP Annals: Manufacturing Technology, 60*(1), 323–326. doi: 10.1016/j.cirp.2011.03.145.

Malhotra, R., Cao, J., Ren, F., Kiridena, V., Cedric Xia, Z., & Reddy, N. V. (2011b). Improvement of geometric accuracy in incremental forming by using a squeezing toolpath strategy with two forming tools. *Journal of Manufacturing Science and Engineering, Transactions of the ASME, 133*(6). doi: 10.1115/1.4005179.

Maqbool, F., & Bambach, M. (2018). Dominant deformation mechanisms in single point incremental forming (SPIF) and their effect on geometrical accuracy. *International Journal of Mechanical Sciences, 136*, 279–292. doi: 10.1016/j.ijmecsci.2017.12.053.

Marabuto, S. R., Afonso, D., Ferreira, J. A. F., Melo, F. Q., Martins, M., & De Alves Sousa, R. J. (2011). Finding the best machine for SPIF operations. A brief discussion. *Key Engineering Materials, 473*, 861–868. doi: 10.4028/www.scientific.net/KEM.473.861.

Meier, H., Buff, B., Laurischkat, R., & Smukala, V. (2009). Increasing the part accuracy in dieless robot-based incremental sheet metal forming. *CIRP Annals: Manufacturing Technology, 58*(1), 233–238. doi: 10.1016/j.cirp.2009.03.056.

Micari, F., Ambrogio, G., & Filice, L. (2007). Shape and dimensional accuracy in single point incremental forming: State of the art and future trends. *Journal of Materials Processing Technology, 191*(1–3), 390–395. doi: 10.1016/j.jmatprotec.2007.03.066.

Najafabady, S. A., & Ghaei, A. (2016). An experimental study on dimensional accuracy, surface quality, and hardness of Ti-6Al-4 V titanium alloy sheet in hot incremental forming. *International Journal of Advanced Manufacturing Technology, 87*(9–12), 3579–3588. doi: 10.1007/s00170-016-8712-3.

Najm, S. M., & Paniti, I. (2020). Study on effecting parameters of flat and hemispherical end tools in spif of aluminium foils. *Tehnicki Vjesnik, 27*(6), 1844–1849. doi: 10.17559/TV-20190513181910.

Oleksik, V., Dobrota, D., Racz, S. G., Rusu, G. P., Popp, M. O., & Avrigean, E. (2021). Experimental research on the behaviour of metal active gas tailor welded blanks during single point incremental forming process. *Metals*, 11, 198.

Peng, W., Ou, H., & Becker, A. (2019). Double-sided incremental forming: A review. *Journal of Manufacturing Science and Engineering, Transactions of the ASME, 141*(5). doi: 10.1115/1.4043173.

Portman, V. T. (2011). Stiffness evaluation of machines and robots: Minimum collinear stiffness value approach. *Journal of Mechanisms and Robotics, 3*(1). doi: 10.1115/1.4003444.

Praveen, K., Lingam, R., & Venkata Reddy, N. (2020). Tool path design system to enhance accuracy during double sided incremental forming: An analytical model to predict compensations for small/large components. *Journal of Manufacturing Processes, 58*, 510–523. doi: 10.1016/j.jmapro.2020.08.014.

Racz, G. S., Breaz, R. E., Oleksik, V. S., Bologa, O. C., & Brîndau, P. D. (2019). Simulated 3-axis versus 5-axis processing toolpaths for single point incremental forming. *IOP Conference Series: Materials Science and Engineering, 564*(1). doi: 10.1088/1757-899X/564/1/012023.

Radu, M. C., & Cristea, I. (2013). Processing metal sheets by SPIF and analysis of parts quality. *Materials and Manufacturing Processes, 28*(3), 287–293. doi: 10.1080/10426914.2012.746702.

Rakesh, L., Amit, S., & Reddy, N. V. (2016). Deflection compensations for tool path to enhance accuracy during double-sided incremental forming. *Journal of Manufacturing Science and Engineering, Transactions of the ASME, 138*(9), 1–11. doi: 10.1115/1.4033956.

Reddy, N.V., Lingam, R., & Cao, J. (2015). Incremental metal forming process in manufacturing. In *Handbook of Manufacturing Engineering and Technology*, Springer-Verlag: London.

Shrivastava, P., Kumar, P., Tandon, P., & Pesin, A. (2018). Improvement in formability and geometrical accuracy of incrementally formed AA1050 sheets by microstructure and texture reformation through preheating, and their FEA and experimental validation. *Journal of the Brazilian Society of Mechanical Sciences and Engineering, 40*(7), 1–15. doi: 10.1007/s40430-018-1255-9.

Skjoedt, M., Bay, N., Endelt, B., & Ingarao, G. (2008). Multi stage strategies for single point incremental forming of a cup. *International Journal of Material Forming, 1*(SUPPL. 1), 1199–1202. doi: 10.1007/s12289-008-0156-3.

Skjoedt, M., Silva, M. B., Martins, P. A. F., & Bay, N. (2010). Strategies and limits in multistage single-point incremental forming. *The Journal of Strain Analysis for Engineering Design*. doi: 10.1243/03093247JSA574.

Trzepieciński, T., Oleksik, V., Pepelnjak, T., Najm, S. M., Paniti, I., & Maji, K. (2021). Emerging trends in single point incremental sheet forming of lightweight metals. *Metals, 11*(8). doi: 10.3390/met11081188.

Verbert, J., Belkassem, B., Henrard, C., Habraken, A. M., Gu, J., Sol, H., … Duflou, J. R. (2008). Multi-step toolpath approach to overcome forming limitations in single point incremental forming. *International Journal of Material Forming, 1*(SUPPL. 1), 1203–1206. doi: 10.1007/s12289-008-0157-2.

Verbert, J., Aerens, R., Vanhove, H., Aertbeliën, E., & Duflou, J. R. (2009). Obtainable accuracies and compensation strategies for robot supported SPIF. *Key Engineering Materials, 410–411*, 679–687. doi: 10.4028/www.scientific.net/KEM.410-411.679.

Wang, Y., Huang, Y., Cao, J., & Reddy, N. V. (2009). Experimental study on a new method of double side incremental forming. *Proceedings of the ASME International Manufacturing Science and Engineering Conference, MSEC2008, 1*, 601–607. doi: 10.1115/MSEC_ICMP2008-72279.

Wang, H., Zhang, R., Zhang, H., Hu, Q., & Chen, J. (2018). Novel strategies to reduce the springback for double-sided incremental forming. *International Journal of Advanced Manufacturing Technology, 96*(1–4), 973–979. doi: 10.1007/s00170-018-1659-9.

Wei, H., Chen, W., & Gao, L. (2011). Spring back investigation on sheet metal incremental formed parts. *World Academy of Science, Engineering and Technology, 79*(7), 285–289. doi: 10.5281/zenodo.1054841.

Yao, Z., Li, Y., Yang, M., Yuan, Q., & Shi, P. (2017). Parameter optimization for deformation energy and forming quality in single point incremental forming process using response surface methodology. *Advances in Mechanical Engineering, 9*(7), 1–15. doi: 10.1177/1687814017710118.

Zhang, Z., Ren, H., Xu, R., Moser, N., Smith, J., Ndip-Agbor, E. & Cao, J. (2015). A mixed double-sided incremental forming toolpath strategy for improved geometric accuracy. *Journal of Manufacturing Science and Engineering, Transactions of the ASME, 137*(5). doi: 10.1115/1.4031092.

Zhang, S., Tang, G. H., Li, Z., Jiang, X., & Li, K. J. (2020). Experimental investigation on the springback of AZ31B Mg alloys in warm incremental sheet forming assisted with oil bath heating. *International Journal of Advanced Manufacturing Technology, 109*(1–2), 535–551. doi: 10.1007/s00170-020-05678-z.

Zhu, H., Wang, H., & Liu, Y. (2019). Tool path generation for the point-pressing-based 5-axis CNC incremental forming. *International Journal of Advanced Manufacturing Technology, 103*(9–12), 3459–3477. doi: 10.1007/s00170-019-03756-5.

# 5 Springback in Tailor Welded Blanks Using a Heuristic Approach

*SS Panicker*
BITS Pilani-K.K. Birla Goa Campus

*H Krishnaswamy*
Indian Institute of Technology Madras

*N Chakraborti*
Czech Technical University in Prague

## CONTENTS

DOI: 10.1201/9781003164241-5

## 5.1   INTRODUCTION

Weight reduction of vehicles for improving the performance and fuel efficiency to comply with emerging zero-emission strategies is an ongoing challenge for the automotive industry. For instance, a 10% reduction in vehicle weight increases the fuel efficiency by 5%–8%, and a reduction of 100 kg mass of a passenger car cuts the $CO_2$ emission by about 12.5 g/km [1]. Weight reduction also helps to reduce the power required for acceleration and braking and in turn enable the design of small-sized engines and power-transmission and braking systems. With the introduction of electric vehicles, the focus on lightweighting strategies continues to remain a priority [2]. One of the major strategies adopted to reduce the vehicle mass is the use of lightweight materials to manufacture the automotive body components, that is, body in white (BIW), outer and inner hoods, floor pan, seat side member, muffler shell, and so on. The geometric complexity of these thin-walled components continuously increases owing to improved aerodynamic design, ergonomics, and aesthetics of the vehicle body. Any lightweighting solutions should be made without compromising the strength of the structure to meet the required safety requirements.

A direct solution for lightweighting is to replace the conventional steel sheets with materials having equivalent strength to weight ratios* such as new-age high strength steels (HSLA, DP980, DP600, TWIP, etc.), aluminium alloys, and magnesium alloys. However, the room-temperature formability of aluminium and magnesium alloy sheets is poor and exhibits high springback while forming, when compared to steel counterparts [3–7]. The cost-effectiveness while using the modern age steel sheets is also a factor that limits its widespread application. Apart from changing the material, significant weight reduction can also be achieved by modifying the design of the component without compromising the required functional requirements. In either of the above approaches, namely, change of material and/or design of components, the thickness of the initial blank is chosen to prevent failure (either strength-based or stiffness-based failure) in critical areas of the component. Such an approach often results in unnecessary material distribution in less critical areas. The ideal case is to design components by distributing just the right amount of material at different locations depending on its service requirements. Theoretically, a mosaic of different materials with varying thicknesses should be used to form sheet metal components (Figure 5.1a).

The concept of tailor welded blank (TWB) is a pragmatic adaptation of this idea where materials of different grades and/or thicknesses are joined prior to the forming process. To acquire overall body weight reduction, the optimization in the selection of sheet metal grade, thickness, and formability is necessary. One such optimization method is the usage of TWBs instead of monolithic sheets [9,10]. Figure 5.1b

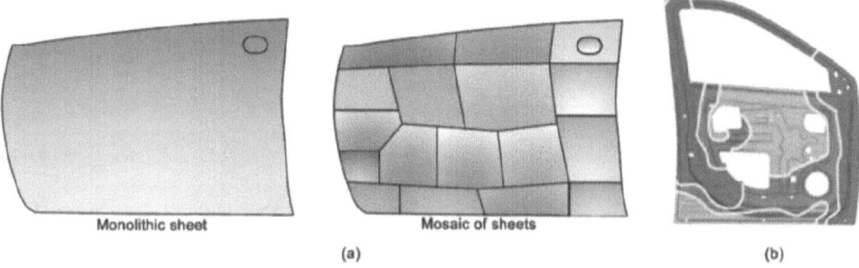

| Monolithic sheet | Mosaic of sheets | |
| --- | --- | --- |
| (a) | | (b) |

**FIGURE 5.1** (a) Schematic to explain the monolithic blank (left) of constant thickness and the tailored blank combining different material grades and thicknesses and (b) topology of an inner door panel obtained through multi-objective optimization. (Reproduced by permission from Sun et al. [8] Copyright© 2018 Elsevier Ltd.)

shows such an attempt for the manufacturing an automotive door panel through multi-objective optimization [8]. However, the suggested theoretical (mosaic) and even the presented optimized blank approaches face practical challenges. The major constraints identified are as follows: (1) sophistication in welding lines directly affects the easiness of manufacturing [11]; (2) the complex welding routes increase the uncertainties of loading, welding, and forming that will significantly affect the performance of TWB structures [12]; and (3) the critical and complicated mechanical behaviours of HAZ between different materials [13,14], affecting the structural performance of TWBs.

### 5.1.1 TAILOR WELDED BLANKS

As the name signifies, TWB is a semi-finished sheet metal blank manufactured by joining two or more blanks together in a plane prior to forming, as shown in Figure 5.2a [8]. Depending upon the requirement, the sheets to be welded (similar or dissimilar materials) can possess different mechanical properties, microstructures, thicknesses, and surface coatings. Such high level customization is attained by applying localized engineered materials, essential only to the distinct locations, rather than allocating them through the entire body panels [9]. Let us consider the door inner[†] of automotive door assembly. Typically, the service requirements of the component are not unique. For instance, the door assembly is supported at hinges in the front pillar of the automotive structure and must be strong and stiff to bear the entire load of the door assembly (Figure 5.3a). The local deflection should not exceed the tolerance limit for smooth opening and closure of doors. Doors are often slammed hard (Figure 5.3b), and the posterior end of the door should be tough during such repeated events. In addition to these, the doors must be resistant to dynamic fatigue loads, side crash (Figure 5.3c), and dent[‡] (Figure 5.3d). Thus, a single sheet component is subjected to different load cycles, and the critical region under each condition varies. A sheet component of monolithic blank cannot meet all the requirements. Currently, the components can be complemented with local stiffeners and other strengthening elements, although this adds to the manufacturing assembly cost and weight.

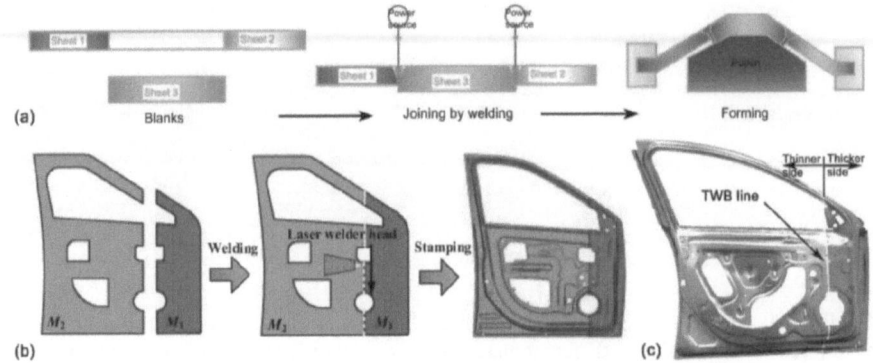

**FIGURE 5.2** (a) Schematic of TWB forming and (b) actual sequence for TWB and stamping of the car door panel. (Reproduced by permission from Sun et al. [8] Copyright© 2018 Elsevier Ltd.) (c) Photograph of an actual stamped inner door panel using TWB. (Courtesy of TWB Company, LLC§.)

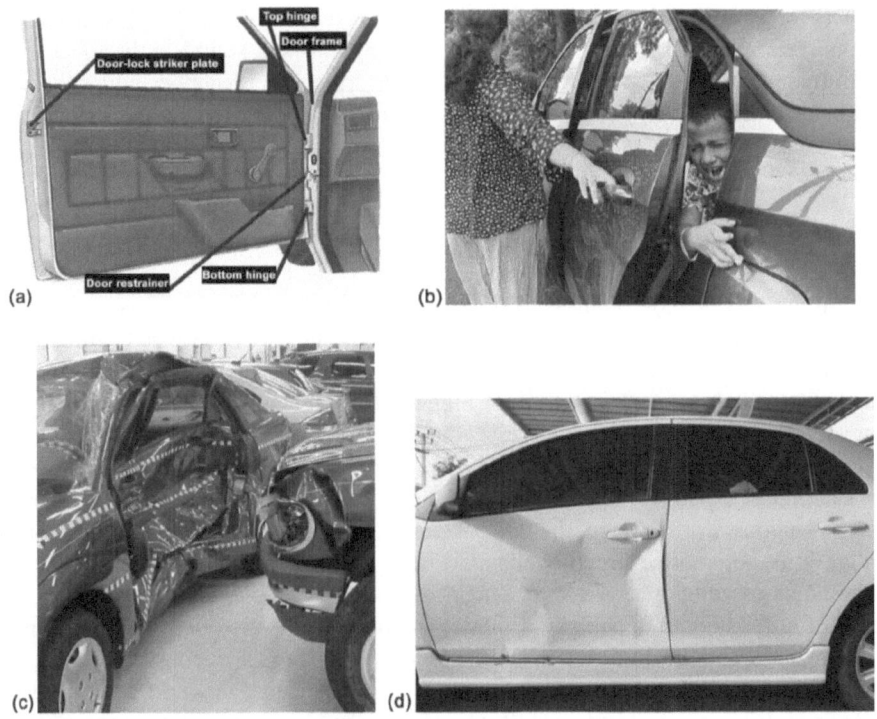

**FIGURE 5.3** (a) Hinges and pins for car door support. (Courtesy of www.howacarworks.com.**) (b) Impact of hard slamming of a car door causing injuries.†† (c) Side crash due to a severe accident. (Courtesy of Wikipedia.‡‡) (d) Large dent in a car door. (Courtesy of center-valleyauto.com.§§)

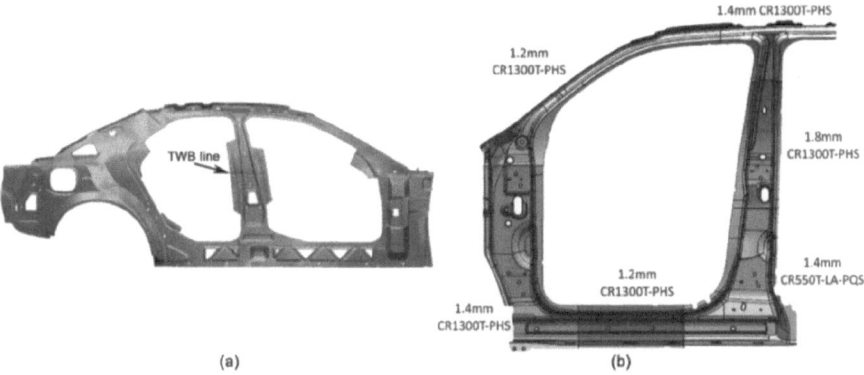

**FIGURE 5.4** (a) Stamped side frame from TWBs (Courtesy of TWB Company, LLC.***) (b) six-pieces laser welded blanks with differential thicknesses and grades used to reduce the overall weight of cars. (Reproduced by permission from Gestamp and Stellantis/FCA [18].)

Alternately, a suitable combination of sheets can be "tailored" to eliminate or at least reduce such supports and stiffeners, improve the crashworthiness, and thereby reduce the weight of the finished part. A typical TWB stamping method is shown in Figure 5.2b. Figure 5.2c shows such an inner door panel made from friction stir welded TWB with the thicker gauge toward the hinge side and the thinner gauge toward the other side. Additionally, an indirect benefit is the improved material utilization. For instance, a conventional car door involves forming a very large rectangular blank that is deformed to shape, where the central hollow region is scrapped. The usage of TWBs as shown in Figure 5.4, where small pieces are joined, will drastically reduce the input blank weight.

The most common methods for joining these small blanks are mash seam welding, laser beam welding, and friction stir welding (FSW) [15]. Other established methods such as induction welding and electron beam welding, although rarely used, can also be considered [16,17]. Table 5.1 outlines the compatibility of different welding techniques for automobile TWB manufacturing. Figure 5.5 shows the schematics of mash-seam, laser, and FSW techniques along with respective weld zone interpretations. Mash-seam welding is a type of resistance welding technique where the metallic sheets are overlapped and joined by moving electrode disks (Figure 5.5a). Appropriate current and voltage applied through the disc along with squeezing action melt the overlaying surfaces. The process has excellent productivity, although it requires a controlled environment to reduce electrode surface pollution, especially for coated steel sheets.

Laser welding is a fusion welding process which uses a focused high intensity laser beam irradiated at the well-prepared square-edged joint (Figure 5.5b). The high energy density by the beam instantly melts and evaporates the interface metals. A keyhole ($10^{-3}$m range) is formed due to recoil pressure of evaporation, and the wall is further heated to act as a heat source. Thus, the joining region is melted further to create the welding process. The process produces narrow deep welds at the joint. Laser welding is highly favoured for manufacturing TWBs as it facilitates remote

**TABLE 5.1**

**Comparative Trade-Off between Different Welding Techniques Adapted in Automobile TWB Manufacturing**

| Weld Methods/Effect of Weld Methods | Electron Beam Welding | Induction Welding | Mash Seam Welding | Laser Beam Welding | Friction Stir Welding |
|---|---|---|---|---|---|
| Appearance | Decent | Not good | Not good | Very good | Decent |
| Rework weld | Not needed | Needed | Cold rolling | Not needed | Minor |
| Coated/uncoated | Possible | Possible | Possible | Possible | Possible |
| Similar and dissimilar metals | Possible | Possible | Possible | Possible | Possible |
| Welding tool/ambience | Electron beam gun/preferred vacuum | Electromagnetic coils/open atmosphere | Seam welding disks/open atmosphere††† | Laser beam gun/flexible*** | Rotating tool/flexible*** |
| Curved welds | Possible | Not possible | Not possible | Possible | Possible |
| High/low strength | Possible | Difficult | Possible | Possible | Possible |
| Changing thicknesses | Possible | Not possible | Difficult | Possible | Possible |
| Edge preparation | Precise | Easy | Medium | Precise | Medium |
| Heat-affected zone | Narrow | Wide | Wide | Narrow | Wide |
| Forming behavior | Very good | Difficult | Good | Very good | Very good |
| Industrial integration | Difficult | Good | Good | Very good | Very good |

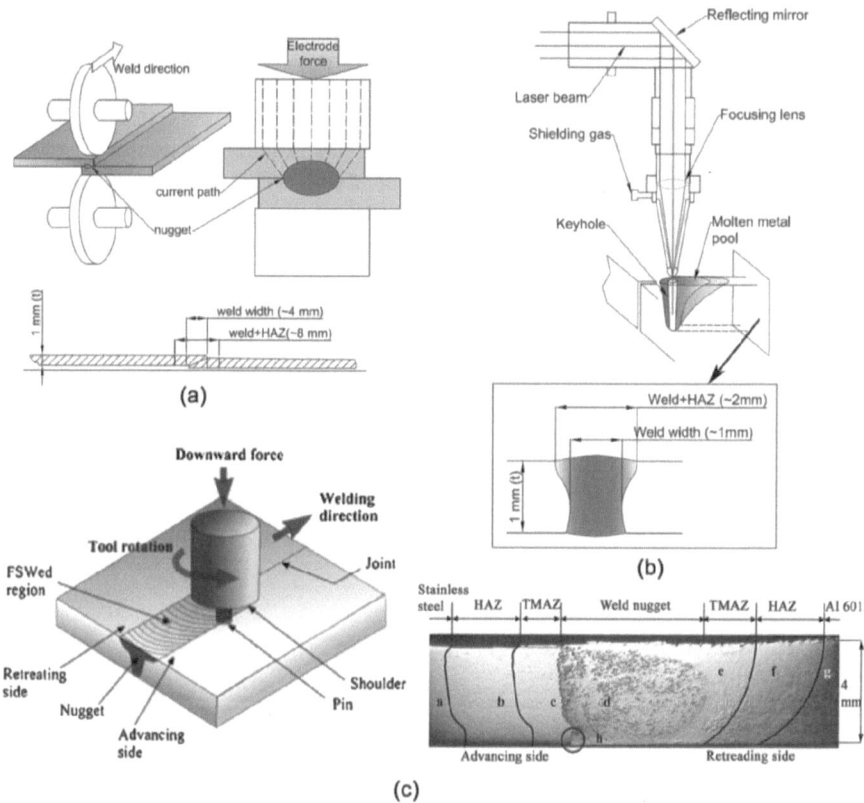

**FIGURE 5.5** Schematic representations of the widely used welding technologies for manufacturing TWBs. (a) Resistance mash seam welding. (b) Laser beam welding. (Reproduced by permission from He et al. [21] Copyright© 2014 Elsevier Ltd.) (c) The typical weld zone of dissimilar steel-aluminum FSW joint. (Reproduced by permission from Uzun et al. [22] Copyright© 2004 Elsevier Ltd.)

welding[‡‡‡] with high accuracy, gives better accessibility to intricate locations, produces a narrow weld zone, provides better finishing and high welding rates, and is highly economical [19]. Hence, laser welding is the only process that can produce complex weld lines in flexible working atmospheres.[§§§] As mentioned earlier, the application of aluminium and/or magnesium TWB is increasingly being considered in automotive applications to further reduce the weight in automotive applications. However, the high reflectivity, low viscosity at the molten state, and the presence of oxide layers of aluminum pose serious hindrance to its weldability by laser. The issues such as hot cracking in the fusion zone, hydrogen-induced porosity, and loss of alloying elements are also common in laser welded aluminium blanks. This leads to poor coupling and reduction of strength [16]. Furthermore, TWB of steel–aluminium and/or steel–magnesium (dissimilar base metals) is difficult to obtain using laser welding due to the formation of brittle inter-metallic phases [20].

The solid-state FSW technique was introduced to join "un-weldable" materials such as aluminium alloys [23]. A rotating tool is used in this process, which is plunged into the interface of two sheets that generates heat due to friction (Figure 5.5c). The heat generated will soften the interface material, and the rotary tool "stirs" the adjoining materials creating intermetallic bonds. The materials do not melt and solidify during the welding process. The FSW technique is cost-effective and possesses many advantages such as low energy input requirement, a relatively short setup and welding time, low temperature, and low distortion compared to other fusion welding processes [22]. Recent advancements also demonstrated the effectiveness of FSW for joining the difficult-to-weld steel-aluminum sheet combinations [10]. Potential applications of TWBs are also found outside automotive industries including aerospace, rail car, alternate energy, shipbuilding, and construction domains. FSW generates a wider weld zone (Figure 5.5c), but little or no post-weld preparation is required and the resultant TWBs show promising deformation behaviours [23].

As an input blank, the TWB seeks to place the right material at the right location. However, it can be a nightmare for tool room engineers to produce dies for forming TWB components. The component and tool design for TWB aims to enhance the formability and reduce the springback [24].

### 5.1.2 FORMABILITY OF TWB

While ductility is a material-dependent property, formability is a combined measure of material and load path on the limit of deformation. In the case of TWB, base materials on each side and the weld region (although very narrow) have distinct forming limit diagrams (FLDs). The FLD of a TWB therefore depends upon the strength ratio of the base metal and weld configurations. The strength ratio in TWB could be due to either different grades of materials or different thicknesses or both (Figure 5.6a) [25]. Korouyeh et al. [25] investigated the formability of TWB with

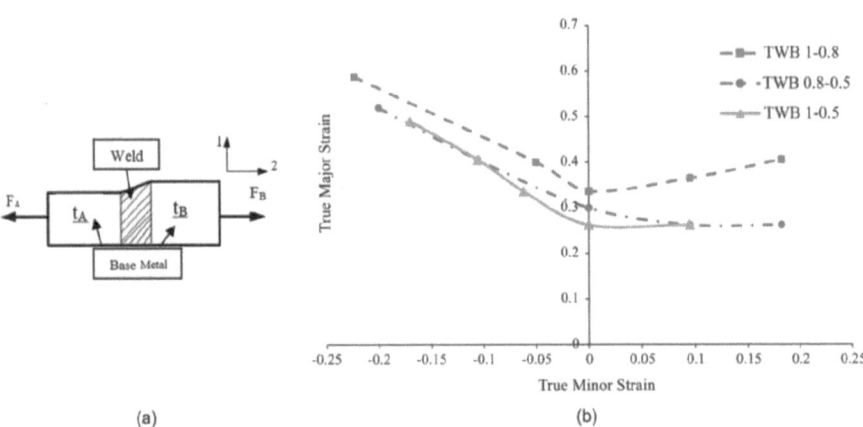

(a)                                                     (b)

**FIGURE 5.6**  (a) Schematic of TWBs of differential thicknesses and influence of TWBs of different thicknesses on the FLDs. (Reproduced by permission from Korouyeh et al. [25] Copyright© 2013 Elsevier Ltd.)

different thickness ratios (1.25–2) using steel sheets and showed that the formability decreased with the increase in thickness ratio (Figure 5.6b). Similar findings can also be found elsewhere [26–28]. Beyond a critical thickness ratio, the thinner side of the TWB could be deformed to its maximum formability limit, while the thicker side might be still under the elastic state. This critical thickness ratio was referred to as the limiting thickness ratio (LTR). It was shown that the formability decreases as the absolute deviation $(\Delta = TR - LTR)$ of actual thickness ratio to LTR increase [25]. The force perpendicular to the weld for unit width is $F = \sigma t$, where $\sigma$ and $t$ are the in-plane tensile stress and thickness, respectively. Imposing force equilibrium, a limiting strength ratio (LSR) can be obtained similar to LTR. The LTR is defined as the ratio of the thinner to thicker sheets. Since the LTR corresponds to the elastic limit of the thicker sheet, the LSR is the ratio of tensile strength (corresponding to the forming limit) of the thinner sheet and yield strength (elastic limit) of the thicker sheet. In a recent report [29], an attempt was made to construct a master curve of ratio of limiting strain and thickness difference irrespective of the thickness ratio. While the particular report [29] is an empirical and preliminary attempt, generic guidelines on formability can be evolved using such approaches.

In addition to the strength ratio, the weld plays an important role in the formability. A study using low carbon steel TWBs of similar grades and thicknesses shows that the FLDs of TWBs are highly sensitive to the weld location and the weld orientation [30]. As can be seen in Figure 5.7a, when the weld is longitudinal, both the blanks are subjected to similar displacement and formability is limited by the minimum limiting strain of the blanks or interface. Usually, the interface or weld region in steel TWBs is stronger and less ductile, and hence the longitudinal welds exhibit lesser formability. On the other hand, with the transverse weld, the weaker side deforms to its maximum limit (the weld line moves toward the stronger side), displaying better formability [28]. In another similar work on steel TWB [31], it is shown that the weld orientation's effect on formability is also dependent upon the strain path. In that study, superior formability under plane stain conditions was obtained in the longitudinally oriented weld line and by the transverse weld line along the uniaxial strain path (Figure 5.7b).

Due to the factors mentioned above, predicting the FLD of TWB is complicated and several attempts to do so have been made in the past. For instance, Safdarian [32] modified the traditionally used limiting strain criterion after Marciniak and Kuczynski (M-K method) to include the bending effect for TWB. Korouyeh et al. [33] analyzed different numerical criteria for formability prediction and found that the second derivative of thinning (SDT) could predict closer to experimental observation. In another related study [34], it was shown that the failure based on maximum punch force predicted the results better. However, it also concluded that none of the methods are accurate enough and the accuracy can be improved only by better modelling of the interface region between the two materials.

Owing to the difference in thickness and material properties between the base metals, the strain path taken by each side would vary when compared to monolithic blanks. This difference in properties gives rise to another important issue of weld line movement (WLM). Since the weld line is relatively brittle, it is advisable to prevent significant shift of the weld line in the plane of the sheet during deformation.**** The difference in formability in both the sides would try to move the

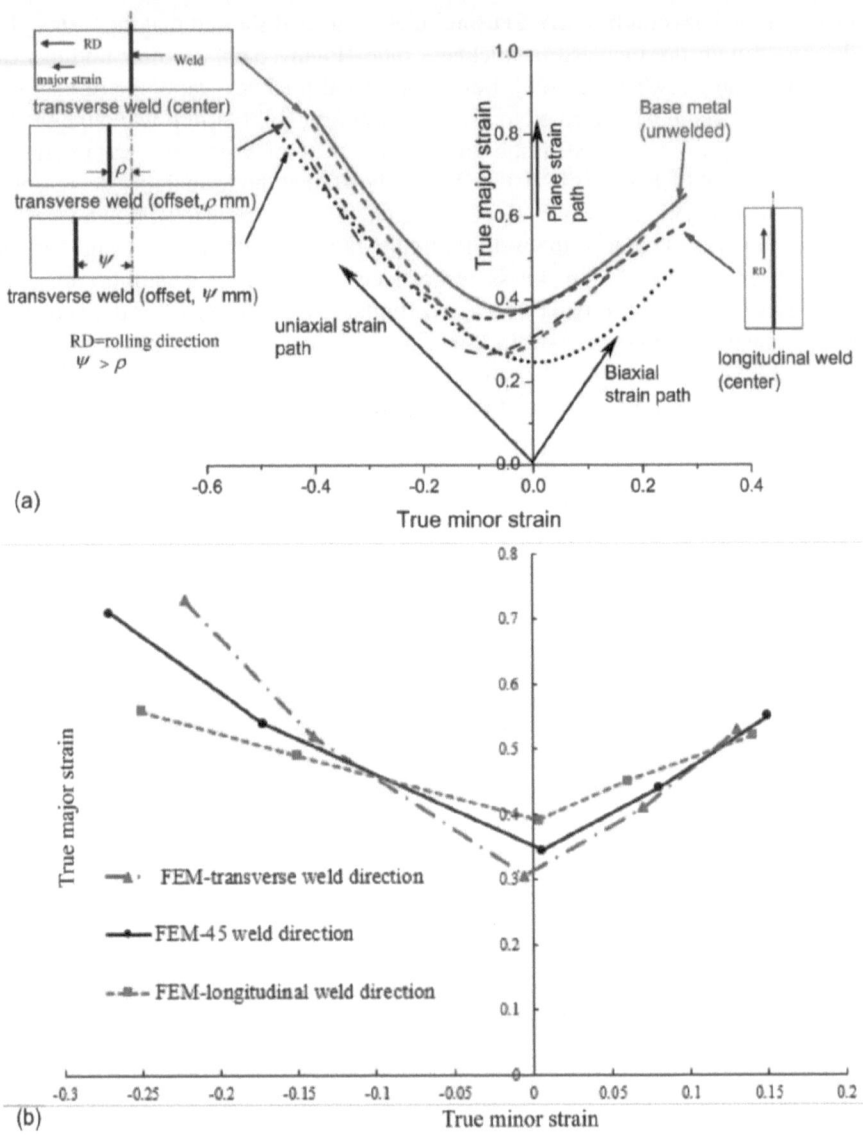

**FIGURE 5.7** Effects of (a) weld location. (Schematic reconstruction based on the results reported in Narayanan and Narasimhan [30].) (b) weld orientation on FLD of low carbon steel TWBs. (Reproduced by permission from Moayedi et al. [31] Copyright© 2020 Elsevier Ltd.)

weld line depending upon the strain induced in both the sides. Non-uniform WLM (see Figure 5.8) can lead to premature failure of the TWB without either of the blanks deforming to its maximum forming limit. Furthermore, the WLM will be dependent on the planar anisotropy that develops non-homogeneous extension of sheets under drawing load (Figures 5.9 and 5.10).

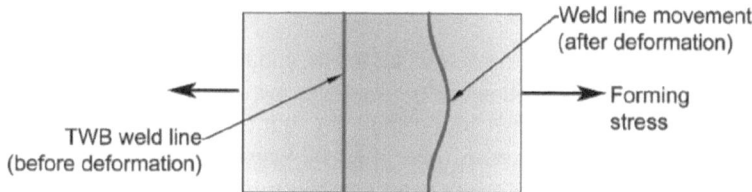

**FIGURE 5.8**    A basic 2-D schematic of non-homogeneous WLM in TWB under simple tensile load.

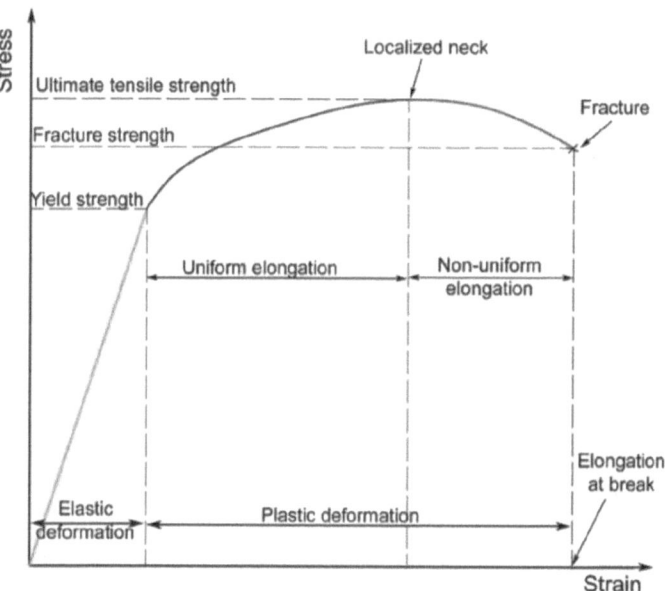

**FIGURE 5.9**    Simple tensile test result.

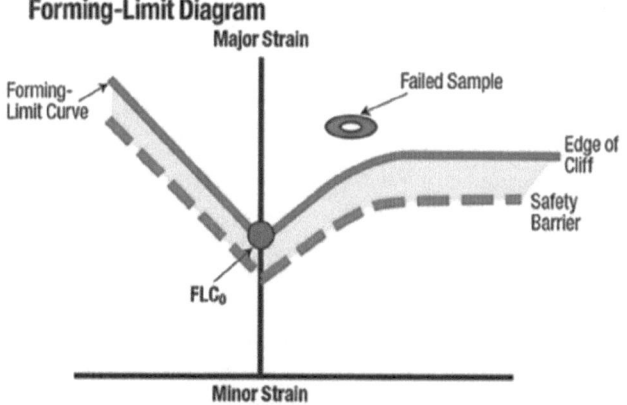

**FIGURE 5.10**    Representation of FLD. (Courtesy of Metal Forming Magazine.[††††])

### 5.1.3  SPRINGBACK IN TWBs

The spring back in TWB is complex due to the combined effect of base metals, the weld region, and the inhomogeneity associated with it. During the deformation of TWBs, the movement of the thicker/high strength region will be less, resulting in reduced planar plastic strain in those regions, which increases the springback. Furthermore, an uneven springback occurs between thicker/stronger material and thinner/weaker material sides. Since springback is dependent on elastic recovery, the section modulus of the components plays an important role [35]. The release of elastic energy during springback can be controlled by locally strain hardening the adjacent region. Similarly, the resistance to deformation varies due to the inherent mechanical anisotropy induced due to crystallographic texture and can affect the springback [35]. The extent of springback in TWB is controlled by the simultaneous influence of inhomogeneous strain distribution in the thickness direction and the elastic–plastic behaviour of the stamped sheet [36].

The weld orientation has a definite influence on the springback behaviour. During V-bending of steel TWBs, it was shown that maximum springback was witnessed in 45° weld orientation [37,38], where the yield strength was maximum. When the blanks are joint along longitudinal orientation, both the blanks are subjected to similar deformation and the springback is high. On the other hand, when the weld line is transverse, part of the strain inhomogeneity is accommodated by the WLM [28]. This can be understood from the experimental observation [39] on springback of TWB U-bending with different thickness (strength) ratios, where the thicker side of transversely welded TWBs was unaffected and only the thinner or heavily deformed side of TWB exhibited different springback characteristics. In V-bending, the springback angle varied inversely with thickness ratio [38]. Furthermore, the mechanical properties of the weld zone, especially HAZ, are also found to be detrimental in the springback of TWBs. Thus, along with the strength and directionality difference, imperfections due to discontinuity of materials within the weld zones also need to be considered while addressing both the formability and springback issues [40]. For instance, Figure 5.11 shows the differential springback that occurred during U-bending operation using TWBs of dissimilar materials. This uneven springback results in a distorted stamping and should be controlled to improve the part quality [15]. Interestingly, springback is influenced by the type of welding process too [41], indirectly through elastic properties.

Usually, the springback compensation is given to the tooling at the early stage of design, for which proper knowledge of the springback behaviour of the materials is necessary. The occurrence of WLM, depending upon the thickness/strength and formability of component sheets, is another serious issue to be considered. A uniform WLM as shown in Figure 5.11 may produce equal material-dependent springback on either side. However, research shows that WLM is not uniform, even for the basic deformation modes – hemispherical dome stretch forming [42–44] and deep drawing [35,45,46]. Hence, the stamping tool design should always consider these two aspects – springback and formability – which further complicate the designing process. Typical methods adopted to control the springback of TWBs are discussed in the next section (Figures 5.12 and 5.13).

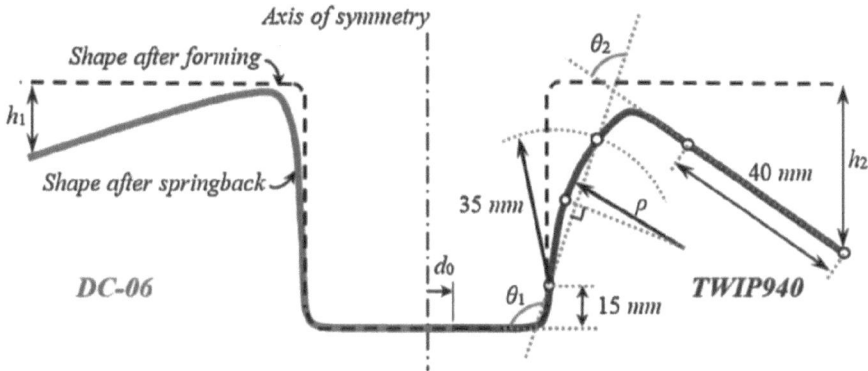

**FIGURE 5.11**   Springback parameters of TWBs of dissimilar materials. (Reproduced by permission from Nguyen et al. [24] Copyright© 2015 WILEY-VCH Verlag GmbH & Co. KGaA, Weinheim.)

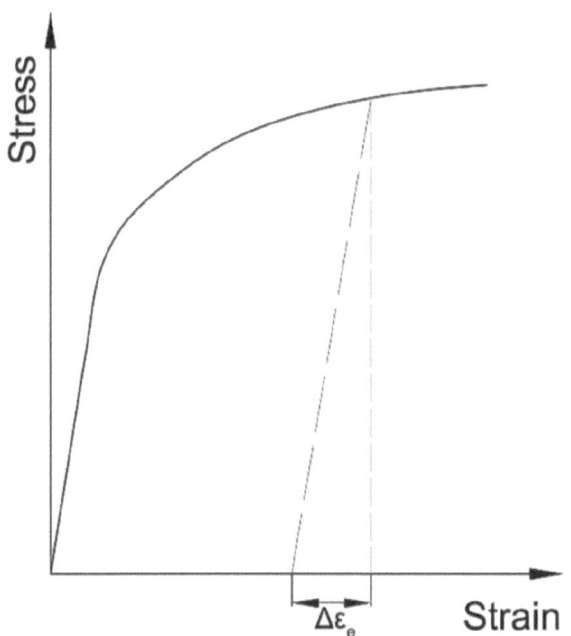

**FIGURE 5.12**   Representation of elastic strain recovery in a simple tensile test.

### 5.1.4 Practical Methods to Control Springback

One of the methods to prevent springback is to induce geometric features that are subjected to large plastic strain. These regions are strain-hardened due to the plastic strain and locally increase the stiffness that prevents the elastic recovery and springback. It can be seen that providing a local geometrical feature on the initial

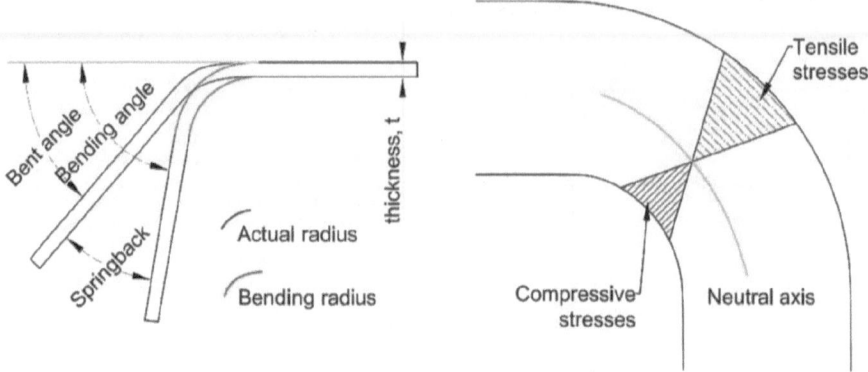

**FIGURE 5.13**  Representation of (a) springback in a simple bend test and (b) stress states in bending.

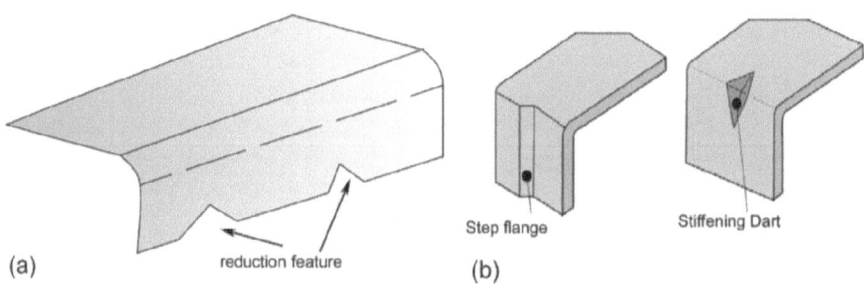

**FIGURE 5.14**  Examples of additional features in a flange to reduce springback: (a) reduction feature and (b) step flange and stiffening dart.

blank, such as a reduction feature at the flange (Figure 5.14a), a step flange, and a dart (stiffening dart) at the radius (Figure 5.14b), can reduce the springback [47]. These features lock the elastic stresses, but the springback reductions are limited in the regions that are at close proximity to these features. Addition of such features is not always advisable as subsequent post-forming deformations, trimming, piercing, punching, heat treatments, and so on may induce unbalanced residual stresses and distortions that significantly affect the quality and strength of the final component. Punch holding time was also found to be a favouring factor for springback reduction [48]. Li et al. [49] reported a dramatic decline of springback in magnesium alloy due to creep and creep recovery when the samples were held in bent states for more than 1 month. The use of servo press has also demonstrated that springback can be reduced by intermittently stopping the punch tool [50,51]. As in overall formability improvement, proper lubrication also aids to some amount of springback reduction [48,52]. Some of the common practical strategies adopted to control the springback are discussed in this section.

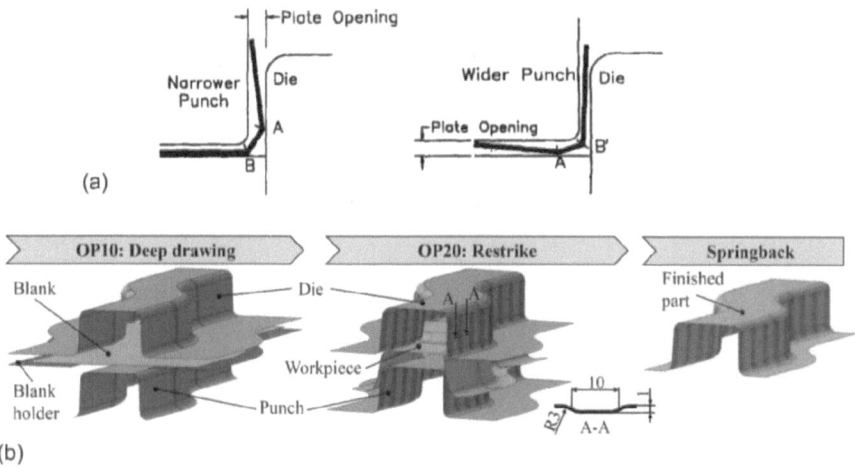

**FIGURE 5.15** Illustrations of (a) basic mechanism of restrike operation. (Reproduced by permission from Shu and Hung [57] Copyright© 1996 Elsevier Ltd.) (b) Restrike (double bend) technique adapted in an S-shaped channel bending operation. (Courtesy of Radonjic and Liewald [53], doi: 10.1088/1757-899X/159/1/012028 is licensed under CC BY-ND 3.0.)

### 5.1.4.1  Restrike Method

A generally used method to compensate for springback is the restrike process, also termed as the 'double bend' technique [53–56]. The restrike method compensates for the bending effects at the corner. A basic representation of the restrike method is shown in Figure 5.15a. The initial bending occurs between B and A. After bending (using a narrower punch), the released part is redeformed using a wider punch which rebends the corner such that the bending occurs between A and B. In practice, the locations A, B, and B' will be indistinguishable and seen as a single bend radius [57]. The process requires additional press working using a compensating punch (as mentioned) or a restrike die. The blank is first formed to near-final shape, followed by the restrike process to obtain the final shape, that is, the shape with reduced springback. An illustration of restrike operation of a typical S-shaped component from a recent research [53] is shown in Figure 5.15b. It is apparent, owing to the material properties, that the tool design for the restrike process is specific to the blank material. Restrike, which may be counted as a multistage process,‡‡‡‡ can cause premature fracture and wrinkle while forming curved whole shapes using hard-to-deform materials [58] like UHSS, high strength aluminum alloys, and magnesium alloys.

### 5.1.4.2  Design Modifications of Tooling

Springback control by providing allowance or compensation on the die is shown in Figure 5.16a [59]. The tooling shape is adjusted according to displacement deviation to compensate for springback [60] rather than controlling it. The modified die/binder shape deforms the sheet to a higher degree and allows the sheet to springback once unloaded, thereby obtaining the nominal shape of the component [61]. As in any sheet forming operation, the severity of springback is also significantly affected by the die

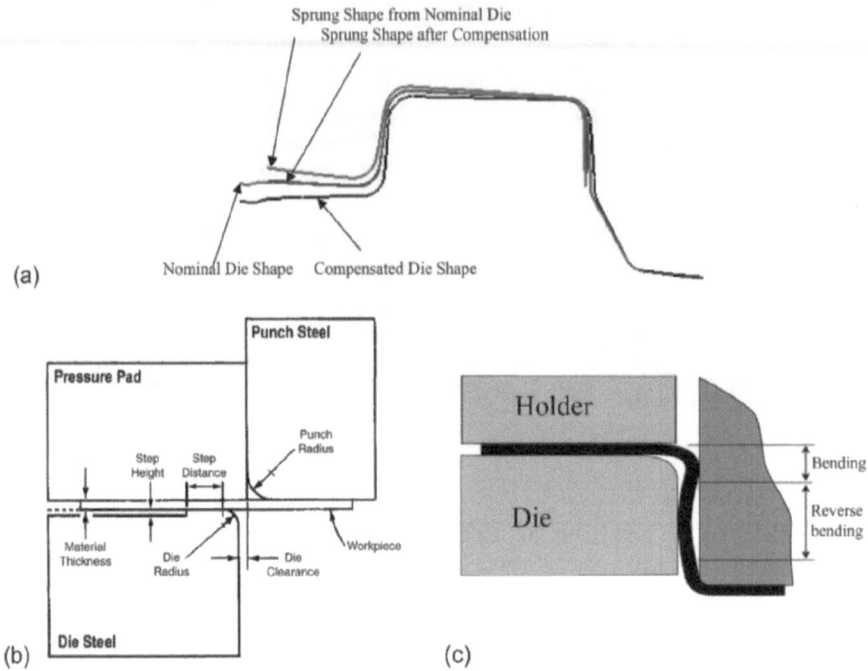

**FIGURE 5.16** Schematics of (a) die allowance given for springback compensation. (Reproduced by permission from Xu et al. [59] Copyright© 2005 AIP Publishing.) (b) a basic die punch binder setup for L-bending. (Reproduced by permission from Ling et al. [62] Copyright© 2005 Elsevier Ltd.) (c) Possible bending in the clearance.

radius and clearance. In a simple example of L-bending (Figure 5.16b), the recommendation is to use the smallest permissible die radius and die clearance [62,63]. The additional bending and unbending of the sheet metal can also be restricted by reducing the die clearance (Figure 5.16c). This also minimizes the angular change while the sheet is coming off from the die radius into the die cavity. However, care should be taken while considering both die radius and clearance [62]. The die radius smaller than the minimum bend radius of the metal causes premature cracking, and reducing the clearance beyond limit causes curvature of bend leg and unnecessary locking (jamming) of materials.

The most common methods used for design compensation for springback are based on displacement adjustment (DA) methods [60,61] and FEA-assisted CAM modelling [64,65]. It is possible using such computer-aided engineering to extract complex process parameters such as thickness variation, effective stresses, effective strains, plastic deformation, and elastic recovery [66,67]. The occurrence of cracks, wrinkles, and under-stretching may be identified easily through digital rendering. The FE predictions invariably depend upon how accurately the material deformation, working parameters, tooling-sheet interactions, and interface properties are included while modelling. Also, the accuracy of FE prediction depends upon the software capability to perform complex calculations [66]. Finally, an experimental validation is necessary to get confidence on the simulated results. Hence, this method demands higher experimental and computational iterative steps even for simple geometries [68] and will become more challenging for non-symmetrical and/or complex geometries.

### 5.1.4.3 Adjusting Forming Severities

Another common method to reduce springback is by providing additional stretching, thereby increasing plastic deformation, or forming severity. Increasing the straining of the sheet by stretching can be a better control of springback. An increase in sheet tension reduces the stress gradient through-thickness, resulting in a reduced bending moment and total springback [69]. It has been demonstrated that a better springback control is possible through intermediate straining, termed as "active" straining, as compared to "passive" straining methods. The latter method uses a constant blank holder force system where the bead force remains unchanged during the forming cycle (hence termed as the passive method). The rigid behaviour of the bead once clamped in the passive method can introduce unwanted high strains in critical locations causing splitting and/or premature failure [70].

The active straining is done at the intermediate stage of deformation, usually allowing the material to deform until the last stage and then locking the sheet at the periphery restricting easy flow of materials,§§§§ thereby inducing more plastic deformation. This process can be done either through variable blank holding force (BHF) or through design modifications of tooling such that lock beads are activated at the targeted punch stroke position. A schematic for the latter method is shown in Figure 5.17a–c, which resulted in significant reduction of springback for an AHSS sheet as shown in Figure 5.17d [47]. Here, the stake beads are manufactured directly on the punch. Locking of the sheet occurs at a targeted punch stroke and final stage deformation occurs under tension, resulting in significant reduction of springback. The usage of retractable stake beads is also in practice. In this method, the beads are located at specially designed slots in the die and are activated to lock the sheet once the target stroke of punch is reached. These methods of using lock beads toward the

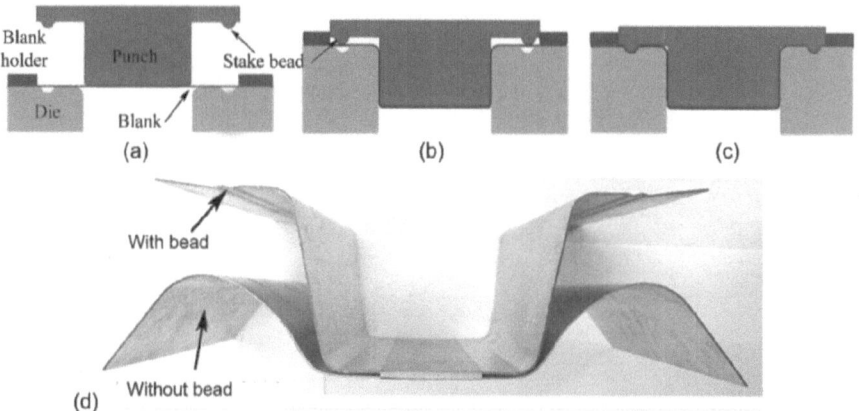

**FIGURE 5.17**  (a–c) Post-form bead method to reduce springback (Courtesy of thefabricator.com.*****) and (d) springback reduction with beads. (Courtesy of Auto/Steel Partnership.†††††)

end of forming are usually termed as post-loading or post-stretch forming strategies [60,71]. Obviously, such modifications are incorporated during tooling design and modifications, following several experimental or numerical try-outs.

While the in-plane stretching methods can control the springback considerably, they require significant tool design modifications depending upon the predetermined dimensional considerations. A larger blank size may be required to locate the stake beads. Additional machining requirements also increase the tool construction costs for retractable stake beads. Adequate structural support is important to counteract lateral thrust forces that can cause catastrophic die failure if the stake beads are located very near to the punch opening. This situation will be prominent while deforming higher strength materials such as AHSS. The lock bead also becomes inefficient while forming a higher strength material with higher work hardening characteristics [72]. In this case, the bead does not impart sufficient locking to stretch the sheet; rather, it will just control the material flow during bending and straightening over the bead profile. For such situations, using a mild steel die might not be sufficient for AHSS materials. Hence, high strength tooling may be required, which further increases the production cost. Similar balancing is also essential in manufacturing multiple C- and/or U-bend sections [60]. Additional complexities also arise for asymmetrical sections and for the manufacturing of hat-shaped (U-bent) and S-shaped components as shown in Figure 5.15b.

### 5.1.4.4  Using Advanced Machine Tools

With the advancement in new machine tools, such as multi-action presses and servo-motor-driven presses, an "active" control of BHF is possible. Using these machines, it is relatively easy to apply variable BHF to give intermediate straining or straining similar to post-stretch forming efficiently. Thus, the complexities in tooling designs with inserts and/or protrusions such as beads and supports can be avoided. A schematic for application of variable BHF is shown in Figure 5.18. Controlling the punch speed was also found to be beneficial in reducing the springback [48,73]. When attempting to increase the sheet tension, care should be exercised to avoid premature failure/splitting, especially for the newer high strength, low-formable materials [69].

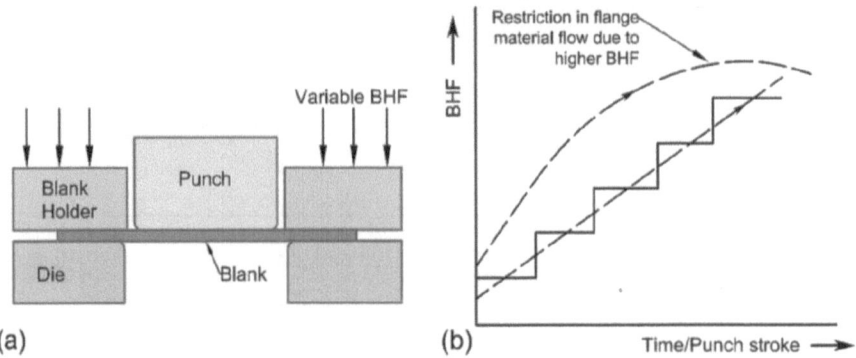

**FIGURE 5.18**  Schematic for variable BHF conditions to control springback. (a) basic principle of BHF in deep drawing process and (b) the relationship between BHF and punch stroke.

## 5.2   SPRINGBACK CONTROL IN TWB FORMING

As seen in the previous section, imparting forming severity maximizing plastic deformation, increasing the straining of the sheet, is the crux to reduce springback. However, the differences in material properties and formability on either side of the weld line increase the complexity of springback control. Additionally, as the weld position, orientation, and weld-zone properties also play a crucial role in the formability (Figure 5.7), these factors will also affect the springback of TWB significantly. Although any of the methods mentioned in Section 5.1.4 can be adopted in springback reduction in TWB forming, this will come with added complexity.

### 5.2.1   BLANK HOLDER FORCE AND TOOL DESIGN MODIFICATION

Three major combinations applied in manufacturing TWBs are (A) similar material–different thickness, (B) dissimilar material–different thickness, and (C) dissimilar material–same thickness. The different thickness combination (A and B) inevitably requires appropriate tool modification for proper accommodation and transfer of BHF as shown in Figure 5.19. The thickness mismatch may be accommodated by modifying the punch. Inherently, even to carry out forming under constant BHF conditions, the clearances on the die and the binder positions should also be modified. Controlling the blank holder force during deformation is essential to control springback in TWBs. Generally, the severity in deformation to reduce springback is achieved through increasing the BHF. The application of BHF can be passive or active.

In the passive method, a constant BHF is given throughout the deformation. Let us consider case C, where TWB is made with different materials having a constant thickness. If the BHF is higher, the material flow of high strength material will be lesser, which may cause the weld line pulling the low strength material to its side. Simultaneously, higher constriction in material flow experienced by the low strength material causes strain localization, resulting in higher possibility of premature fracture. A lower BHF generates a situation similar to unconstrained material flow that causes material flow into the die cavity, producing WLM toward the stronger

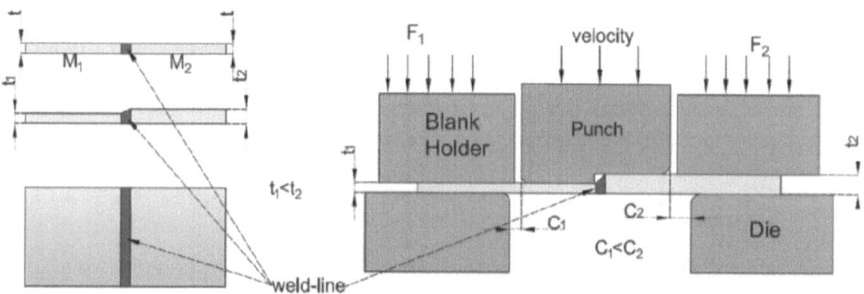

**FIGURE 5.19**   Schematic of different TWB combinations and a basic tool design modification required for TWB forming.

material (Figure 5.8). This phenomenon causes higher thinning and/or splitting of the low strength material when deformed [74].

Usually, for TWBs of different thicknesses (cases A and B), a pocket is placed on the die and/or punch at a precalculated distance near the weld line to avoid unnecessary locking (jam) of the material during deformation [24]. Here, unconstrained material flow in the low strength (thinner) material tends to fail due to buckling/wrinkling [75]. An over-constrained condition may also occur, resulting in WLM toward the thinner material leading to tearing failure near the weld line [75]. The WLM toward the stronger/thicker material also represents a measure of strain localization in the weaker or thinner base metal [74]. In effect, a correlation exists between the WLM and springback. Hence, minimizing the WLM can effectively improve the formability and reduce springback. Inevitably, the BHF for either side of the material needs to be controlled for better control of material flow, which confirms lower WLM, improved formability, and reduced springback. Practically, the usage of a split blank holder with differential holder force at either side depending upon the material strength and thickness may be used. The same gap between the die and binder may be maintained for case C, whereas different gaps according to the thickness difference at either side may be used for cases A and B.

The active method can be applied for better control of springback with proper adjustment of BHF that varies with the punch stroke. Theoretically, the BHF should also vary with the thickness on each side of the TWB, which can be accomplished only by splitting the blank holder. An improvement in formability may also be achieved through proper weld placement and BHF adjustments [46,76,77]. The following are the major issues to be addressed: (1) design and modification of split BH confirming to available press, (2) evaluating the maximum possible BHF levels for each material, and (3) the application of differential BHFs confirming maximum formability, lesser WLM, and minimum springback.

It is apparent that the complete understanding of deformation mechanisms such as surface and thickness strain variations, forming and residual stress variations, changes in the stress state due to friction at the sheet-tooling interface, die, punch, and binder contouring is necessary to accommodate for the springback compensation even by simple tool and die design modifications. Furthermore, the formability of the material should not be compromised.

Due to the complexities in TWB, control of springback based on strategies listed for monolithic blanks will be difficult. Instead, the springback compensation can be adapted to reduce springback in TWBs [38,78,79]. The compensation involves springback prediction and tool modification by adjusting the tool surface proportional to the displacement during springback. The accuracy of the procedure is purely dependent upon the accuracy of predicting springback.

### 5.2.2 Mathematical Modelling of Springback

This section focuses on the constitutive material models and the complexity associated with them in predicting springback numerically using finite element (FE) simulation. Accurate description of the material's elasto-plastic behaviour is necessary for predicting springback. Most of the sheet metals studied are isotropic in the elastic region, and hence the generalized 3D Hooke's law can be used to describe the stress–strain relation of the material.

$$\sigma_{ij} = \frac{E}{1+v}\,\varepsilon_{ij} + \frac{vE}{(1+v)(1-2v)}\,\delta_{ij}\varepsilon_{kk} \tag{5.1}$$

where $E$ & $v$ represent the elastic constants and $\sigma_{ij}$ and $\varepsilon_{ij}$ represent the stress and strain tensors, respectively. The stress–strain relation in the plastic region is not unique. The complete description of the mechanical behavior subjected to plastic deformation involves a yield criterion, a flow rule, and a hardening rule. The yield criterion provides a mathematical condition for plastic deformation,

$$\sigma_{eq} = \phi(\sigma_{ij}) \tag{5.2}$$

where $\sigma_{eq}$ is the effective stress and $\phi$ is the yield function for a given stress state of deformation. The $\sigma_{eq}$ compares the yield condition under the general stress state with that of uniaxial flow stress. The comparison is made by imposing the work hardening hypothesis [80] wherein the yield condition is evaluated for a constant amount of plastic work. The most commonly used yield criterion for isotropic materials is based on von Mises [80] and is given by

$$2\sigma_{eq}^2 = \left(\sigma_{xx} - \sigma_{yy}\right)^2 + \left(\sigma_{yy} - \sigma_{zz}\right)^2 + \left(\sigma_{zz} - \sigma_{xx}\right)^2 + 3\left(\tau_{xy}^2 + \tau_{yz}^2 + \tau_{xz}^2\right). \tag{5.3}$$

Under uniaxial deformation, the equivalent stress reduces to uniaxial yield strength. Typical sheet metals exhibit anisotropy of mechanical properties due to the crystallographic texture (preferred orientation of grains) induced by prior thermomechanical processing. The yield criterion must be modified (Figure 5.20) to accommodate the anisotropic behavior. The earliest attempt to include anisotropy is by Hill [81], who in 1948 modified the von Mises yield criterion as follows:

$$\sigma_{eq}^2 = H\left(\sigma_{xx} - \sigma_{yy}\right)^2 + F\left(\sigma_{yy} - \sigma_{zz}\right)^2 + G\left(\sigma_{zz} - \sigma_{xx}\right)^2$$
$$+ 2N\tau_{xy}^2 + 2L\tau_{yz}^2 + 2M\tau_{xz}^2. \tag{5.4}$$

The Hill48 criterion is very popular [83] in commercial sheet metal forming analysis software, especially for the application of low carbon steel. Several advanced yield criteria [84–88] have been proposed since Hill48, especially to model advanced high strength steels (AHSSs) and aluminum and magnesium alloys. A detailed review on the anisotropic yield criteria can be found in Refs. [89,90]. The plastic strain increment is proportional to the deviatoric stress and its direction normal to the yield surface. The plastic strain can be estimated using the flow rule,

$$d\varepsilon_{ij} = d\lambda\frac{\partial\phi}{\partial\sigma_{ij}} \tag{5.5}$$

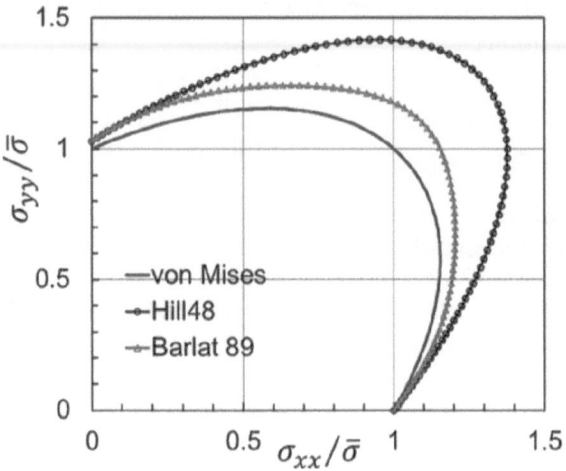

**FIGURE 5.20** Comparison of different anisotropic yield criteria for mild steel. (Material data from Kuwabara et al. [82].)

The above description uses the yield function to derive the plastic strain increment, and hence the flow rule is referred to as the associative flow rule [91]. Alternately, the normality of strain increment can be relaxed and an independent function, $\psi \neq \phi$, can be used to describe the strain increment. Such treatment is referred to as the non-associative flow rule [92,93]. Although the non-associated flow rule can improve the flexibility of the constitutive relation, they are less commonly employed in forming simulations. The plastic deformation at low homologous temperatures witnesses an increase in strength due to strain hardening. The strain hardening during uniaxial tensile test can be described using a one-dimensional model of the form $\sigma(\varepsilon)$. For instance, the Swift hardening model indicates

$$\sigma = K\left(\varepsilon_0 + \varepsilon_p\right)^n \qquad (5.6)$$

Many similar equations can be found in the textbooks [80,90,91]. When extending it to the general stress state, we have already discussed that yield loci are constructed for constant plastic work. During strain hardening, the plastic work increases and so is the size of the yield locus. If the yield locus expands uniformly in all the directions, then the plastic work can be replaced with equivalent or effective plastic strain. The simple one-dimensional equation can be used to predict the size of the yield locus. Such behavior of uniform expansion of yield locus is referred to as isotropic hardening (Figure 5.21). The isotropic hardening assumption is suitable when the loading is proportional throughout. However, if the loading is non-proportional, then the change in strain path after a pre-strain could induce the Bauschinger effect.‡‡‡‡‡ Under such cases, the strain hardening description should be non-isotropic; that is, with deformation, the expansion of yield locus is not uniform. Combined isotropic-kinematic hardening (Figure 5.22) is one of the many approaches to account for this deviation from isotropic hardening. In the combined hardening models, the Bauschinger effect

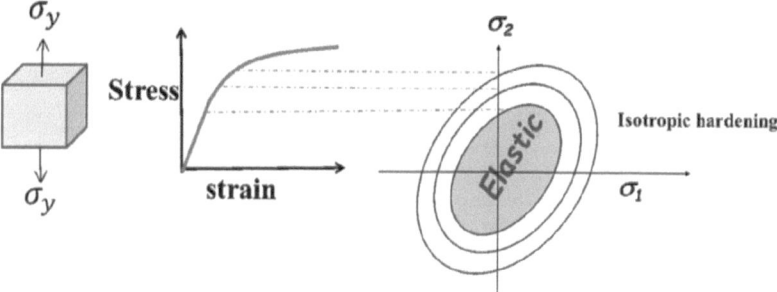

**FIGURE 5.21**  Schematic of isotropic hardening.

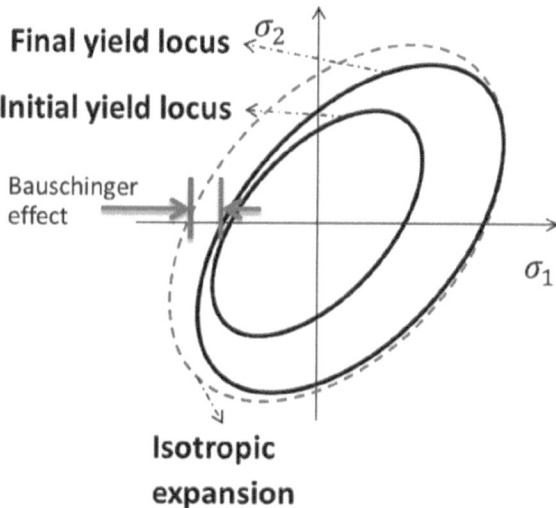

**FIGURE 5.22**  Schematic of combined isotropic-kinematic hardening.

is modelled using a back stress tensor that evolves with strain. In its simplest form, assuming the von Mises criterion for its isotropic component, the yield condition with back stress can be written as

$$F\left(\sigma, \varepsilon_p\right) = \left|\sigma_{\text{eff}} - x\right| - \sigma_y = 0 \tag{5.7}$$

where $x$ is the back stress and $\sigma_y$ is the initial yield stress. In the absence of back stress (proportional loading), only the isotropic component with uniform expansion of yield locus is sufficient with $\sigma_y = \sigma_0 + Q\left(1 - e^{b\varepsilon_p}\right)$; $Q$ and $b$ are material parameters for isotropic strain hardening. The back stress components to model the kinematic component are given as

$$x = \sum_{k=1}^{3} x_k \tag{5.8}$$

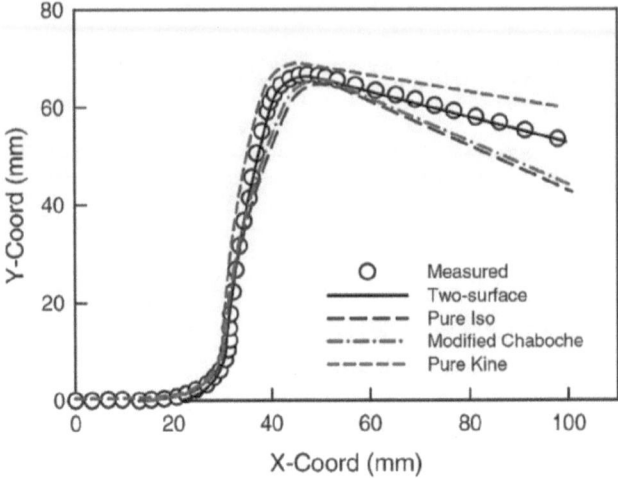

**FIGURE 5.23** Influence of different hardening models on the springback prediction. (Reproduced by permission from Lee et al. [98] Copyright© 2007 Elsevier Ltd.)

where $k$ refers to individual back stress components that assume a general form, $x_k = \dfrac{C_k}{\gamma_k}\left(1 - e^{-\gamma_k\,\varepsilon_p}\right)$; $C$ and $\gamma$ are material coefficients for kinematic hardening.

The springback simulation is performed by constraining a few nodes to arrest the rigid body motion of the component, and the shape change is predicted based on the elastic recovery. Naturally, the accuracy of springback depends upon the accuracy of the stress state [94] estimated prior to springback and the choice of location for displacement constraints. The constraints can be evolved based on practical understanding. The accuracy of the stress state at the end of forming is clearly influenced by the choice of yield criterion and hardening models. The springback behavior involves unloading or change in strain path which results in the Bauschinger effect in the material. As a result, the springback behavior is strongly influenced by the kinematic hardening component too (Figure 5.23). The kinematic hardening takes into account the path dependency of plastic behaviour only. Path dependency exists in the elastic region too; the elastic modulus changes continuously during unloading. This time-dependent variation of elastic modulus is known as the Chord modulus. The Chord modulus influences the elastic recovery and residual stress post springback [69,95]. In addition to the complexities in material behaviour explained above, the accuracy of the stresses at the end of forming is further influenced by the friction model. Traditionally, adhesive theory of friction after Coulomb with a constant friction coefficient is used to model the interface friction during metal forming. However, during the typical sheet forming process, the friction coefficient is known to vary with the applied pressure and local velocity [96]. Unless the variation of friction coefficient is modelled correctly, the shear stress due to friction and therefore the overall stress tensor is incorrectly accounted for. This leads to error in springback prediction. Lee et al. [97] have made a summary on the relative influence of these factors in the simulation of U-draw bending. As expected, the accuracy in springback improved

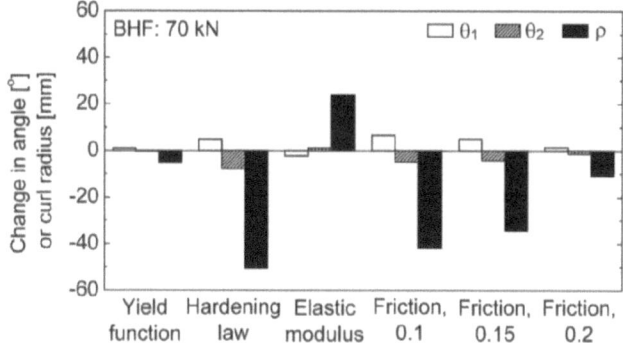

**FIGURE 5.24** Relative influence of different factors on the accuracy of springback. (Reproduced by permission from Lee et al. [97] Copyright© 2015 Elsevier Ltd.)

when anisotropy in yield locus, the kinematic component of hardening, chord modulus variation, and variable friction coefficient were implemented. The study also showed another interesting insight in which the relative influence of these factors is not monotonic (Figure 5.24); for instance, the effect of yield function and hardening law is opposite to that of unloading elastic modulus on the accuracy of springback. Therefore, all the material effects of hardening and friction have to be simultaneously implemented. A summary of advanced issues in springback prediction from the constitutive description has been compiled in an excellent review by Wagoner et al. [69].

The numerical scheme utilized in simulation also has an influence on the forming stress predictions. Most of the complex sheet forming simulations are performed using explicit solver. Unlike the implicit method, explicit analysis does not ensure equilibrium. The time step used should be small so that the dynamic equilibrium achieved in the explicit analysis predicts the forming stresses fairly well. With the continuous research efforts by the software industry, it may be safe to point out that the explicit analysis in most of commercially available FE software predicts the forming stresses within the allowable tolerance limit.

The discussion so far has been focused on the issues of modelling springback using monolithic blanks. The geometry of the actual component and multistage forming process further increases the mathematical complexity of numerical predictions. The springback prediction in TWBs [99] is further complicated by the interaction between the mechanical behaviour of blanks and the weld zone. For practical applications of TWBs involving real life components, detailed modelling of constitutive behavior for accurate springback prediction is difficult and time consuming. Thus, alternate heuristic methods on data-based learning could be adopted to predict springback in TWBs.

## 5.3 DATA-DRIVEN EVOLUTIONARY MODELLING

The comprehensive physical models for the TWBs, though tend to be accurate, are often quite difficult to build as they are generally complex to handle and, in most cases, computationally prohibitive. In this situation, surrogate models can be more conveniently built, if necessary, even using a lesser number of parameters. These models are easier to be utilized in computation and will not lead to any significant error or loss of accuracy if properly constructed. Also, when the data-driven models correctly predict the observed trends, they are also expected to capture the pertinent physics of the system. However, to come up with the right model in most cases is a challenging task, and the evolutionary intelligent modelling strategies are capable of handling it quite effectively. Intelligent evolutionary algorithms [100] create nature-inspired analogs for modelling and optimization for a physical process that may not have any connection with the real biology, and in recent times, many such algorithms were proposed and, as will be explored here, in the case of TWBs, surrogate models were effectively constructed in an evolutionary way, based upon some carefully designed limited amount of FE simulations. This reduces the computational requirement for simulation and optimization based upon the FE method alone. The proposed EvoNN algorithm (Evolutionary Neural Net) is described in the following section.

## 5.3.1 THE EVONN ALGORITHM

Any reliable data-driven modelling approach need to address the issues of overfitting and underfitting problems in their structure. In a typical situation of overfitting, the fitted data tend to capture almost all the trends and fluctuations that are present in its learning data set. This leads to an excellent correlation coefficient. However, any amount of random noise that is likely to be present in the parent data gets captured in the model as well. Since the noise is of random nature, it would be different for the same system at other instances; thus the overfitted model would fail to predict anything meaningful. The opposite of this is the underfitted model, where, through a very conservative strategy, nearly all fluctuations in the training data are smoothened drastically. Such models might succeed in avoiding the fluctuations brought in through random noise but at the same time would fail to capture any physical trends that would be present in the data and thus would be of very little use in all practical purposes.

The EvoNN strategy was designed to apply a bi-objective evolutionary algorithm on a population of neural networks [101–103]. The idea was to come up with the optimum trade-off between the accuracy and complexity of the models in order to develop a set of Pareto optimum models, each having its own architecture, from which one is selected by applying an information criterion [100]. Increasing complexity means using a greater number of parameters, and this would cater for the tendency of overfitting by increasing training accuracy. On the other hand, increasing the training error would reduce the number of parameters but render the model more prone toward underfitting. Since EvoNN picks up a model that is an optimum trade-off between those conflicting tendencies, it intelligently avoids overfitting and underfitting as well.

The EvoNN algorithm has two modules. The first module analyzes the input data (usually noisy) and correlates with the output to generate objective functions. The objective functions obtained are optimized in the second module to determine the Pareto optimality between them. In the first module, the optimization is performed between the two conflicting objectives of complexity and accuracy. The first module has to be executed for each objective function. The optimization in the second module is performed between these objective functions. A predator-prey genetic algorithm is used in both the modules. However, in principle, many other suitable evolutionary algorithms can be used for this purpose. The training module allows *testing with overlapping data sets* in addition to a *single variable response (SVR)* feature, which can assess the impact of any particular decision variables on its corresponding objectives.

## 5.3.2  U-Bending of TWB: A Case Study

In the pursuit of reducing the weight of components in automotive applications, new-age materials including AHSSs aim to replace the conventional low carbon steel sheets. However, AHSSs exhibit very high springback compared to conventional steels due to profound residual stress in deformed parts. The weldabilities of both these materials are comparable and can be easily done through laser welding with a very narrow weld zone. Studies show that such narrow HAZ among steel sheets has negligible effects on the formability of TWBs [74,104].§§§§§

Among the various new-age AHSS materials, the twinning-induced plasticity (TWIP) steels are of particular interest in manufacturing TWB as they possess a combination of high strength and ductility [72]. A TWIP steel–mild steel TWB combination has a high strength ratio of ~4.0, which is very high compared to many previously reported TWB combinations [72,105,106]. In actual forming, the plastic deformation significantly enhances the strength ratio of this TWB combination due to the high strain hardening property of TWIP steel. An early fracture at the drawbead location in the low carbon steel region is expected due to extreme restraining forces of the blank holder. Therefore, different restraining forces are required at either side of the TWB, and the variable blank holder force method was proposed [72,76,107]. Referring to Figure 5.11, it is clear that the springback behaviour of TWB made of TWIP (an AHSS) and mild steel is highly complex compared to the constituent monolithic sheets. Therefore, the U-bending of TWB made of such a material combination would constitute an effective case study.

The benchmark U-draw bending problem established in NUMISHEET 1993 was selected to study the springback of TWB [24]. The selection of this problem is supported by general research documentations on understanding and controlling the elastic recovery of monolithic blanks. This research could obtain acceptable accuracy in size, shape, and design of stamping components from monolithic blanks with controlled springback. Iterative methods for design and modification for optimized tooling geometry were used using FE methods. Application of advanced material models could also aid to attain a certain level of accuracy in prediction of forming with springback. However, investigations on springback of TWB, especially of such a high strength ratio, are sparse [24,35] due to the involvement of complex interdependent factors as discussed in previous sections. The constituent materials of TWB selected in this study are TWIP-940 and DC-06 sheets of a constant thickness. The system of non-constant BHF at different punch strokes was applied to reduce the springback. FE models were used to model the U-bending process, predict the springback, and finally update the springback compensation.

The main objective of this study is to minimize the springback through stretching, exploring the possibility of achieving a near-uniform stress state in the thickness direction throughout the sheet during forming. However, stretching should not surpass the surface damage due to excessive thinning and/or failure limit of the side wall material. Therefore, a proper balance of both these conflicting objectives, (1) reduction in springback and (2) reduction in forming severity by BHF control, is necessary. As there is a requirement of optimizing two inter-dependent, yet conflicting objectives, the routine application of iterative experimental and FE model-based methods to obtain an optimal forming window will be cumbersome. As this typical problem represents a bi-objective Pareto-optimization problem, a genetic algorithm (GA)-based multi-objective optimization technique EvoNN is used. The present Pareto problem is solved using a data-driven evolutionary approach following the two basic steps – (1) constructing metamodels using evolutionary methods and (2) optimizing the problem using EvoNN.

### 5.3.2.1 Setup Description and Springback Measurement

Figure 5.25a represents the schematic of TWB and tooling arrangements used for this study. TWB consisted of DC-06 (mild steel), which is on the left side, and TWIP-940 (AHSS) is on the right side of the weld line with initial dimensions of 35 mm width, 350 mm length, and 1.5 mm thickness. Two discretely controlled split blank holders (BHs), namely, BH1 and BH2, are used to apply different BHFs at either side of the TWB. It should be noted that compared to the benchmark design, the dimensions of the tooling were modified in this case to accommodate the thicker blank [24]. Figure 5.11 gives the springback terminologies of the U-draw bend TWB, where springback on either side and the side wall curl of U-bend TWB is quantified by two angles ($\theta_1$ and $\theta_2$) and $\rho$, respectively. However, to simplify the problem, a single term '$h$' is used in this study to quantify the aggregate effects of springback and side wall curl. Here, $h$ is the vertical distance of the edges at either side of TWB occurring post forming and after springback at either edge after springback. It was

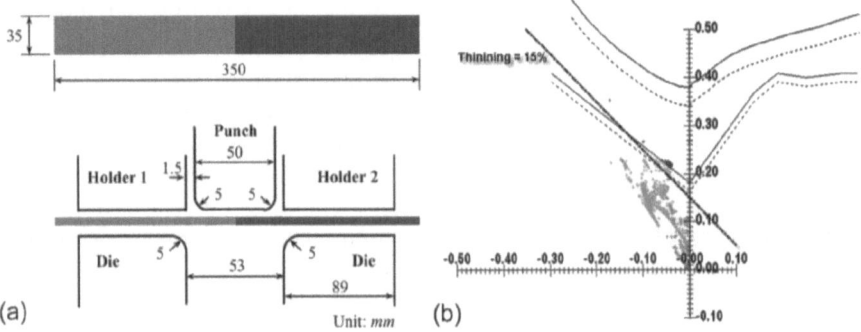

**FIGURE 5.25** (a) Schematic and dimensions of TWB and tool arrangement. DC-06 on the left and TWIP-940 on the right of the weld line and (b) representation of formability of parent metals in the FLD plot [108,109] and thinning criteria. (Reproduced by permission from Nguyen et al. [24] Copyright© 2015 WILEY-VCH Verlag GmbH & Co. KGaA, Weinheim.)

discussed that quantifying the WLM gives two-fold benefits: (1) it can be a representation of the forming quality of TWB and (2) it gives information for tool design modification (e.g., placing pockets, to avoid the material jam considering thickness differences of TWB). Thus, an additional term $d_0$ was also introduced in this case to mention the WLM of TWB.

### 5.3.2.2   FE Modelling and Springback Prediction

The FE models and simulations to predict springback of the U-draw bending process are carried out using the commercially available PAM-STAMP software. Considering the symmetry, and to reduce the overall computational time, a half-model is developed by dividing the setup along the length normal to the weld line of the TWB. The deformable blank is discretized using 7372 Belytschko–Tsay shell elements of 1 mm size with 9 through-thickness integration points. All the tools are modelled using rigid surface elements with surface-to-surface contact at the tool–blank interfaces. A constant Coulomb's friction coefficient value of 0.1 is given at the interface. An active method of BHF control with punch stroke (detailed in the next section) is applied to control the springback. The maximum punch displacement given is 70 mm, and the velocity is set to 10 m/s.

The formability and springback are highly dependent on the constitutive description of the material. Here, the yielding and hardening behaviour of the deforming sheets are incorporated into the FE model using the plane stress anisotropic Yld2000-2D yield function (Eq. 5.8) [84] and Swift's isotropic hardening rule (Eq. 5.9) [110]. The generalized Yld2000-2D function consists of two convex functions as shown below [84,110].

$$\phi\left(\sigma_{ij}\right) = \phi' + \phi'' = 2\bar{\sigma}^a$$

swhere                                                                                   (5.9)

$$\phi' = \left|\tilde{S}_1' - \tilde{S}_2'\right|^a ; \phi'' = \left|2\tilde{S}_1'' + \tilde{S}_2''\right|^a + \left|2\tilde{S}_2'' + \tilde{S}_1''\right|^a$$

$\bar{\sigma}$ follows a typical hardening law such as Swift (Eq. 5.6). Experimental true stress–strain data of TWIP940 and DC06 from Chung et al. [108] and Butuc et al. [109], respectively, are fitted using Swift's equation (Eq. 5.6). The corresponding Yld2000-2D yield function coefficients and constants of Swift's law for material modelling are described in previous literature studies [108,109].

The accuracy of FE prediction is also influenced by the type of failure models defined while modeling. A mostly adopted method to obtain the failure step is the application of forming limit diagram (FLD). FLD provides with the limiting strains at which the material fails due to necking. The particular formability characteristics of both the parent materials are also obtained from previous literature studies [108,109] as shown in Figure 5.25b. The FLD plot reveals that TWIP 940 sheets have much higher formability than DC06 sheets. Being a relatively low formable material, DC06 is prone for early failure during U-draw bending operation. It is noteworthy that unlike in monolithic sheet forming, the definition of forming limits should simultaneously satisfy the forming limit condition of both the sheets in TWB forming. Thus, a maximum thinning criterion considering the low-formable DC06 sheet is considered in this study. It is clear from Figure 5.25b that the straight line corresponding to 15% thinning can provide a reasonable formability characteristic of DC06 material. The placement of the weld line along the major strain direction is shown to reduce the formability due to early failure at the weld region and HAZ near the weld. However, this unfavorable effect is avoided in the present study by modeling the weld line perpendicular to the major strain direction. Moreover, a very narrow weld and HAZ as in laser welding of steel have a negligible impact on the formability of steel TWBs [74,104]. Therefore, instead of explicitly simulating the weld region, the joint of TWB is modelled by sharing the common nodes of materials modelled at either side of the weld.

### 5.3.2.3   Variable BHF Method to Control Springback

As discussed earlier, the effectiveness of BHF in U draw bending cases is obtained by increasing the forming severity to reduce springback, simultaneously taking care of the non-occurrence of early failure. While considering the TWB with such a high strength ratio, the application of the same BHF throughout is prone to generate poor control of these contradicting objectives. Owing to higher strength, the WLM will be more toward the TWIP 940 side. Therefore, split BHs – BH1 and BH2 – are applied for DC06 and TWIP 940 sides. Additionally, a varying BHF is given according to the punch stroke to induce an 'active' method of forming severity and springback control (refer to Section 5.2.1). Hence, in this study, a separated BH – active BHF control strategy – is adopted to obtain highly favourable formability and springback control conditions. The differential BHF applied by BH1 and BH2 is given according to five distinct intervals of punch strokes, 0%–20%, 20%–40%, 40%–60%, 60%–80%, and 80%–100%, as demonstrated in Figure 5.26. The corresponding BHF ranges given are represented as A, B, C, D, and E for BH1 and F, G, H, I, and J for BH2.

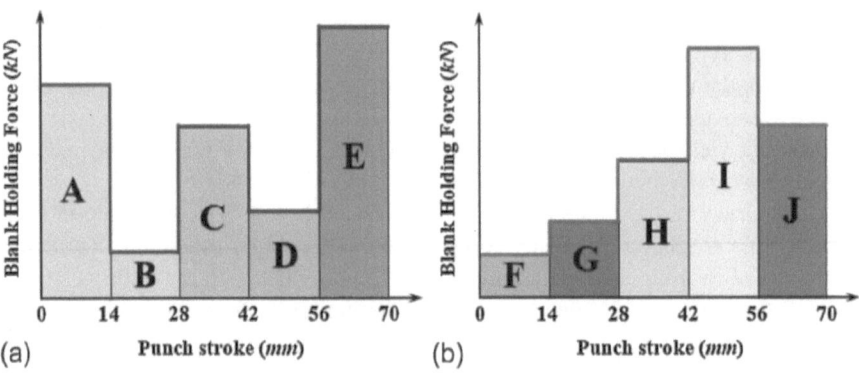

**FIGURE 5.26**   Schematic of conceptual variable BHF for (a) BH1-DC06 side and (b) BH2-TWIP 940 side. (Reproduced by permission from Nguyen et al. [24] Copyright© 2015 WILEY-VCH Verlag GmbH & Co. KGaA, Weinheim.)

### 5.3.3 Predator-Prey EvoNN GA

The present case study requires optimization of two opposing objectives (simultaneous reduction in springback and forming severity) arising during the adjustments of variable split BHF conditions. This is a peculiar bi-objective optimization problem demanding a trade-off between the two opposing objectives, and the Predator-Prey EvoNN algorithm can be used for obtaining the optimal Pareto frontier. The Predator-Prey EvoNN algorithm is one among the nature-inspired GA methods that have been successfully implemented in several engineering optimization problems [101,111–121]. As the name indicates, the Predator-Prey EvoNN method mimics the natural evolutionary process of *prey* and *predators* arbitrarily populated in a forest. Accordingly, a Moore-type two-dimensional computational lattice of unique geometric symmetry with random population of distinct entities – *prey* and *predators* – is generated, and the basic configuration is shown in Figure 5.27. The computing lattice can be visualized as the toroidal surface with uniform distribution of nodes with randomly populated predators and prey, shown with light grey squares and black circles, respectively. A non-empty node in the lattice initially accommodates either a prey or a predator, and the adjacent nodes act as its neighbourhood. Further documentation on the topic and details into similar strategic approaches may be found elsewhere [116,122,123].

Among the initial random population of prey and predators generated by the GA, the prey denotes a set of viable solutions. The predators prune the prey population according to a weighted *fitness function* $\xi$ (Eq. 5.10), which is related to the objective

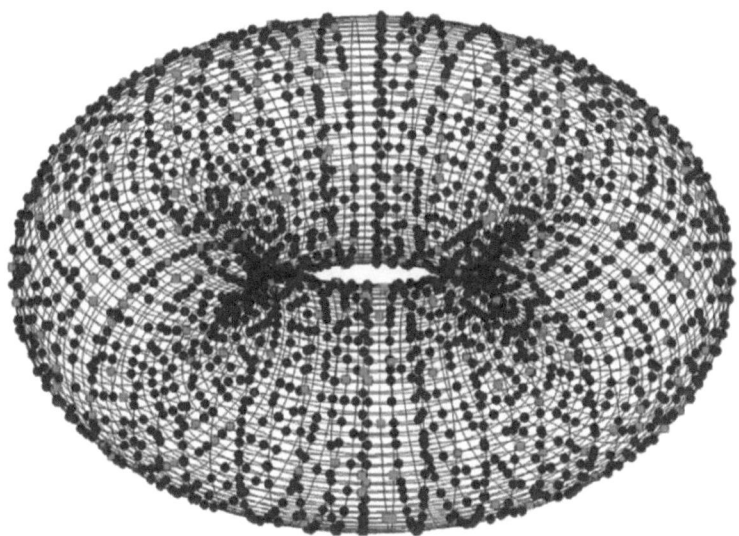

**FIGURE 5.27**   Basic configuration of the computational lattice generated with Predator-Prey random population (light grey square denotes predators, and black circles denotes prey). (Reproduced by permission from Nguyen et al. [24] Copyright© 2015 WILEY-VCH Verlag GmbH & Co. KGaA, Weinheim.)

function values related to each prey, $F_{1,i}$ and $F_{2,i}$. A unique weight value $\omega_j$ for each predator is generated using a uniform random number.

$$\xi_{ij} = \omega_j F_{1,i} + \left(1 - \omega_j\right) F_{2,i}; 0 \le \omega_j \le 1 \qquad (5.10)$$

The surviving prey population after a predetermined number of generations of predator activity based on Eq. (5.10) is then ranked using the Fonseca criterion [124]. The best among these estimates the Pareto frontier. Hence, the next important step is to define the objective functions giving appropriate constraints that restrict the limits of objectives.

### 5.3.3.1    Objective and Constraint Functions

In this study, the variable-BHF strategy at either side of the TWB as explained in Section 5.3.2.3 is adopted for springback control. As already mentioned, the variable-BHF strategy leads to the necessity of optimizing the problem considering two conflicting objectives, namely, to simultaneously minimize the springback effect $f_1(x)$ and forming severity, $f_2(x)$. The objective functions defined are dependent on the control parameters, that is, the variable BHFs provided at different levels of punch strokes, represented as $A, B, C, D, E, F, G, H, I,$ and $J$ (Figure 5.26). These parameters are selected as the *design variables* such that the final best results of the variable-BHF profile will be $\chi_{\text{opt.}} = \left(A_{\text{opt}}, B_{\text{opt}}, C_{\text{opt}}, D_{\text{opt}}, E_{\text{opt}}, F_{\text{opt}}, G_{\text{opt}}, H_{\text{opt}}, I_{\text{opt.}}, J_{\text{opt.}}\right)$.

Revisiting Figure 5.11, the stronger material TWIP940 is more sensitive to produce a higher springback, $h_2$. Accordingly, the first objective is to minimize the springback measure in the TWIP940 side, or

$$f_1(\chi) = h_2 \to \min \qquad (5.11)$$

As detailed in Section 5.3.2.2, the forming severity and tearing fracture are highly dependent on the maximum thinning condition of TWB, especially in the DC06 side. Hence, the second objective, considering $t_i$ as the instantaneous thickness and $t_0$ as the initial thickness, is defined as

$$f_2(\chi) = \left[ \max\left\{ \frac{t_0 - t_i}{t_0} \right\}, i = \overline{1, N} \right] \to \min \qquad (5.12)$$

where $N$ is the number of elements under investigation, counted along the length of the one-element-width strip of the TWB chosen at the symmetry plane in the FE model.

Referring to FLD in Figure 5.25b, the maximum allowable thinning of the weaker DC-06 material is limited to 0.15 (equivalent to 15% thinning criterion in the FLD). As seen in Figure 5.19, the WLM of dissimilar thick TWB is restricted by tool modifications to avoid the possible issue of locking/jamming of the material during forming. A 5 mm WLD is allowed in this study. The limits given for forming severity by maximum thinning of DC-06 mild steel and the allowable WLM are considered as the *constraint functions* $g_1(x)$ and $g_2(x)$ shown in Eqs. (5.13) and (5.14), respectively. Usually, the feasible region of the design parameters within the prescribed limit in surrogate models is selected from a multi-dimensional *convex hull* [125]. The conditions for design feasibility are further met by describing a set of functional inequalities that need to be fulfilled for solving the problem. In this case, a minimum of 2 kN BHF should be applied to suppress the reaction forces by the blank. Otherwise, it will cause detrimental upward movement of the binder while forming. Such inequality conditions are directly expressed as $g_k(x)$ (Eq. 5.15).

$$g_1(\chi) = \max \left\{ \frac{t_0 - t_i}{t_0} \right\} < 0.15, i = \overline{1, N} \tag{5.13}$$

$$g_2(\chi) = |d_0| \le 5\,\text{mm} \tag{5.14}$$

$$g_k(\chi) \le 0, k = \overline{1, M} \tag{5.15}$$

All these constraints set the lower and upper bounds of the variable design parameters as $X_j^L \le X_j \le X_j^U, j = \overline{1, 10}$. It is noteworthy that any conditions other than given upper and lower bounds are considered as special cases of the generalized form of Eq. (5.15). After setting the objective and constraint functions, the optimal conditions of BHFs are obtained through (1) metamodel construction and (2) solving the bi-objective problem using Predator-Prey GA.

### 5.3.3.2   Metamodel Construction and Optimization

The concept of metamodeling is adopted to simultaneously improve the efficiency of GA and reduce the number of FE simulations. The basic data for evaluating the objective functions within the defined constraints are generated through FE simulations. These data are then called for solving the function evaluation in the bi-objective optimization problem. The construction and training of the metamodel are done using judiciously designed samples collected from FE simulation output data. This approach is termed as data-driven metamodeling. All the inputs for optimization calculations are done through suitable design of experiment (DOE) methods [126]. The main aim of the DOE is to cover the entire sampling space using suitable sampling points. In this study, a stratified sampling technique known as *Latin hypercube sampling method* [127] is used, where all the combinations of design parameters are denoted by $\chi = (A, B, C, D, E, F, G, H, I, J)$. The selection of these design parameters is done such that it is uniformly scattered within the design space.

As mentioned before, the FE simulations of U-bending are done to generate approximate data for metamodeling. Basic details of the FE model are discussed in Section 5.3.2.2. The variable BHFs for two sides of TWBs are given as input according to the sampling $\chi$. Initially, 200 sampling points are simulated to generate an early approximation. A two-stage FE simulation strategy is adopted for obtaining the sampling points. The first step in *Stage 1* is to evaluate the formability as per the defined constraints in terms of FLD and/or above-mentioned constraint functions.

Then, the stress and strain in one-element-width strips are recorded. Further, the second objective function ($f_2(\chi)$) and constraint functions ($g_1(\chi)$, $g_2(\chi)$) are calculated. In *Stage 2*, the springback simulations are done to calculate springback in terms of $h_1$ and $h_2$. It should be noted that the *Stage 1* and *Stage 2* simulations are done using explicit and implicit solvers of the FE package, respectively.

The construction of optimum metamodels with accuracy and complexity as their objectives is achieved by selecting statistically significant sampling points from FE simulation results. The metamodels thus constructed are used to evaluate the output for any number of inputs, which essentially replaces the computationally intensive FE simulations. The application of metamodels significantly reduces the function calculation time and thus improves the GA performance. The desired outputs corresponding to a number of statistically significant combinations of DOE sampling steps are obtained by feeding the respective data as input to FE simulations. An in-house code built exclusively for this study automated the pre-processing and post-processing tasks of the FE simulations. A dataset is generated, which contains all the post-processing output data obtained for all the sampling sets. The feasible design space was continuously updated to eliminate anomaly and non-viable cases. If necessary, this updating may also be used to enhance the effectiveness of the metamodel while generating probable additional sampling units.

Finally, the EvoNN technique using Predator-Prey GA is applied to solve the specific multi-objective optimization problem. It is noteworthy that employing the EvoNN optimal neural networks for function evaluation significantly reduced the computationally intensive FE simulations.

## 5.4   RESULTS AND DISCUSSION

### 5.4.1   INFLUENCE OF UNIFORM BHF

The influence of uniform BHF on either side of the TWB during forming is analyzed. The shape after U-bending and thinning contour obtained under uniform BHF conditions is shown in Figure 5.28. The respective springback measure on both sides ($h_1$ for DC-06 and $h_2$ for the TWIP 940 side) and maximum thinning for DC-06 steel after each constant BHF condition is plotted in Figure 5.29. The higher strength obviously resulted in increased springback, and lower thinning behavior of TWIP-940 during U-bending. Minimal springback measurement ($h_2$) and significant thinning are observed in the softer material side of DC-06 steel. Thus, the magnitude of maximum achievable uniform BHF up to tearing failure as per the maximum thinning limit in FLD (0.15) is obtained. The maximum BHF obtained is 90 kN. Springback in either side is found to gradually increase with a small increase in BHF and reduced further. Similar observation is also found for monolithic sheet forming [128]. The higher BHF uniformly distributes the through-thickness stress component in the stretching direction, which is predominantly in the tensile mode. However, in comparison, the reduction in springback is very less in this study, and a small increase of BHF above 90 kN caused thinning beyond the allowable limit.

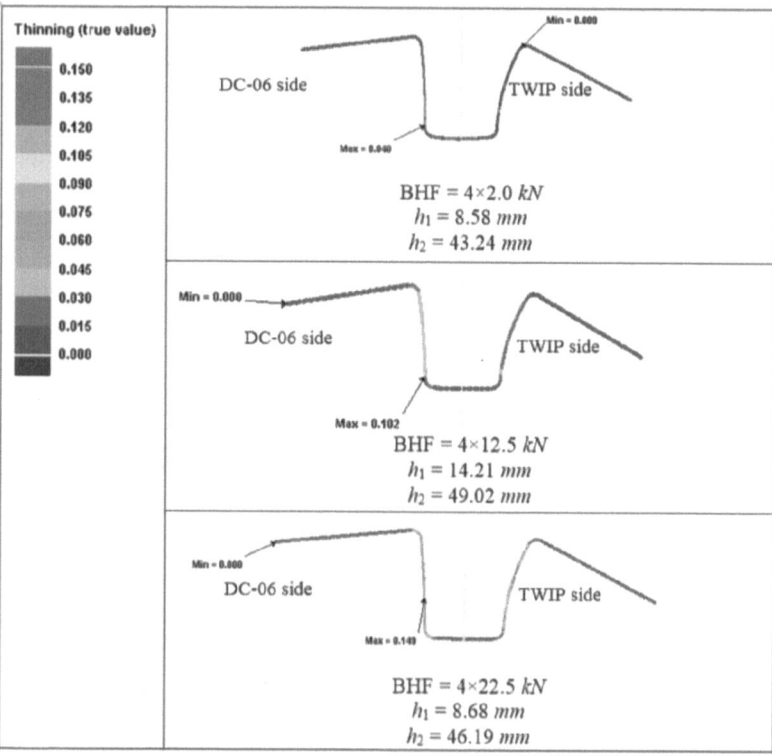

**FIGURE 5.28** Shape after springback and thinning contour of U-bend TWB under identical BHF conditions. (Reproduced by permission from Nguyen et al. [24] Copyright© 2015 WILEY-VCH Verlag GmbH & Co. KGaA, Weinheim.)

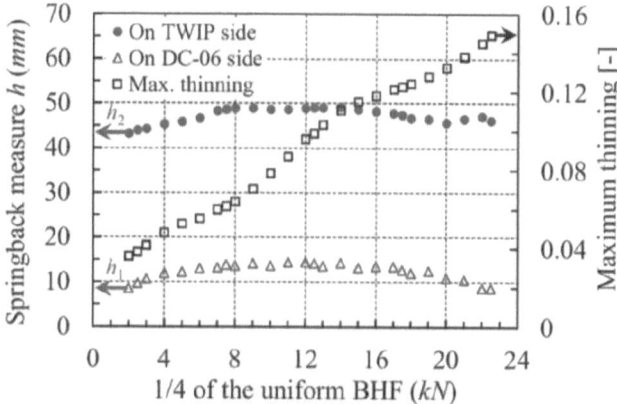

**FIGURE 5.29** Plot of springback measurement of U-bent TWB and maximum thinning in DC-06 side under different levels of uniform BHF. (Reproduced by permission from Nguyen et al. [24] Copyright© 2015 WILEY-VCH Verlag GmbH & Co. KGaA, Weinheim.)

### 5.4.2 Obtaining the Feasible Region

It is suggestive from Figure 5.29 that the main objective functions, namely, springback behavior ($f_1$) and forming severity ($f_2$), have opposing responses to the changes in design variables. For instance, a reduction in BHF decreases the forming severity, $f_2$, and vice versa. However, the low range of BHF induces lesser springback in either side of TWB, whereas a higher springback is observed with a reduction in BHF within the mid to high range. Moreover, an increase in BHF significantly increases the forming severity mainly in the DC-06 side due to excessive thinning. Hence, it is apparent that the lowest magnitude of BHF is an allowable range for an optimal solution as it reduces both forming severity and springback measurement. The mid to high value of BHF can also be a feasible regime as it reduces the springback with moderate thinning. Therefore, the simultaneous optimization of these two objectives results in a conflict while controlling the design variables. Additional complication will also arise while dealing with the constraint function, $g_2$. Then, a correlation study between the individual design variables and defined objectives cannot yield to an optimal solution.

The feasible region in design space is obtained after eliminating the samples that violate the constraints. This region consists of only the allowable solutions, which are approximated by a 10-D *convex hull* within the space of the design variables. The feasible region thus obtained is then used as constraints of the problem. The final allowable solutions must be inside all the 2-D convex hulls so that the bad off-springs after GA operations are eliminated. Hence, the bad solutions on the Pareto frontier obtained are removed in subsequent steps.

### 5.4.3 Pareto Frontier

Figure 5.30 shows the Pareto frontier of the present optimization problem. The function space in the figure represents the objective function $f_1$ plotted as a

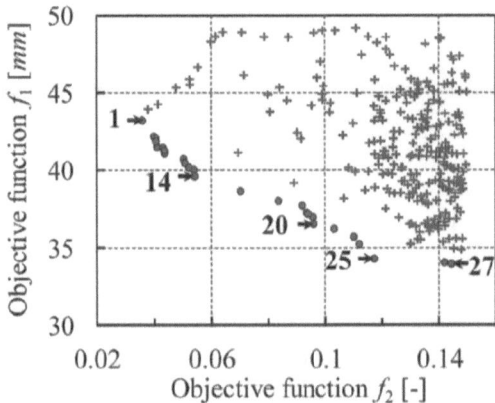

**FIGURE 5.30** The Pareto frontier and overall dataset obtained for the present bi-objective optimization problem. (Reproduced by permission from Nguyen et al. [24] Copyright© 2015 WILEY-VCH Verlag GmbH & Co. KGaA, Weinheim.)

function of second objective function $f_2$. All the data points corresponding to a pair of BHF profiles on either side of TWBs are also shown. The filled circles, numbered 1, 14, 20, 25, and 27, represent the optimized solution points considering the two objectives. Essentially, the Pareto frontier is a bi-objective functional space that separates the feasible and infeasible solutions. The loci of Pareto optimal points are the ones that represent solutions of opposing objectives such that betterment of one objective will result in weakening of the other. In other words, each of these points represents the best compromise, that is, the best trade-off points obtained from solutions of the two conflicting objectives. It is not possible to produce any further improvement of a Pareto optimal solution. It is seen that the Pareto frontier is positioned at a lower functional space than the cloud of the original dataset. Using this method, it is possible to generate improvised pairs of BHF profiles compared to any of the sampling points used in building the dataset.

In this study, the validation of EvoNN-predicted optimal solutions is cross-checked using the data obtained from FE simulations. The Pareto frontier method, which is based on the principle of optimality, is used to depict the trade-off behavior of the opposing objectives. The optimal solutions of the BHF profiles on either side of TWB according to the Pareto frontier are enlisted in Table 5.2. The corresponding values of springback measurements ($h_1$ and $h_2$), maximum thinning ($t_{max}$), and WLM predicted using the EvoNN and FE simulations are also given in the table. Even though there is good agreement between the trends predicted through EvoNN metamodel optimization and FE simulation, discrepancy in magnitude of objective functions still existed. This discrepancy might be due to the high spatial dimensions that require a large number of sampling points to generate better metamodels. The Pareto optimal front provides a list of optimal trade-offs, all of which are equally optimal.

The BHF profiles on either side of TWBs at selected Pareto front points 14, 20, 25, and 27 are, respectively, shown in Figure 5.31a–d. The springback measurement decreases, however, with a simultaneous increase in $t_{max}$ (i.e., the forming severity). The Pareto front obtained through data-driven Predator-Prey EvoNN provides a number of equivalent optimal solutions from which a suitable selection can be made. Usually, an external factor considering practicality and actual constrictions is invoked to select a feasible solution from the Pareto frontier. For instance, in view of better process controlling, the occurrence of abrupt change in the BHF profile is usually avoided. It can be noted from Figure 5.31 that the BHF combinations obtained from the Pareto front are rather complicated, and the classical optimization techniques usually fail to provide such feasible combinations. Another important aspect to note is that while U-bend forming of TWB, the highest BHF value in the TWIP-940 side is obtained well before the final step and is relatively lesser during the final step. However, due to its high strength, the higher work hardening of TWIP-940 steel will produce more resistance to deformation during the final forming step. Hence, it is advisable to apply a restraining force during the final step which also reduces the WLM from low strength DC-06 steel to the stronger TWIP-940 side.

**TABLE 5.2**

**Optimal Solutions Obtained from Pareto Frontier and Corresponding Springback Measurement, $t_{max}$, and WLM**

| Point | BHP Profile on DC-06 Side | | | | | BHP Profile on TWIP Side | | | | | $h_1$ (mm) | $h_2$ (mm) | Max. Thinning | Weldline Dliisp. (mm) |
|---|---|---|---|---|---|---|---|---|---|---|---|---|---|---|
| | A | B | C | D | E | F | G | H | I | J | | | | |
| | 2.00 | 2.00 | 2.00 | 2.00 | 2.00 | 2.00 | 2.00 | 2.00 | 2.00 | 2.00 | 8.58 | 43.24 | 0.036 | 1.11 |
| 2 | 2.81 | 2.17 | 2.28 | 2.38 | 2.47 | 2.47 | 2.65 | 9.54 | 2.46 | 2.33 | 9.48 | 42.20 | 0.040 | 1.23 |
| 3 | 2.73 | 2.17 | 2.41 | 2.46 | 2.47 | 2.47 | 2.65 | 9.89 | 2.36 | 2.33 | 9.69 | 42.09 | 0.041 | 1.25 |
| 4 | 2.95 | 2.17 | 2.34 | 2.51 | 2.47 | 2.47 | 3.05 | 11.49 | 3.15 | 2.33 | 9.07 | 41.90 | 0.041 | 1.20 |
| 5 | 2.82 | 2.17 | 2.33 | 2.57 | 2.47 | 2.47 | 2.99 | 10.80 | 2.92 | 2.33 | 9.74 | 41.54 | 0.041 | 1.43 |
| 6 | 2.82 | 2.17 | 2.40 | 2.48 | 2.47 | 2.48 | 3.19 | 11.41 | 3.23 | 2.33 | 9.21 | 41.48 | 0.043 | 1.58 |
| 7 | 2.67 | 2.22 | 2.38 | 2.58 | 2.47 | 2.47 | 2.77 | 12.13 | 2.61 | 2.38 | 9.09 | 41.31 | 0.044 | 1.67 |
| 8 | 2.79 | 2.17 | 2.39 | 2.57 | 2.47 | 2.48 | 2.89 | 11.55 | 2.96 | 2.33 | 9.14 | 41.26 | 0.044 | 1.60 |
| 9 | 2.86 | 2.17 | 2.40 | 2.53 | 2.47 | 2.48 | 3.03 | 11.82 | 2.33 | 2.35 | 9.04 | 41.10 | 0.044 | 1.61 |
| 10 | 3.17 | 2.26 | 2.52 | 2.80 | 2.56 | 2.68 | 4.17 | 15.22 | 2.33 | 2.36 | 9.00 | 40.75 | 0.050 | 2.04 |
| 11 | 2.96 | 2.17 | 2.99 | 2.88 | 2.68 | 2.47 | 3.76 | 15.80 | 2.35 | 2.33 | 9.64 | 40.51 | 0.051 | 2.02 |
| 12 | 3.06 | 2.17 | 2.48 | 2.81 | 2.57 | 2.47 | 3.86 | 15.19 | 2.33 | 3.30 | 8.83 | 40.21 | 0.052 | 2.23 |
| 13 | 2.79 | 2.17 | 2.58 | 2.78 | 2.56 | 2.47 | 3.88 | 15.14 | 2.33 | 2.33 | 9.11 | 40.05 | 0.054 | 2.93 |
| 14 | 2.68 | 2.17 | 2.56 | 2.79 | 2.59 | 2.47 | 3.89 | 15.24 | 2.33 | 2.33 | 8.53 | 39.64 | 0.054 | 3.05 |
| 15 | 3.97 | 2.17 | 4.70 | 3.86 | 2.76 | 2.66 | 3.23 | 18.02 | 2.33 | 2.73 | 10.48 | 38.66 | 0.070 | 2.24 |
| 16 | 6.87 | 6.97 | 8.21 | 5.84 | 6.41 | 2.70 | 19.37 | 22.43 | 11.40 | 2.33 | 11.95 | 38.04 | 0.083 | 1.32 |
| 17 | 6.68 | 6.57 | 8.17 | 6.08 | 6.43 | 2.47 | 18.99 | 22.35 | 11.75 | 3.06 | 12.63 | 37.75 | 0.092 | 2.07 |
| 18 | 7.02 | 6.67 | 8.19 | 5.72 | 6.14 | 2.70 | 19.29 | 22.37 | 11.08 | 2.33 | 12.75 | 37.30 | 0.093 | 2.10 |
| 19 | 11.84 | 9.45 | 8.56 | 11.11 | 10.10 | 4.04 | 22.68 | 22.67 | 22.60 | 3.10 | 12.08 | 37.22 | 0.094 | -0.55 |
| 20 | 7.43 | 6.95 | 8.38 | 8.62 | 5.57 | 2.94 | 19.65 | 22.68 | 14.38 | 2.33 | 12.18 | 36.99 | 0.096 | 2.06 |
| 21 | 7.41 | 6.45 | 8.41 | 6.46 | 4.56 | 2.58 | 19.10 | 22.68 | 13.44 | 2.33 | 12.33 | 36.57 | 0.096 | 2.13 |
| 22 | 8.88 | 8.79 | 8.54 | 7.03 | 7.87 | 4.13 | 21.76 | 22.68 | 17.06 | 3.02 | 11.06 | 36.24 | 0.103 | 2.78 |
| 23 | 9.79 | 9.37 | 8.82 | 7.18 | 8.54 | 3.55 | 22.64 | 22.68 | 17.65 | 3.03 | 11.36 | 35.72 | 0.110 | 2.75 |
| 24 | 10.01 | 9.13 | 8.03 | 6.88 | 8.02 | 3.84 | 22.41 | 21.69 | 18.74 | 2.86 | 11.13 | 35.25 | 0.112 | 2.97 |
| 25 | 10.96 | 9.36 | 8.41 | 8.25 | 7.06 | 3.85 | 22.68 | 22.64 | 20.24 | 2.81 | 11.18 | 34.30 | 0.117 | 3.02 |
| 26 | 18.88 | 12.20 | 21.53 | 8.80 | 19.90 | 6.38 | 10.56 | 29.04 | 11.86 | 29.08 | 8.82 | 34.05 | 0.142 | -1.33 |
| 27 | 21.34 | 9.34 | 8.52 | 18.18 | 8.47 | 4.18 | 22.68 | 22.68 | 22.64 | 2.81 | 9.70 | 33.96 | 0.144 | 1.66 |

*Source:* Reproduced by permission from Nguyen et al. [24] Copyright© 2015 WILEY-VCH Verlag GmbH & Co. KGaA, Weinheim.

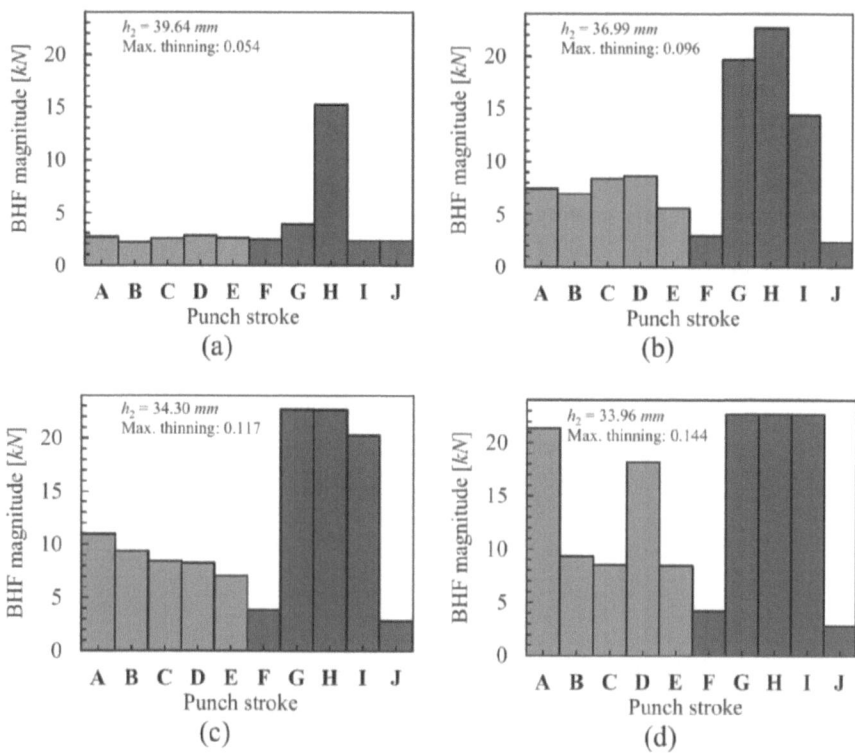

**FIGURE 5.31** BHFs for the specific Pareto front solution points: (a) point 14, (b) point 20, (c) point 25, and (d) point 27. (Reproduced by permission from Nguyen et al. [24] Copyright© 2015 WILEY-VCH Verlag GmbH & Co. KGaA, Weinheim.)

## 5.5  SUMMARY

A general overview of the need, applicability, and formability aspects of TWB, especially in the passenger vehicle body manufacturing, is briefly covered. The mostly used practical methods to reduce springback and their feasibility in TWB stamping are discussed in detail. The basics on the occurrence and analytical modeling of springback through numerical predictions are also elaborated. Essentially, all the efforts to obtain a feasible optimum condition for springback prediction have to satisfy two opposing objectives, that is, the simultaneous reduction of springback and forming severity. As the materials on either side of TWB behave differently, the complexity associated with springback prediction of TWB also increases. Thus, the selection of feasible design parameters considering these complications is a challenging task. It is found that the optimization of these conflicting objectives through the FE simulation methods adopting classical optimization methods will be computationally prohibitive. Hence, it is proposed to use a data-driven EvoNN metamodeling method.

Here, the EvoNN-based Predator-Prey algorithm is used for optimizing the conflicting bi-objective problem. In this regard, a case study of U-bend forming of a TWB manufactured using low strength DC-06 and TWIP-940 high strength steel is demonstrated in detail. The complications of high strength ratio difference, difference in thicknesses, and WLM are considered while defining the optimization problem. The variable BHF strategy on either side of the TWB is used for better control of the forming behavior, which further adds the complexity of the problem. The objective functions are defined to obtain a feasible range of BHFs that generate the best possible solutions of the bi-objective problem through Pareto frontier. The metamodeling method significantly reduced the number of FE simulations to obtain the feasible region of optimized solutions. Also, this multi-objective GA method generated the sets of BHF combinations that are otherwise impossible to obtain through classical methods.

The EvoNN method can extensively predict new information by efficiently capturing the existing data. The whole process can be automated once the key input factors are decided, eliminating possible human error while handling numerous data. Therefore, a better control in the process sequence and product quality can be assured with a minimum number of computational simulations and cumbersome experimentations.

## NOTES

\*    Note that substituting materials with less density is often not the right approach. It is essential to replace materials with superior strength to weight ratio.

†    Automotive door assembly consists of two main sheet metal panels, the outer smooth panel and an inner structural panel joined to form a hollow and stiff door assembly.

‡    In principle, dent resistance is a requirement for door outer and not door inner. Nevertheless, the point tries to address the multitude of functional requirements in a sheet component.

§    https://www.twbcompany.com/applications/#door-inner-with-header (accessed date 19 Feb 2022).

\*\*    https://www.howacarworks.com/bodywork/renewing-hinge-pins-and-hinges (dated 19 Feb 2022).

††    Illustrative photograph taken by the first author at BITS Pilani-K. K. Birla Goa campus (dated 20 Aug 2022).

‡‡    Crash test between a 1996 Ford Explorer and 2000 Ford Focus photographed at the Insurance Institute for Highway Safety Vehicle Research Center by Brady Holt. https://en.wikipedia.org/wiki/Side_collision#/media/File:Ford_Focus_versus_Ford_Explorer_crash_test_IIHS.jpg (dated 20 Aug 2022).

§§    https://centervalleyauto.com/how-to-fix-a-large-dent-in-a-car-door (dated 19 Feb 2022).

\*\*\*    https://www.twbcompany.com/applications/#door-inner-with-header (dated 19 Feb 2022).

†††    Recently, special attentions are being given to under-water using laser and solid-state friction stir welding methods [129].

‡‡‡    The welding tool (power source for welding) is not directly in contact with the blanks to be joined. With the narrow beam concentration, intricate spaces, which are usually unapproachable can be welded using this technique.

§§§    Unlike electron beam welding (EBW), the laser beam welding does not require any vacuum. The laser beam welding also produces comparative weld zone dimensions as in EBW. Even under-water laser welding is also being considered for special circumstances [129].

****        In addition to formability issues, weld line movement is a bigger problem for die design. If the TWB includes sheets of different thickness, the movement of weld line has to be accommodated in the die design, which is complex. The above discussion however deals with the weld line movement only from the formability point of view.

††††      https://www.metalformingmagazine.com/article/?/quality-control/material-testing/forming-limit-diagrams-then-and-now (dated 20 Feb 2022).

‡‡‡‡      Inhomogeneity in surface and thickness strains during multi-stage processes are inherent, which may also enhance the springback tendencies. This phenomenon occurs for materials having low ductility with relatively higher yield strength. The deformation in a multistage process potentially reduces the formability and traverses through different strain paths [130].

§§§§     Analogous to drawbead action in stretch-deep draw operation.

*****    https://www.thefabricator.com/stampingjournal/article/stamping/r-d-update-reducing-springback-using-poststretching-with-stake-beads (dated: 21 Feb 2022).

†††††    https://ahssinsights.org/forming/springback/correcting-springback/ (dated: 21 Feb 2022).

‡‡‡‡‡    Bauschinger effect refers to the flow stress softening with change in strain path after a certain pre-strain. Most commonly used experiment is a tension-compression test, wherein the yielding under compression after a tensile prestrain occurs at a lower stress than that of tension.

§§§§§    The HAZ in laser welded steel TWBs is stronger than the parent metals and shows higher elastic modulus. The higher yield strength tends to increase the springback, whereas higher elastic modulus reduces the springback. These two contradicting factors tend to produce little effect on the springback behaviour of steel TWBs [104]. On the contrary, the Young's modulus of FSW aluminum TWB is comparable, and yield strength is lower than that of parent metals. Hence, it would be important to consider the HAZ of FSW aluminum TWBs [131].

## REFERENCES

1. Mallick PK, editor. *Materials, Design and Manufacturing for Lightweight Vehicles.* Elsevier: Amsterdam, Netherlands, 2010.
2. Czerwinski F. Current trends in automotive lightweighting strategies and materials. *Materials (Basel)* 2021;14. doi: 10.3390/ma14216631.
3. Panicker SS, Panda SK. Investigations into improvement in formability of AA5754 and AA6082 sheets at elevated temperatures. *J Mater Eng Perform* 2019;28:2967–82. doi: 10.1007/s11665-019-04030-1.
4. Lee YS, Kim MC, Kim SW, Kwon YN, Choi SW, Lee JH. Experimental and analytical studies for forming limit of AZ31 alloy on warm sheet metal forming. *J Mater Process Technol* 2007;187–8:103–7. doi: 10.1016/j.jmatprotec.2006.11.118.
5. Hua L, Meng F, Song Y, Liu J, Qin X, Suo L. A constitutive model of 6111-T4 aluminum alloy sheet based on the warm tensile test. *J Mater Eng Perform* 2014;23:1107–13. doi: 10.1007/s11665-013-0834-2.
6. Van Sy L, Nam NT. Hot incremental forming of magnesium and aluminum alloy sheets by using direct heating system. *Proc Inst Mech Eng Part B J Eng Manuf* 2013;227:1099–110. doi: 10.1177/0954405413484014.
7. Chen FK, Huang TB. Formability of stamping magnesium-alloy AZ31 sheets. *J Mater Process Technol* 2003;142:643–7. doi: 10.1016/S0924-0136(03)00684-8.
8. Sun G, Tan D, Lv X, Yan X, Li Q, Huang X. Multi-objective topology optimization of a vehicle door using multiple material tailor-welded blank (TWB) technology. *Adv Eng Softw* 2018;124:1–9. doi: 10.1016/j.advengsoft.2018.06.014.

9. Hovanski Y, Upadhyay P, Carsley J, Luzanski T, Carlson B, Eisenmenger M, et al. High-speed friction-stir welding to enable aluminum tailor-welded blanks. *JOM* 2015;67:1045–53. doi: 10.1007/s11837-015-1384-x.

10. Hussein SA, Tahir ASM, Hadzley AB. Characteristics of aluminum-to-steel joint made by friction stir welding: A review. *Mater Today Commun* 2015;5:32–49. doi: 10.1016/j.mtcomm.2015.09.004.

11. Li G, Xu F, Huang X, Sun G. Topology optimization of an automotive tailor-welded blank door. *J Mech Des Trans ASME* 2015;137:1–8. doi: 10.1115/1.4028704.

12. Song X, Sun G, Li Q. Sensitivity analysis and reliability based design optimization for high-strength steel tailor welded thin-walled structures under crashworthiness. *Thin-Walled Struct* 2016;109:132–42. doi: 10.1016/j.tws.2016.09.003.

13. Sun G, Xu F, Li G, Huang X, Li Q. Determination of mechanical properties of the weld line by combining micro-indentation with inverse modeling. *Comput Mater Sci* 2014;85:347–62. doi: 10.1016/j.commatsci.2014.01.006.

14. Li G, Xu F, Sun G, Li Q. Identification of mechanical properties of the weld line by combining 3D digital image correlation with inverse modeling procedure. *Int J Adv Manuf Technol* 2014;74:893–905. doi: 10.1007/s00170-014-6034-x.

15. Kinsey BL. Tailor welded blanks for the automotive industry. In: *Tailor Welded Blanks for Advanced Manufacturing*. Woodhead Publishing: Sawston, Cambridge, 2011. doi: 10.1016/B978-1-84569-704-4.50007-1.

16. Merklein M, Johannes M, Lechner M, Kuppert A. A review on tailored blanks: Production, applications and evaluation. *J Mater Process Technol* 2014;214:151–64. doi: 10.1016/j.jmatprotec.2013.08.015.

17. Saunders FI, Wagoner RH. Forming of tailor-welded blanks. *Metall Mater Trans A Phys Metall Mater Sci* 1996;27:2605–16. doi: 10.1007/BF02652354.

18. Reed D, Belanger P. Hot stamped steel one-piece door ring in the all-new 2019 Ram 1500. Present 2018 Gt Des Steel 2018.

19. El-Batahgy A, Kutsuna M. Laser beam welding of AA5052, AA5083, and AA6061 aluminum alloys. *Adv Mater Sci Eng* 2009;2009. doi: 10.1155/2009/974182.

20. Hong KM, Shin YC. Prospects of laser welding technology in the automotive industry: A review. *J Mater Process Technol* 2017;245:46–69. doi: 10.1016/j.jmatprotec.2017.02.008.

21. He X, Gu F, Ball A. A review of numerical analysis of friction stir welding. *Prog Mater Sci* 2014;65:1–66. doi: 10.1016/j.pmatsci.2014.03.003.

22. Uzun H, Dalle Donne C, Argagnotto A, Ghidini T, Gambaro C. Friction stir welding of dissimilar Al 6013-T4 To X5CrNi18-10 stainless steel. *Mater Des* 2005;26:41–6. doi: 10.1016/j.matdes.2004.04.002.

23. Buffa G, Fratini L, Merklein M, Staud D. Investigations on the mechanical properties and formability of friction stir welded tailored blanks. *Key Eng Mater* 2007;344:143–50. doi: 10.4028/www.scientific.net/KEM.344.143.

24. Nguyen NT, Hariharan K, Chakraborti N, Barlat F, Lee MG. Springback reduction in tailor welded blank with high strength differential by using multi-objective evolutionary and genetic algorithms. *Steel Res Int* 2015;86:1391–402. doi: 10.1002/srin.201400263.

25. Korouyeh RS, Moslemi Naeini H, Torkamany MJ, Liaghat G. Experimental and theoretical investigation of thickness ratio effect on the formability of tailor welded blank. *Opt Laser Technol* 2013;51:24–31. doi: 10.1016/j.optlastec.2013.02.016.

26. Bhaskar VV, Narayanan RG, Narasimhan K. Effect of thickness ratio on formability of tailor welded blanks (TWB). *AIP Conf Proc* 2004;712:863–8. doi: 10.1063/1.1766635.

27. Chan LC, Chan SM, Cheng CH, Lee TC. Formability and weld zone analysis of tailor-welded blanks for various thickness ratios. *J Eng Mater Technol Trans ASME* 2005;127:179–85. doi: 10.1115/1.1857936.

28. Panda SK, Kumar DR. Study of formability of tailor-welded blanks in plane-strain stretch forming. *Int J Adv Manuf Technol* 2009;44:675–85. doi: 10.1007/s00170-008-1888-4.

29. Wu J, Hovanski Y, Miles M. Investigation of the thickness differential on the formability of aluminum tailor welded blanks. *Metals (Basel)* 2021;11:1–10. doi: 10.3390/met11060875.

30. Narayanan RG, Narasimhan K. Influence of the weld conditions on the forming-limit strains of tailor-welded blanks. *J Strain Anal Eng Des* 2008;43:217–27. doi: 10.1243/03093247JSA344.

31. Moayedi H, Darabi R, Ghabussi A, Habibi M, Foong LK. Weld orientation effects on the formability of tailor welded thin steel sheets. *Thin-Walled Struct* 2020;149:106669. doi: 10.1016/j.tws.2020.106669.

32. Safdarian R. Forming limit diagram prediction of tailor welded blank by modified M-K model. *Mech Res Commun* 2015;67:47–57. doi: 10.1016/j.mechrescom.2015.05.004.

33. Korouyeh RS, Naeini HM, Liaghat G. Forming limit diagram prediction of tailor-welded blank using experimental and numerical methods. *J Mater Eng Perform* 2012;21:2053–61. doi: 10.1007/s11665-012-0156-9.

34. Safdarian R, Jorge RMN, Santos AD, Naeini HM, Parente MPL. A comparative study of forming limit diagram prediction of tailor welded blanks. *Int J Mater Form* 2015;8:293–304. doi: 10.1007/s12289-014-1168-9.

35. Padmanabhan R, Oliveira MC, Laurent H, Alves JL, Menezes LF. Study on springback in deep drawn tailor welded blanks. *Int J Mater Form* 2009;2:829–32. doi: 10.1007/s12289-009-0566-x.

36. Marretta L, Ingarao G, Di Lorenzo R. Design of sheet stamping operations to control springback and thinning: A multi-objective stochastic optimization approach. *Int J Mech Sci* 2010;52:914–27. doi: 10.1016/j.ijmecsci.2010.03.008.

37. Gautam V, Raut VM, Kumar DR. Analytical prediction of springback in bending of tailor-welded blanks incorporating effect of anisotropy and weld zone properties. *Proc Inst Mech Eng Part L J Mater Des Appl* 2018;232:294–306. doi: 10.1177/1464420715624261.

38. Gautam V, Kumar DR. Experimental and numerical investigations on springback in V-bending of tailor-welded blanks of interstitial free steel. *Proc Inst Mech Eng Part B J Eng Manuf* 2018;232:2178–91. doi: 10.1177/0954405416687146.

39. Seo DG, Chang SH, Heo YM. Springback characteristics of tailor-welded strips in U-draw bending. *Met Mater Int* 2003;9:571–6. doi: 10.1007/BF03027257.

40. Zadpoor AA, Sinke J, Benedictus R. Finite element modeling and failure prediction of friction stir welded blanks. *Mater Des* 2009;30:1423–34. doi: 10.1016/j.matdes.2008.08.018.

41. Darmawan AS, Anggono AD, Nugroho S. Springback phenomenon analysis of tailor welded blank of mild steel in U-bending process. *AIP Conf Proc* 2019;2014:030021. doi: 10.1063/1.5112425.

42. Panda SK, Kumar DR. Improvement in formability of tailor welded blanks by application of counter pressure in biaxial stretch forming. *J Mater Process Technol* 2008;204:70–9. doi: 10.1016/j.jmatprotec.2007.10.076.

43. Abbasi M, Hamzeloo SR, Ketabchi M, Shafaat MA, Bagheri B. Analytical method for prediction of weld line movement during stretch forming of tailor-welded blanks. *Int J Adv Manuf Technol* 2014;73:999–1009. doi: 10.1007/s00170-014-5850-3.

44. Bandyopadhyay K, Panda SK, Saha P. Investigations into the influence of weld zone on formability of fiber laser-welded advanced high strength steel. *J Mater Eng Perform* 2014;23:1465–79. doi: 10.1007/s11665-014-0881-3.

45. Bandyopadhyay K, Lee M-G, Panda SK, Saha P, Lee J. Formability assessment and failure prediction of laser welded dual phase steel blanks using anisotropic plastic properties. *Int J Mech Sci* 2017;126:203–21. doi: 10.1016/j.ijmecsci.2017.03.022.

46. Bandyopadhyay K, Panda SK, Saha P, Padmanabham G. Limiting drawing ratio and deep drawing behaviour of dual phase steel tailor welded blanks: FE simulation and experimental validation. *J Mater Process Technol* 2015;217:48–64. doi: 10.1016/j.jmatprotec.2014.10.022.

47. Keeler S, Kimchi M, Mooney PJ. *Advanced High-Strength Steels Guidelines Version 6.0*. WorldAutoSteel: Brussels, Belgium, 2017, 314.

48. Choudhury IA, Ghomi V. Springback reduction of aluminum sheet in V-bending dies. *Proc Inst Mech Eng Part B J Eng Manuf* 2014;228:917–26. doi: 10.1177/0954405413514225.

49. Li B, McClelland Z, Horstemeyer SJ, Aslam I, Wang PT, Horstemeyer MF. Time dependent springback of a magnesium alloy. *Mater Des* 2015;66:575–80. doi: 10.1016/j.matdes.2014.03.035.

50. Majidi O, Barlat F, Lee MG. Effect of slide motion on springback in 2-D draw bending for AHSS. *Int J Mater Form* 2016;9:313–26. doi: 10.1007/s12289-014-1214-7.

51. Majidi O, Barlat F, Lee MG, Kim DJ. Formability of AHSS under an attach-detach forming mode. *Steel Res Int* 2015;86:98–109. doi: 10.1002/srin.201400001.

52. Liao J, Xue X. Experimental investigation of deep drawing of C-rail benchmark using various lubricants. *MATEC Web Conf* 2015;21:04011. doi: 10.1051/matecconf/20152104011.

53. Radonjic R, Liewald M. Approaches for springback reduction when forming ultra high-strength sheet metals. *IOP Conf Ser Mater Sci Eng* 2016;159:0–11. doi: 10.1088/1757-899X/159/1/012028.

54. Mamutov AV, Golovashchenko SF, Bessonov NM, Mamutov VS. Electrohydraulic forming of low volume and prototype parts: Process design and practical examples. *J Manuf Mater Process* 2021;5. doi: 10.3390/jmmp5020047.

55. Gösling M, Güner A, Burchitz I, Thülig T, Carleer B. Effect of coining on springback behaviour. *IOP Conf Ser Mater Sci Eng* 2018;418:0–7. doi: 10.1088/1757-899X/418/1/012106.

56. Chou IN, Hung C. Finite element analysis and optimization on springback reduction. *Int J Mach Tools Manuf* 1999;39:517–36. doi: 10.1016/S0890-6955(98)00031-5.

57. Shu JS, Hung C. Finite element analysis and optimization of springback reduction: The "double-bend" technique. *Int J Mach Tools Manuf* 1996;36:423–34. doi: 10.1016/0890-6955(95)00072-0.

58. Shinmiya T, Urabe M, Fujii Y. Development of press forming technologies for ultra-high strength steel sheets utilizing computer aided engineering. *JFE Tech Rep* 2019;24:39–45.

59. Xu S. Springback prediction, compensation and correlation for automotive stamping. *AIP Conf Proc* 2005;778:345–52. doi: 10.1063/1.2011244.

60. Liao J, Xue X, Zhou C, Barlat F, Gracio JJ. A springback compensation strategy and applications to bending cases. *Steel Res Int* 2013;84:463–72. doi: 10.1002/srin.201200220.

61. Yang XA, Ruan F. A die design method for springback compensation based on displacement adjustment. *Int J Mech Sci* 2011;53:399–406. doi: 10.1016/j.ijmecsci.2011.03.002.

62. Ling YE, Lee HP, Cheok BT. Finite element analysis of springback in L-bending of sheet metal. *J Mater Process Technol* 2005;168:296–302. doi: 10.1016/j.jmatprotec.2005.02.236.

63. Cho JR, Moon SJ, Moon YH, Kang SS. Finite element investigation on spring-back characteristics in sheet metal U-bending process. *J Mater Process Technol* 2003;141:109–16. doi: 10.1016/S0924-0136(03)00163-8.

64. Rosochowski A. Die compensation procedure to negate die deflection and component springback. *J Mater Process Technol* 2001;115:187–91. doi: 10.1016/S0924-0136(01)00805-6.

65. Livatyali H, Wu HC, Altan T. Prediction and elimination of springback in straight flanging using computer-aided design methods Part 1: Experimental investigations. *J Mater Process Technol* 2002;120:348–54. doi: 10.1016/S0924-0136(01)01161-X.

66. Silveira VL, Haase OC, Stemler PMA, Viana RAM, Duarte AS. FEM stress state analysis on springback reduction methods: Variable blank holder force and stake bead. *IOP Conf Ser Mater Sci Eng* 2018;418. doi: 10.1088/1757-899X/418/1/012109.

67. Kitayama S, Huang S, Yamazaki K. Optimization of variable blank holder force trajectory for springback reduction via sequential approximate optimization with radial basis function network. *Struct Multidiscip Optim* 2013;47:289–300. doi: 10.1007/s00158-012-0824-2.

68. Ren H, Xie J, Liao S, Leem D, Ehmann K, Cao J. In-situ springback compensation in incremental sheet forming. *CIRP Ann* 2019;68:317–20. doi: 10.1016/j.cirp.2019.04.042.

69. Wagoner RH, Lim H, Lee MG. Advanced issues in springback. *Int J Plast* 2013;45:3–20. doi: 10.1016/j.ijplas.2012.08.006.

70. Liu YC. The effect of restraining force on shape deviations in flanged channels. *J Eng Mater Technol* 1988;110:389–94. doi: 10.1115/1.3226067.

71. Sunseri M, Cao J, Karafillis AP, Boyce MC. Accommodation of springback error in channel forming using active binder force control: Numerical simulations and experiments. *J Eng Mater Technol Trans ASME* 1996;118:426–35. doi: 10.1115/1.2806830.

72. Hariharan K, Nguyen NT, Chakraborti N, Lee MG, Barlat F. Multi-objective genetic algorithm to optimize variable drawbead geometry for tailor welded blanks made of dissimilar steels. *Steel Res Int* 2014;85:1597–607. doi: 10.1002/srin.201300471.

73. Choi MK, Huh H. Effect of punch speed on amount of springback in U-bending process of auto-body steel sheets. *Procedia Eng* 2014;81:963–8. doi: 10.1016/j.proeng.2014.10.125.

74. Zadpoor AA, Sinke J, Benedictus R. Mechanics of tailor welded blanks: An overview. *Key Eng Mater* 2007;344:373–82. doi: 10.4028/www.scientific.net/kem.344.373.

75. Kinsey BL, Cao J. An analytical model for tailor welded blank forming. *J Manuf Sci Eng Trans ASME* 2003;125:344–51. doi: 10.1115/1.1537261.

76. Kinsey B, Liu Z, Cao J. Novel forming technology for tailor-welded blanks. *J Mater Process Technol* 2000;99:145–53. doi: 10.1016/S0924-0136(99)00412-4.

77. Panda SK, Li J, Hernandez VHB, Zhou Y, Goodwin F. Effect of weld location, orientation, and strain path on forming behaviour of AHSS tailor welded blanks. *J Eng Mater Technol Trans ASME* 2010;132:1–11. doi: 10.1115/1.4001965.

78. Wang H, Zhou J, Zhao TS, Liu LZ, Liang Q. Multiple-iteration springback compensation of tailor welded blanks during stamping forming process. *Mater Des* 2016;102:247–54. doi: 10.1016/j.matdes.2016.04.032.

79. Liu XJ, Zhang M, Huo XH, Ge XJ, Liu B. Research on the spring-back compensation method of tailor welded blank U-shaped part. *Appl Mech Mater* 2013;373–375:1970–4. doi: 10.4028/www.scientific.net/AMM.373-375.1970.

80. Marciniak Z, Duncan JL, Hu SJ. *Mechanics of Sheet Metal Forming,* Second Edition. Butterworth-Heinemann: Oxford, 2002.

81. Hill R. A theory of the yielding and plastic flow of anisotropic metals. *Proc R Soc London A Math Phys Eng Sci* 1948;193:281–97.

82. Kuwabara T, Kuroda M, Tvergaard V, Nomura K. Use of abrupt strain path change for determining subsequent yield surface: Experimental study with metal sheets. *Acta Mater* 2000;48:2071–9. doi: 10.1016/S1359-6454(00)00048-3.

83. Hariharan K, Prakash RV, Sathya Prasad M. Influence of yield criteria in the prediction of strain distribution and residual stress distribution in sheet metal formability analysis for a commercial steel. *Mater Manuf Process* 2010;25:828–36. doi: 10.1080/10426910903496847.

84. Barlat F, Brem JC, Yoon JW, Chung K, Dick RE, Lege DJ, et al. Plane stress yield function for aluminum alloy sheets - Part 1: Theory. *Int J Plast* 2003;19:1297–319. doi: 10.1016/S0749-6419(02)00019-0.

85. Hill R. A user-friendly theory of orthotropic plasticity in sheet metals. *Int J Mech Sci* 1993;35:19–25. doi: 10.1016/0020-7403(93)90061-X.

86. Aretz H, Barlat F. New convex yield functions for orthotropic metal plasticity. *Int J Non Linear Mech* 2013;51:97–111. doi: 10.1016/j.ijnonlinmec.2012.12.007.

87. Vegter H, Van Den Boogaard AH. A plane stress yield function for anisotropic sheet material by interpolation of biaxial stress states. *Int J Plast* 2006;22:557–80. doi: 10.1016/j.ijplas.2005.04.009.

88. Banabic D, Kuwabara T, Balan T, Comsa DS, Julean D. Non-quadratic yield criterion for orthotropic sheet metals under plane-stress conditions. *Int J Mech Sci* 2003;45:797–811. doi: 10.1016/S0020-7403(03)00139-5.

89. Banabic D, Barlat F, Cazacu O, Kuwabara T. Advances in anisotropy of plastic behaviour and formability of sheet metals. *Int J Mater Form* 2020;13:749–87. doi: 10.1007/s12289-020-01580-x.

90. Banabic D. Sheet metal forming processes: Constitutive modelling and numerical simulation, 2010. doi: 10.1007/978-3-540-88113-1.

91. Hosford WF, Caddell RM. *Metal Forming: Mechanics and Metallurgy,* Third Edtion. Cambridge University Press: New York, 2007. doi: 10.1016/0378-3804(85)90124-X.

92. Stoughton TB, Yoon JW. Review of Drucker's postulate and the issue of plastic stability in metal forming. *Int J Plast* 2006;22:391–433. doi: 10.1016/j.ijplas.2005.03.002.

93. Taherizadeh A, Green DE, Ghaei A, Yoon JW. A non-associated constitutive model with mixed iso-kinematic hardening for finite element simulation of sheet metal forming. *Int J Plast* 2010;26:288–309. doi: 10.1016/j.ijplas.2009.07.003.

94. Lu R, Liu X, Xu Z, Hu X, Shao Y, Liu L. Simulation of springback variation in the U-bending of tailor rolled blanks. *J Brazilian Soc Mech Sci Eng* 2017;39:4633–47. doi: 10.1007/s40430-017-0778-9.

95. Zang SL, Lee MG, Hoon Kim J. Evaluating the significance of hardening behaviour and unloading modulus under strain reversal in sheet springback prediction. *Int J Mech Sci* 2013;77:194–204. doi: 10.1016/j.ijmecsci.2013.09.033.

96. Kim C, Lee JU, Barlat F, Lee MG. Frictional behaviours of a mild steel and a TRIP780 steel under a wide range of contact stress and sliding speed. *J Tribol* 2014;136:1–7. doi: 10.1115/1.4026346.

97. Lee JY, Barlat F, Lee MG. Constitutive and friction modeling for accurate springback analysis of advanced high strength steel sheets. *Int J Plast* 2015;71:113–35. doi: 10.1016/j.ijplas.2015.04.005.

98. Lee MG, Kim D, Kim C, Wenner ML, Wagoner RH, Chung K. A practical two-surface plasticity model and its application to spring-back prediction. *Int J Plast* 2007;23:1189–212. doi: 10.1016/j.ijplas.2006.10.011.

99. Kim J, Lee W, Chung KH, Kim D, Kim C, Okamoto K, et al. Springback evaluation of friction stir welded TWB automotive sheets. *Met Mater Int* 2011;17:83–98. doi: 10.1007/s12540-011-0212-2.

100. Chakraborti N. *Data-Driven Evolutionary Modeling in Materials Technology* (First Edition). CRC Press: Boca Raton,2022. doi: 10.1201/9781003201045.

101. Pettersson F, Chakraborti N, Saxén H. A genetic algorithms based multi-objective neural net applied to noisy blast furnace data. *Appl Soft Comput J* 2007;7:387–97. doi: 10.1016/j.asoc.2005.09.001.

102. Chakraborti N. *Evolutionary Data-Driven Modeling.* Elsevier Inc.: Amsterdam, Netherlands, 2013. doi: 10.1016/B978-0-12-394399-6.00005-9.

103. Chakraborti N. Strategies for evolutionary data driven modeling in chemical and metallurgical systems. In: J. Valadi, P. Siarry (Eds.), *Appl Metaheuristics Process Engineering.* Springer International Publishing: Cham, 2014, pp. 89–122. doi: 10.1007/978-3-319-06508-3_4.

104. Zhao KM, Chun BK, Lee JK. Finite element analysis of tailor-welded blanks. *Finite Elem Anal Des* 2001;37:117–30. doi: 10.1016/S0168-874X(00)00026-3.

105. Sharma RS, Molian P. Yb:YAG laser welding of TRIP780 steel with dual phase and mild steels for use in tailor welded blanks. *Mater Des* 2009;30:4146–55. doi: 10.1016/j.matdes.2009.04.033.

106. Shao H, Gould J, Albright C. Laser blank welding high-strength steels. *Metall Mater Trans B Process* 2007;38:321–31. doi: 10.1007/s11663-007-9026-5.
107. Chen W, Lin GS, Hu SJ. A comparison study on the effectiveness of stepped binder and weld line clamping pins on formability improvement for tailor-welded blanks. *J Mater Process Technol* 2008;207:204–10. doi: 10.1016/j.jmatprotec.2007.12.100.
108. Chung K, Ahn K, Yoo DH, Chung KH, Seo MH, Park SH. Formability of TWIP (twinning induced plasticity) automotive sheets. *Int J Plast* 2011;27:52–81. doi: 10.1016/j.ijplas.2010.03.006.
109. Butuc MC, Teodosiu C, Barlat F, Gracio JJ. Analysis of sheet metal formability through isotropic and kinematic hardening models. *Eur J Mech A/Solids* 2011;30:532–46. doi: 10.1016/j.euromechsol.2011.03.005.
110. Basak S, Panda SK, Lee MG. Formability and fracture in deep drawing sheet metals: Extended studies for pre-strained anisotropic thin sheets. *Int J Mech Sci* 2020;170:105346. doi: 10.1016/j.ijmecsci.2019.105346.
111. Kovačič M, Rožej U, Brezočnik M. Genetic algorithm rolling mill layout optimization. *Mater Manuf Process* 2013;28:783–7. doi: 10.1080/10426914.2012.718475.
112. Ulutan D, Özel T. Multiobjective optimization of experimental and simulated residual stresses in turning of Nickel-Alloy IN100. *Mater Manuf Process* 2013;28:835–41. doi: 10.1080/10426914.2012.718474.
113. Mitra T, Pettersson F, Saxén H, Chakraborti N. Blast furnace charging optimization using multi-objective evolutionary and genetic algorithms. *Mater Manuf Process* 2017;32:1179–88. doi: 10.1080/10426914.2016.1257133.
114. Klancnik S, Brezocnik M, Balic J, Karabegovic I. Programming of CNC milling machines using particle swarm optimization. *Mater Manuf Process* 2013;28:811–5. doi: 10.1080/10426914.2012.718473.
115. Datta S, Zhang Q, Sultana N, Mahfouf M. Optimal design of titanium alloys for prosthetic applications using a multiobjective evolutionary algorithm. *Mater Manuf Process* 2013;28:741–5. doi: 10.1080/10426914.2013.773020.
116. Giri BK, Hakanen J, Miettinen K, Chakraborti N. Genetic programming through bi-objective genetic algorithms with a study of a simulated moving bed process involving multiple objectives. *Appl Soft Comput J* 2013;13:2613–23. doi: 10.1016/j.asoc.2012.11.025.
117. Nandi AK, Deb K, Datta S. Genetic algorithm-based design and development of particle-reinforced silicone rubber for soft tooling process. *Mater Manuf Process* 2013;28:753–60. doi: 10.1080/10426914.2013.773022.
118. Salgotra R, Singh U, Singh S, Mittal N. A hybridized multi-algorithm strategy for engineering optimization problems. *Knowledge-Based Syst* 2021;217:106790. doi: 10.1016/j.knosys.2021.106790.
119. Tutum CC, Deb K, Hattel JH. Multi-criteria optimization in friction stir welding using a thermal model with prescribed material flow. *Mater Manuf Process* 2013;28:816–22. doi: 10.1080/10426914.2012.736654.
120. Tutum CC, Deb K, Baran I. Constrained efficient global optimization for pultrusion process. *Mater Manuf Process* 2015;30:538–51. doi: 10.1080/10426914.2014.994752.
121. Giri BK, Pettersson F, Saxén H, Chakraborti N. Genetic programming evolved through bi-objective genetic algorithms applied to a blast furnace. *Mater Manuf Process* 2013;28:776–82. doi: 10.1080/10426914.2013.763953.
122. Li X. A real-coded predator-prey genetic algorithm for multiobjective optimization. *Proceeding of Evolutionary Multi-Criterion Optimization, Second International Conference, EMO 2003*, Faro, Port. April 8–11, 2003, pp. 207–21. doi: 10.1007/3-540-36970-8_15.
123. Deb K. *Mutliobjective Optimization Using Evolutionary Algorithm*, vol. 44, First Edition. John Wiley & Sons, Ltd: Hoboken, NJ, 2001.
124. Fonseca CMM. *Multiobjective Genetic Algorithms with Application to Control Engineering Problems*. The University of Sheffield: Sheffield, 1995.

125. Siegel A. A parallel algorithm for understanding design spaces and performing convex hull computations. *J Comput Math Data Sci* 2022;2:100021. doi: 10.1016/j.jcmds.2021.100021.

126. Kleijnen JPC. Design and analysis of simulation experiments. In: *Statistics and Simulation*. Springer International Publishing: Cham, 2018, pp. 3–22. doi: 10.1007/978-3-319-76035-3_1.

127. McKay MD, Beckman RJ, Conover WJ. A comparison of three methods for selecting values of input variables in the analysis of output from a computer code. *Technometrics* 2000;42:55–61. doi: 10.1080/00401706.2000.10485979.

128. Papeleux L, Ponthot JP. Finite element simulation of springback in sheet metal forming. *J Mater Process Technol* 2002;125–126:785–91. doi: 10.1016/S0924-0136(02)00393-X.

129. Fu Y, Guo N, Cheng Q, Zhang D, Feng J. Underwater laser welding for 304 stainless steel with filler wire. *J Mater Res Technol* 2020;9:15648–61. doi: 10.1016/j.jmrt.2020.11.029.

130. Bandyopadhyay K, Basak S, Panda SK, Saha P. Use of stress based forming limit diagram to predict formability in two-stage forming of tailor welded blanks. *Mater Des* 2015;67:558–70. doi: 10.1016/j.matdes.2014.10.089.

131. Liu S, Chao YJ. Determination of global mechanical response of friction stir welded plates using local constitutive properties. *Model Simul Mater Sci Eng* 2005;13:1–15. doi: 10.1088/0965-0393/13/1/001.

# 6 Investigation of Twist Springback Pattern of the AA6061 Strip

*MN Nashrudin and AB Abdullah*
Universiti Sains Malaysia

## CONTENTS

## 6.1 INTRODUCTION

The use of lightweight components in modern transportation is becoming popular, although there are many challenges due to safety and comfort issues. Over the years, rising consumer demands have resulted in an increase in vehicle weights due to the addition of safety and comfort-related components (Allwood, 2012). The main purpose of using light materials is to reduce the total body weight of vehicles. In addition to using light materials, different design strategies, such as the design and manufacturing technology, can also be utilized to obtain lightweight construction including plates and tubes (Chatti et al., 2012; Zhan et al., 2016). Metal forming is one of the manufacturing technologies that is intensely used in the automotive industry, where many components are produced through plastic deformation processes. In the process, the raw material in the form of a metal sheet or bulk is deformed plastically to obtain the required shape and size.

One of the most important factors for consideration in forming processes is the formability of the material. Forming processes are influenced by many factors, including the material properties, geometry (such as thickness), and complexity of the formed part. If remained uncontrolled, these factors can lead to many defects such as springback, which causes undesirable shape changes and difficulties in assembly. There are three types of springbacks, namely, wall springback, flange springback and twist springback (Pacak et al., 2019; Eggertsen and Mattiasson, 2012). Compared to others, twist springback is the most critical and difficult to be predicted. Twist springback is the effect of torsion moments within the cross section of the automotive part, and the twist is visible over the product area (Dezelak et al., 2014). This springback is the tendency of the twisted part to return to its original shape due to elastic recovery of the twist angle on the removal of applied torque and typically occurs after twist forming and bending a section (Zhang et al., 2020; Kashfi et al., 2017; Abdullah and Samad, 2014; Dwivedi et al., 2002). The behaviour of the formed part with twist springback may be different from and more complex than two-dimensional springback in flanges and wall springback (Xue et al., 2015). From a mechanics point of view, twist springback is the result of torsional moments in the cross section of the workpiece and the different rotations of two cross sections along the axis (Li et al., 2011; Xue et al., 2016). The unbalanced elastic–plastic deformation and residual stress released in the deformed part produce torsional moments that cause twist springback (Xue et al., 2016; Liao et al., 2017; Xue et al., 2014; Lal et al., 2017). This commonly occurs on the unsymmetrical part after the forming operation because of the non-uniform deformation gradients. It can also appear in symmetrical parts due to the sheet metal position, non-uniform lubrication, and damaged or poorly designed draw beads. These problems may result in unequal material flow and unpredictable twist springback patterns (Dezelak et al., 2014).

In terms of applications, mostly automotive parts have been presented, for example, hat channel by Geka et al. (2013), U-shape rails by Takamura et al. (2011), twist rail (Li et al., 2011) and other parts like propeller blade by Abdullah et al. (2012). Table 6.1 lists most of the applications in the twist springback studies. In general, there are not many parts have been used as case studies on the twist springback. The U channel and twist rail are the most used.

From the perspective of material, various materials have been studied and various findings can be observed. Table 6.2 lists the material being used in the twist springback studies. In terms of material, HSS is the most common as it is typically found on automotive parts.

## 6.2   LITERATURE REVIEW

Recently, there have been many simulation-based investigations conducted on the twist springback behaviour of metals. Chen et al. (2019) investigated the prestrain effect of DP500 using complex nonlinear elastoplastic behaviours and twist

**TABLE 6.1**

**Some of the Case Studies Carried Out on the Twist Springback**

| References | Name of the Part | Photo |
|---|---|---|
| Geka et al. (2013), Xue et al. (2016), Takamura et al. (2011), and Dezelak et al. (2014) | Hat channel/U-shape rails | 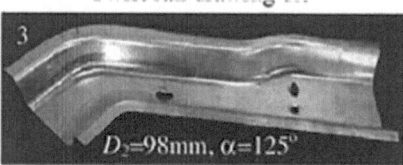 |
| Li et al. (2011) | Twist rail/ S-rail | 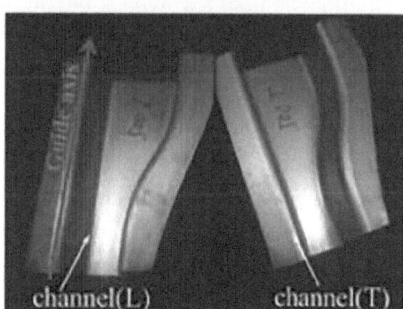 |
| Liao et al. (2020) | L/T channel | 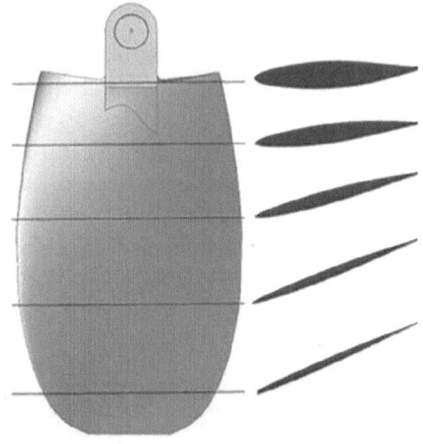 |
| Abdullah et al. (2012) | Propeller blade | |

**TABLE 6.2**
**List of materials and profiles that have been studied**

| References | Material | Profile |
|---|---|---|
| Abdullah et al. (2012) | AA6061-T6 | Sheet |
| Xue et al. (2016) | Dual Phase steel DP-500 | Sheet |
| Xue et al. (2015) | AA6060-T4 | Square tube |
| Liao et al. (2020) | Advanced High-Strength Steel DP-500 | Sheet |
| Feifei et al. (2013) | Ti-3Al-2.5V | Pipe |

springback under two-step loading paths. They found that pre-straining not only influences the residual stress but also affects the elastic modulus distribution. In another work, Xue et al. (2018) simulated an aluminium alloy extruded tube after going through the mandrel-rotary draw bending to observe the twist springback. The modelling of material behaviours has used a non-quadratic anisotropic yield criterion integrated with combined isotropic and kinematic hardening to describe the strain–stress behaviours including anisotropy, Bauschinger effects and permanent softening. The simulation results indicate that the mandrel nose placement mainly affects the longitudinal angular springback but not twist deformation in the circumferential sections. Xie et al. (2018) have compared three different hardening models (including Ziegler, Johnson-Cook and combined hardening models) in the investigation of twist springback of a double C rail made of transformation-induced plasticity 780 (TRIP 780) steel. Liao et al. (2020) considered the effect of pre-strain on the twist springback. Furthermore, Wang et al. (2020) conducted a 3D finite element model of the roll-bending process to study the pattern of twist defects of aluminium alloy Z-section profiles with large cross sections. The results indicated and verified that the effective control of the twist defects of the profile could be realized by adjusting the side roller so that the exit guide roll is higher than the one at the entrance (the side rolls presented an asymmetric loading mode with respect to the main rolls) and increasing the radius of the upper roll. Similarly, Lin et al. (2018) performed a numerical simulation analysis of the bending process with multipoint dies for the aluminium profile. The results showed that the springback of the aluminium profile could reach a minimum when the number of die units was 25 under the precondition of saving cost and ensuring the quality of forming parts. Jia et al. (2018) evaluated and compared the twist springback pattern of an advanced high-strength steel using an unbalanced post-stretching configuration. The simulation results indicated that the twist springback can be effectively controlled by reducing the post-stretching proximate to the asymmetric part area. For tube bending–twisting, Zhang and Wu (2016) considered few parameters including bending radii, twisting angles, residual radii, and residual angles. In another work, Jiang et al. (2013) considered thickness and die radius in the bending of diamond shape perforated lattice truss panels. They found that the springback is decreased with the friction coefficient increase, while the springback is increased with the die radius increase. As the plate thickness increases, the springback decreases firstly and then increases and suggested 1.3 mm as optimum thickness, in combination with a larger punching friction coefficient and smaller die radius.

In recent years, the demand for aluminium alloy in automotive manufacturing industries for lightweight vehicle construction is increasing due to its high strength and good formability (Hirsch, 2014). Many approaches have been developed to address the problem of springback on uniform sections by simulations and experimentation, and a number of springback studies on uniform sections are available in the literature. However, research on non-uniform thickness sections is still lacking. Thus, this is the aim of the present study to evaluate the twist springback pattern of a non-uniform thickness section by both numerical and experimentation and compared it with the twist springback pattern of a uniform profile.

## 6.3   METHODOLOGY

The study begins with the simulation of twist forming of aluminium alloy strip using finite element modelling (FEM), followed by specimen preparation and material data obtained from tensile test. Next, the experimental validation will be carried out. The measurement of the twist springback on each of the specimen profiles will be then elaborated. The same measurement methods will be performed for both finite element (FE) and experimental procedures.

## 6.4   FINITE ELEMENT MODELLING (FEM)

The FEM method has been used to verify the manufacturability of sheet metals and can be utilized to predict the springback response effectively. It is a numerical technique used to minimize error function and create a stable solution (Senthil-Anbazhagan and Dev-Anand, 2016). In the present study, the evaluation is divided into two major stages: FE simulation and experimental validation. The first stage involved the analysis of the twist-forming process by using FE simulation. In the second stage, experimentations are conducted using a semiautomatic torsion machine to validate the results of the simulation. ANSYS Workbench software was used to simulate the twist-forming process of the specimens modelled using SolidWorks CAD software. Input data, such as the stress–strain curve and material properties of AA6061, were obtained using results from tensile testing. In the simulation, twist angles at the loading and unloading steps were determined under two conditions to obtain the twist springback values. At the loading step, a specified moment was defined to perform the twist-forming operation. The instantaneous release at the unloading step was crucial to the twist springback. Appropriate meshing and convergence tests were conducted to estimate an appropriate meshing number and size of 0.9 mm for 7956 elements. The moments and boundary conditions were applied to the deformed geometry to the twist-forming process. The fixed support and moment were defined on each side of the specimen to represent the actual situation of the twist-forming operation, as shown in Figure 6.1. The specimens were twisted in the longitudinal direction and the twist springback was measured in the transverse direction. The construction geometry path was applied on the edge of the specimen profile to measure the deflection during the loading and unloading processes on the y-axis. The calculation method of the springback on simulation is discussed in twist springback measurement sections.

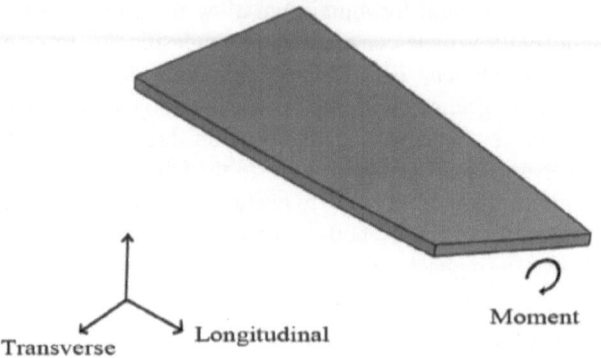

**FIGURE 6.1**   Fixed support and moment defined on the specimen.

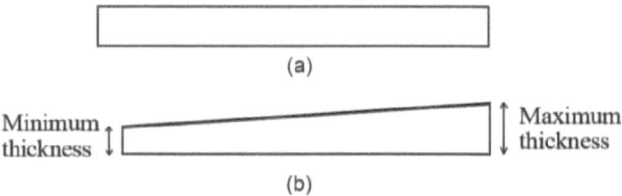

**FIGURE 6.2**   The developed model of the profiles: (a) flat and (b) tapered.

## 6.5  SPECIMEN PROFILES

Five profiles were modelled to represent the uniform thickness and non-uniform thickness sections, as shown in Figure 6.2. A tapered profile described the condition where the thickness changes linearly, as shown in Figure 6.2b. These profiles were modelled in 3D because of the presence of persistent anticlastic curvature during twist forming. Twist forming is unsymmetrical, which is different from simple bending. Large discrepancies between the simulations and experiments on persistent anticlastic curvature can develop when the sheet tension approaches the yield stress; consequently, this discrepancy can affect the rotation of the strip along the longitudinal and transverse directions (Li et al., 2002).

## 6.6  MATERIAL DATA

Formability is an important factor in sheet metal forming. To obtain the input data necessary for material properties related to formability, tensile testing was conducted in accordance with ASTM E8M to obtain the stress–strain curve of the material. Figure 6.3 shows the geometry of the tensile test specimen prepared using wire cut EDM. The gauge length, width, and overall length were 34.13, 15, and 100 mm, respectively. The test was carried out at room temperature using a universal testing machine equipped with an extensometer at 1 mm/min speed.

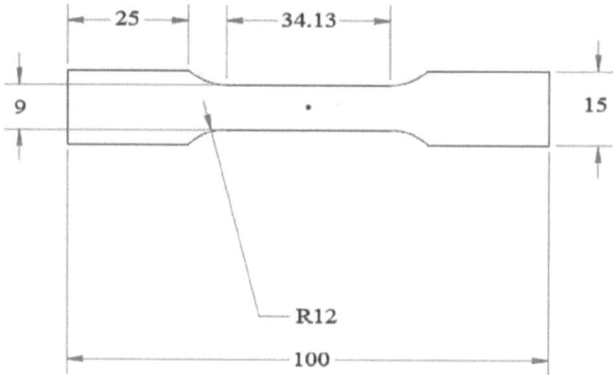

**FIGURE 6.3**   Tensile test specimen dimensions (unit in mm).

---

**TABLE 6.3**
**Twist-Forming Parameters for Experiment**
**for a Non-Uniform Thickness Section**

| Materials | AA6061 |
|---|---|
| Thickness ratio | 0.8, 0.85, 0.88, 1.0 |
| Twist angle | 10°, 20°, 30°, 40° |

---

## 6.7   VALIDATION

Experimental validation of the simulation is conducted by firstly identifying the twist-forming parameters, namely, the twist angle and thickness ratio. Both flat and tapered profiles were used for validation.

## 6.8   TWIST-FORMING PARAMETERS

Two twist-forming parameters, namely, twist angle and thickness ratio, were evaluated. Other parameters such as twisting speed and torque values were not considered. The thickness ratio was set between 0.8 and 1.0. The twist angle was set from 10° to 40° on an aluminium alloy (AA6061) material, as summarized in Table 6.3.

## 6.9   SPECIMEN PREPARATION

Typically, a non-uniform thickness section can be prepared by cold working or rolling processes. In this study, the non-uniform profile was cold-formed using a 100-ton mechanical press machine. Figure 6.4 shows the die set for the preparation of the non-uniform profile. The formed specimens were then cut into strips with dimensions of 70 mm × 30 mm.

**FIGURE 6.4**   Die set for producing a non-uniform profile.

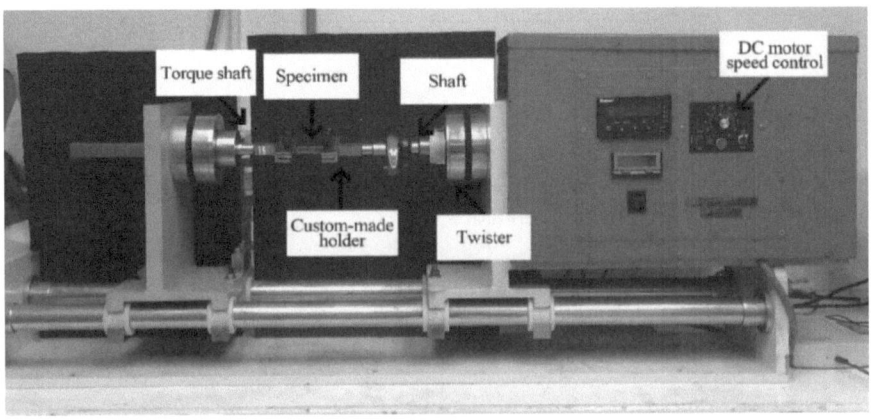

**FIGURE 6.5**   Semiautomatic torsion machine.

## 6.10   TORSION TEST MACHINE SETUP

The twist-forming experiment was conducted in accordance with ASTM A938 using a semiautomatic torsion machine, as shown in Figure 6.5. Modifications were made to the load release mechanism for unloading purposes. In the experiment, observations were made at different twist angles, and the torque values were recorded. The

(a)                                                      (b)

**FIGURE 6.6**    (a) Load cell and (b) arc length on the twister.

---

**TABLE 6.4**
**Arc Length for each Twist Angle**

| Twist Angle (°) | Arc Length (mm) |
|---|---|
| 10 | 4.2 |
| 20 | 8.4 |
| 30 | 12.6 |
| 40 | 16.8 |

---

twist angle was measured using the arc length of the twister mounted on the torsion machine, and the torque was measured using the load cell, as shown in Figure 6.6a. The arc length of the twist angle, shown in Figure 6.6b, was calculated using the following formula:

$$S = r\theta,\qquad(6.1)$$

where $S$ is the arc length, $r$ is the radius of the twister, and $\theta$ is the twist angle. The arc length values for the twist angle are summarized in Table 6.4. The semiautomatic torsion machine was powered with a direct current (DC) motor and speed-reducing gearbox with a ratio of 1:60. The twister speed can be adjusted using the DC motor speed control to a maximum revolution of 1750 rpm. To measure the torque, a tension/compression load cell of the $S$ beam type with a maximum capacity of 300 kg was used. The flat T-shaped specimen holder was customized with a hexagonal end to mount a rectangular cross-sectional strip, as shown in Figure 6.7.

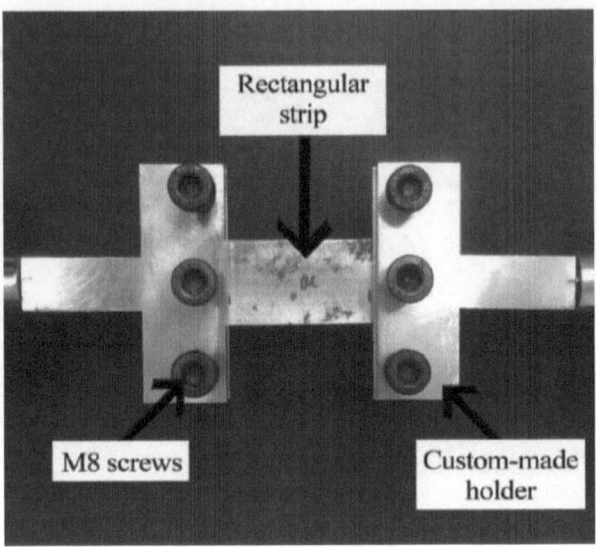

**FIGURE 6.7**    Strip mounted on the holder and fix with six screws.

As the parameters of the rolling direction and twisting speed were not studied in this research, the rolling direction for the specimen strips tested was assumed at 90°, and the twisting speed of the process was kept constant at 15 rpm. The specimens were assumed ideally fixed in the customized holder and no sliding occurred during the loading and unloading process.

## 6.11   TWIST SPRINGBACK MEASUREMENT

In the measurement of the twist springback, two major angles should be determined: the angle after forming, i.e., the angle observed after loading completion, and the angle after unloading, i.e., after the load was removed from the twisted part. In this case, obtaining the profile after unloading is the most challenging task. The profile may be affected if the twisted strip is removed from the twister arm. Therefore, online measurement is the most reliable method, which can be done without removing the specimen from the machine. An online measurement method was employed to capture the image of the profile, which was subsequently processed using MATLAB® software. The image was captured from the transverse direction of the specimen, as shown in Figure 6.8. The twist direction was considered in line with the longitudinal axis of the specimen. The standard definition of springback is the difference between the angles after loading ($\theta_o$) and the angle after unloading ($\theta_f$), as illustrated in Figure 6.9. In the simulation, the deflection on loading and unloading was measured on the profile extracted from the software. In the experimentation, the twist springback angle was measured using a similar approach using MATLAB software. The twist springback angles for the flat and tapered profiles were measured in the transverse direction as the twisted strip image can only be captured from that direction.

**FIGURE 6.8** Captured image during twist forming from the transverse direction.

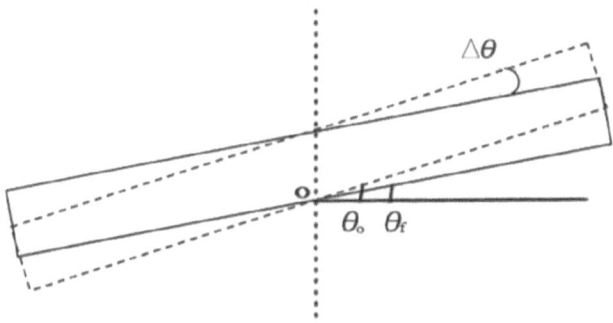

**FIGURE 6.9** Standard definition of springback, i.e., angle after loading, $\theta_o$ and after unloading, $\theta_f$.

## 6.12 FLAT PROFILE

The angles after loading and after unloading were measured from the x-axis of the top surface of the strip. Given that the thickness for the flat profile was uniform, the angles for the top, middle, and bottom of the strip were assumed to be the same. The flat profile and determination of both angles from the transverse axis are illustrated in Figure 6.10. The $\theta_o$ and $\theta_f$ for the flat profile can be directly calculated using the Pythagoras theorem. The lengths of x and y were determined from the simulation and MATLAB, while the twist springback ($\Delta\theta$) was obtained from the difference between $\theta_o$ and $\theta_f$.

## 6.13 TAPERED PROFILE

The calculation of the springback for the strip with a tapered profile differed from the calculations of the flat profile due to the non-uniform thickness. Figure 6.11 illustrates the tapered profile from the longitudinal axis. The $\theta_o$ and $\theta_f$ for the tapered

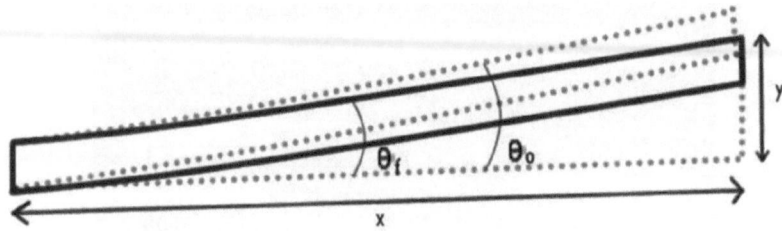

**FIGURE 6.10** Illustration of the flat profile during twist forming. The solid line represents the profile after unloading and the dotted line represents the profile after loading.

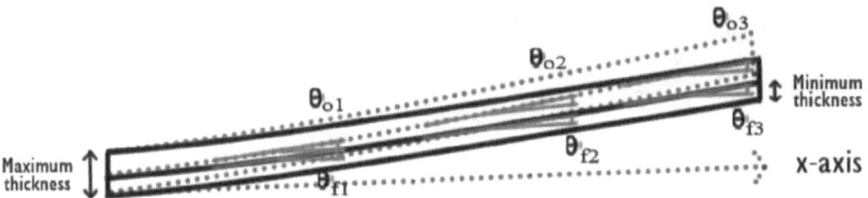

**FIGURE 6.11** View from the transverse direction of the tapered profile. The solid line represents the profile after unloading and the dotted line represents the profile after loading.

profile were measured from the transverse direction. The top surface of the side with minimal thickness will produce a curvature after the twist forming due to the uneven thickness distribution. Therefore, the angle cannot be directly measured from the lines. Here, both $\theta_o$ and $\theta_f$ were obtained using a tangent line, as shown in Figure 6.13. The average of the angles from the number of tangency points on the profiles was recorded to obtain the springback $\Delta\theta$. The number of points of the tangent line would depend on the strip length and curvature radius.

## 6.14   RESULTS AND DISCUSSION

The discussion is divided into two sections for the simulation and experimental validation. In the simulation, the twist-forming process was simulated on the ANSYS Workbench software to evaluate the effect of twist angle and the thickness ratio on the twist springback by using static structural analysis. The forming process was carried out in two steps, namely, the loading and unloading steps, to determine the twist springback. The simulation was conducted for both the flat and tapered profiles with different thickness ratios and twist angles, and the results of the simulation are shown in Table 6.5.

## 6.15   EFFECTS OF TWIST ANGLE

The twist angle in the flat profile was significant to the twist springback angle. Increasing the twist angle of the AA6061 strip increases the twist springback. The results also showed that the springback increased at a low rate when the twist angle

**TABLE 6.5**
**Result of Simulation for Flat and Tapered Profiles**

| | Twist Angle | | | |
|---|---|---|---|---|
| **Thickness** | **10°** | **20°** | **30°** | **40°** |
| **Ratio** | | Average Springback (°) | | |
| 0.80 | 2.32 | 4.55 | 4.68 | 3.9 |
| 0.85 | 2.3 | 4.5 | 4.62 | 3.86 |
| 0.88 | 2.29 | 4.47 | 4.56 | 3.82 |
| 1.0 (flat profile) | 1.90 | 3.39 | 3.93 | 4.05 |

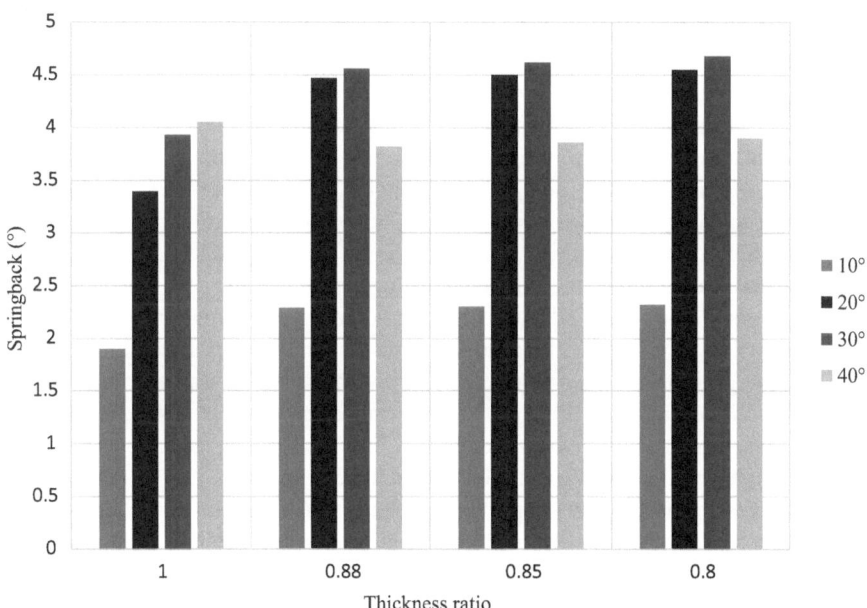

**FIGURE 6.12**  Twist springback for different thickness ratios for AA6061.

increased at a specific value because the recoverable percentage was low. When the twist angle is further increased, the elastic phase became a non-recoverable plastic with decreased total deformation. The elastic phase then became negligible and resulted in a constant twist springback constant beyond a certain value of twist angle. Twist springback is related to the unbalanced residual stresses in the specimen.

## 6.16  EFFECTS OF THE THICKNESS RATIO

Thickness is one of the most significant parameters that should be considered in predicting twist springback. As shown in Figure 6.12, the thickness ratio of 1.0 (flat profile) resulted in a lower twist springback than that with the low thickness ratio for the tapered profile. This result can be attributed to the difference in thicknesses. When the thickness is high, the twist springback would decrease. Furthermore, the flow stress of the material increases with decreasing thickness. When the thickness ratio is low, which is represented as a high inclination of the thickness between both sides, the material flow stress would increase, resulting in an increase in the twist springback.

## 6.17   VALIDATION

For the experimental validation, the twist-forming process was performed using a semiautomatic torsion machine. The parameters considered were the twist angle and thickness ratio. The experimental results are shown in Table 6.6 for the flat and tapered profiles.

As shown in Figure 6.13, the twist springback of the flat profile for AA6061 was in good agreement with both the simulation and experimental results. The percentage error was between 1% and 17%, which was considered low. In contrast, the simulation results for the tapered profiles showed a large deviation value from the experimental results, possibly due to the strain hardening effect (Figure 6.14). During the preparation of the specimen, the tapered profile for the experimental method underwent a cold forming process to reduce the thickness. Strain hardening (also called work hardening or cold working) is a condition where the hardness of the metal is increased and strengthened through plastic deformation. During cold working, if the strain hardening is not relieved, then flow stress increases continuously with deformation. The flow stress of the material increases due to the work hardening characteristics. The grain-to-grain variations in the flow stress within the specimen would also increase. Thus, the flow stress and twist springback amount would both increase. There is a large difference between the results of the twist springback angles between the simulation and the experimentation for the tapered profiles, which may be due to the changes in mechanical properties as a result of the strain hardening effect. The specimen with a thin profile due to the strain hardening exhibited a higher yield strength than that of the thicker specimen. The low thickness ratio also resulted in increased twist springback because of the high yield strength. The flow stress in the material has also increased, which contributed to the increase in elastic stress. This

## TABLE 6.6
### Result of Experiment for Flat and Tapered Profiles

|  | Twist Angle | | | |
|---|---|---|---|---|
|  | 10° | 20° | 30° | 40° |
| Thickness Ratio | Average Springback (°) | | | |
| 0.8 | 1.40 | 4.03 | 5.25 | 5.96 |
| 0.85 | 1.56 | 4.27 | 5.41 | 5.50 |
| 0.88 | 1.32 | 3.87 | 5.55 | 5.53 |
| 1.0 | 1.92 | 3.62 | 4.1 | 4.06 |

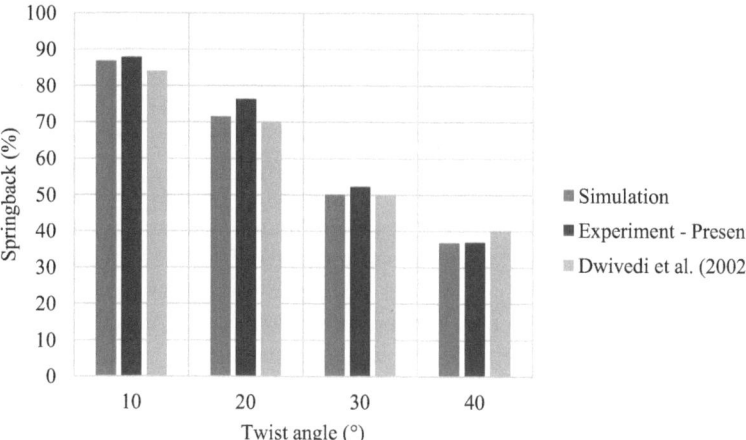

**FIGURE 6.13**   Twist springback of the flat profile for simulation and experimental methods.

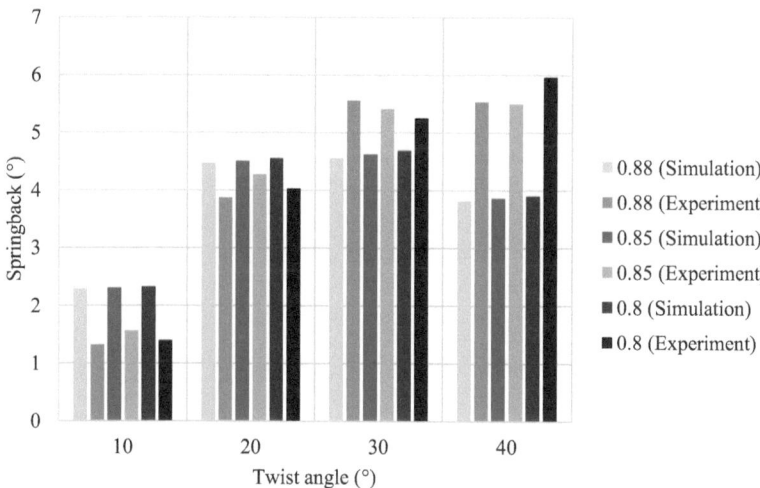

**FIGURE 6.14**   Twist springback of the tapered profile for simulation and experimental methods at different thickness ratios.

elastic stress allowed the specimen to return to its original shape after unloading. In conclusion, it was observed that the springback decreased with an increasing thickness ratio. As discovered in the experiment, for the flat specimen, the twist springback decreased with the increase of the twist angle. With deviation <10% between experiment and simulation. In addition, this is in agreement with the pattern in the obtained result by Dwivedi et al. (2002). However, for the tapered specimen, different and inconsistent patterns were observed. Such a decrease occurred because the specimen was still in the elastic stress phase even at a high twist angle in the experiment due to the high yield strength.

## 6.18   CONCLUSIONS

A twist-forming process was successfully modelled using ANSYS Workbench soft-ware to evaluate the twist springback of a non-uniform thickness section by using static structural analysis. Two profiles, namely, flat and tapered profiles, were evalu-ated. The twist springback was calculated by evaluating the twist during loading and unloading. The springback measurements were made from the transverse direc-tion to the profile that is twisted in the longitudinal direction. The results showed that the twist springback increases up to a certain twist angle and then subsequently decreases upon further twisting. A low recoverable percentage was recorded as the twist angle increased as the elastic phase entered into the non-recoverable plastic phase, thus decreasing the total deformation. For the tapered profile, the low-thick-ness-ratio specimens have a higher twist springback angle as compared to speci-mens with a high thickness ratio. This result can be attributed to the increased flow stress of the material as the thickness ratio decreases, increasing the twist spring-back response. For the flat-profile specimens, the simulation results were in good agreement with the results from the experimentations. However, for the tapered specimens, there are large differences between the simulation and experimentation results. These differences can be attributed to the strain hardening effects during the preparation of the tapered specimens, which changes the properties of the mate-rial. This strain hardening effect also increases the twist springback as the yield strength of the material is increased. Future work can consider further evaluation of the material such as the hardness distribution in the simulation, and to determine its influence on springback.

## ACKNOWLEDGEMENT

This grant was funded by the Ministry of Higher Education Malaysia (203/PMEKANIK/6071308).

## REFERENCES

Abdullah, A. B.,, Sapuan, S. M.,, Samad, Z., Khaleed, H. M. T., & Aziz, N. A., (2012). Geometrical error analysis of cold forged AUV propeller blade using optical measure-ment method. *Advanced Materials Research*, 383–390, 7117–7121.

Abdullah, A. B., & Samad, Z. (2014). Measurement of twist springback on AA6061-T6 alumi-num alloy strip - a preliminary result. *Applied Mechanics and Materials, 699*, 44–48.

Allwood, J. M. (2012). *Sustainable materials – with both eyes open. University of Cambridge - The Future in Practice*, United Kingdom.

Chatti, S., Pietzka, D., Khalifa, N. B., Jäger, A., Selvaggio, A., & Tekkaya, A. E. (2012). Lightweight construction by means of profiles. *Key Engineering Materials, 504–506*, 369–374.

Chen, S., Liao, J., Xiang, H., Xue, X. and Pereira, A. B. (2021). Pre-strain effect on twist springback of a 3D P-channel in deep drawing, *Journal of Materials Processing Technology, 287*, 116224.

Dezelak, M., Stepisnik, A., Pahole, I., & Ficko, M. (2014). Evaluation of twist springback pre-diction after an AHSS forming process. *International Journal of Simulation Modelling, 13*(2), 171–182.

Dwivedi, J. P., Shah, S. K., Upadhyay, P. C., & Das Talukder, N. K. (2002). Springback analysis of thin rectangular bars of non-linear work-hardening materials under torsional loading. *International Journal of Mechanical Sciences, 44*(7), 1505–1519.

Eggertsen P.A. & Mattiasson K. (2012). Experiences from experimental and numerical springback studies of a semi-industrial forming tool. *International Journal of Material Forming, 5*, 341–59.

Feifei, S., Yang, H., Li, H., Zhan, M. & Li, G. (2013). Springback prediction of thick-walled high-strength titanium tube bending. *Chinese Journal of Aeronautics*, 26, 1336–1345. doi: 10.1016/j.cja.2013.07.039

Geka, T., Asakura, M., Kiso, T., Sugiyama, T., Takamura, M. and Asakawa, M. (2013). Reduction of springback in hat channel with high-strength steel sheet by stroke returning deep drawing, *Key Engineering Materials*, 554–557, 1320–1330.

Hirsch, J. (2014). Recent development in aluminium for automotive applications. *Transactions of Nonferrous Metals Society of China (English Edition), 24*(7), 1995–2002.

Jia, Y., Pu, C., Zhu, F., Zhou, D. et al. (2018). Numerical Study of Twist Spring-back Control with an Unbalanced Post-stretching Approach for Advanced High Strength Steel, *SAE Technical Paper* 2018–01–0806.

Jiang, W., Yang, B., Guan, X and Luo, Y. (2013). Bending and Twisting Springback Prediction in the Punching of the Core for a Lattice Truss Sandwich Structure, *Acta Metall. Sin. (Engl. Lett.)*, 26(3) 241–246.

Kashfi, M., Bakhtiyari, D., Ghavamian, A., Kashfi, M., Kahal, P. (2017). A numerical-experimental investigation on spring back in rectangular section thin wall hollows and extracting a relation to estimate twisted angle by considering strain harden behavior. *Iranian Journal of Manufacturing Engineering*, 4(1), 38–43.

Lal, R. K., Dwivedi, J. P. and Singh, V. P. (2017). Springback analysis of hollow rectangular bar of linear work-hardening material, *Emerging Materials Research, 6*(2), 404–412

Li, H., Sun, G., Li, G., Gong, Z., Liu, D., & Li, Q. (2011). On twist springback in advanced high-strength steels. *Materials and Design, 32*(6), 3272–3279.

Li, K. P., Carden, W. P., & Wagoner, R. H. (2002). Simulation of springback. *International Journal of Mechanical Sciences, 44*(1), 103–122.

Liao, J., Xue, X., Lee, M.-G., Barlat, F., Vincze, G., & Pereira, A. B. (2017). Constitutive modeling for path-dependent behavior and its influence on twist springback. *International Journal of Plasticity*, 1–25.

Liao, J., Chen, S., Xue, X. and Xiang, H. (2020). On twist springback of a curved channel with pre-strain effect, *International Journal of Lightweight Materials and Manufacture, 3*(2), 108–112.

Lin, X., Li, Y., Cai, Z., Liang, J., Liang, C., Yu, K., Li, X., Wang, A. and Liao, Y. (2018). Effect of Flexible 3D Multipoint Stretch Bending Dies on the Shape Accuracy and the Optimal Design, *Advances in Materials Science Engineering*, Article ID 1095398

Pacak, T., Taticek, F. and Vales, M. (2019). Compensation of Springback in Large Sheet Metal Forming. *Acta Polytechnica*, 59(5) 483–489.

Senthil-Anbazhagan, A. M., & Dev-Anand, M. (2016). Design and Crack Analysis of Pressure Vessel Saddles Using Finite Element Method. *Indian Journal of Science and Technology, 9*(21), 1–12.

Takamura, M. Fukui, A, Hama, T, Sunaga, H., Makinouchi, A and Asakawa, M. (2011). Investigation on twisting and side wall opening occurring in curved hat channel products made of high strength steel sheets. The 8th International Conference and Workshop on Numerical Simulation of 3D Sheet Metal Forming Processes, AIP Conf. Proc., 1383 pp. 887–894

Wang, A., Xue, H., Bayraktar, E., Yang, Y., Saud S. and Chen, P. (2020). Analysis and Control of Twist Defects of Aluminum Profiles with Large Z-Section in Roll Bending Process, *Metals, 10*(1), 31

Xie, Y. M., Huang, R. Y., Tang, W., Pan B. B. & Zhang, F. (2018). An Experimental and Numerical Investigation on the Twist Springback of Transformation Induced Plasticity 780 Steel Based on Different Hardening Models, *International Journal of Precision Engineering and Manufacturing, 19*, 513–520

Xue, X., Liao, J., Vincze, G., & Barlat, F. (2015). Twist springback characteristics of dual-phase steel sheet after non-axisymmetric deep drawing. *International Journal of Material Forming, 1*, 1–12.

Xue, X., Liao, J., Vincze, G., & Gracio, J. (2014). Modelling of mandrel rotary draw bending for accurate twist springback prediction of an asymmetric thin-walled tube. *Procedia Engineering, 216*, 405–417.

Xue, X., Liao, J., Vincze G. & Pereira, A. B. (2018). Control strategy of twist springback for aluminium alloy hybrid thin-walled tube under mandrel-rotary draw bending, *International Journal of Material Forming, 11*, 311–323

Xue, X., Liao, J., Vincze, G., Sousa, J., Barlat, F., & Gracio, J. (2016). Modelling and sensitivity analysis of twist springback in deep drawing of dual-phase steel. *Materials and Design, 90*, 204–217.

Zhan, M., Wang, Y., Yang, H. and Long, H. (2016). An analytic model for tube bending springback considering different parameter variations of Ti-alloy tubes. *Journal of Materials Processing Technology, 236*. 123–137.

Zhang, S., & Wu, J. (2016). Springback prediction of three-dimensional variable curvature tube bending. *Advances in Mechanical Engineering, 8*(3). 1–13. doi:10.1177/ 1687814016637327

Zhang, Z., Wu, J., Liang, B., Yang, J., & Wang, M. (2020). Investigation to the torsion generation of spatial tubes in bending-twisting process. *International Journal of Advanced Manufacturing Technology, 107*(3/4), 1191–1203.

# 7 Springback Study of Non-Uniform Thickness Aluminium Blanks

*MF Adnan and AB Abdullah*
Universiti Sains Malaysia

*MF Jamaludin*
Universiti Malaya

## CONTENTS

## 7.1   RESEARCH BACKGROUND

The world's transportation system is heavily dependent on petroleum-based fuels and products. It accounts for about 40% of the world's oil consumption of nearly 75 million barrels of oil per day (Mayyas et al., 2012). As non-renewable sources of energy are depleting, there is an urgent need for increased efficiency to reduce energy consumption. Eco-friendly solutions in transportation and fuel efficiency

DOI: 10.1201/9781003164241-7

will also reduce the detrimental impact on the environment, especially air pollution. Apart from advancement in propulsion technology, alternative fuel types and hybrid vehicles, another initiative to increase fuel efficiency and reduce emissions across the board is by vehicle lightweighting strategy. This strategy attempts to utilize alternative materials and optimized component design to reduce the overall mass of the vehicle.

A lighter vehicle requires less energy from the engine to overcome inertial forces during acceleration and deceleration, thus requiring less work to move the vehicle. For a typical combustion engine, more than 85% of the energy in the fuel is lost to thermal and mechanical inefficiency, while the remaining 12%–15% is actually used for motion, which is directly affected by the vehicle's weight (Joost, 2012). Weight reduction is still the most cost-effective means to reduce fuel consumption and greenhouse gases from the transportation sector, with an improvement of 7% for every 10% reduction in the vehicle's total weight (Ghassemieh, 2010).

Until very recently, the average sizes and weights of vehicles sold worldwide have climbed steadily over the decades following the oil crisis of the 1970s. The increase in weights was due to improved safety and comfort features such as added entertainment, anti-lock brake systems, airbags and added safety body structures, all of which contribute to the vehicles' weight gains (Prawoto et al., 2013). It has been proposed that mass reduction technology with minimal additional manufacturing cost could achieve up to 20% reduction in the weight of new vehicles (Air et al., 2010). Historical trends in vehicle design, new vehicle designs and concepts for future model redesigns have included considerations of mass reductions. Mass reduction can occur in smaller increments, for example, by reducing the mass of individual components, or through a total revamp of the vehicle's overall design (Davis, 2010). However, this weight reduction must not compromise the safety aspects of the body-in-white (BIW) parts. Thus, parts that are redesigned to be thinner should also satisfy the mechanical properties required to comply with Crash Test Norma (Abdulhay et al., 2011).

There are two important automotive safety concepts which should be considered, namely, crashworthiness and penetration resistance. Crashworthiness is the ability to absorb the impact energy and be survivable for the passengers. It can also be defined as the potential to absorb energy through controlled failure modes and mechanisms, providing a gradual decay in the load profile during absorption. On the other hand, penetration resistance is related to the total absorption capacity without allowing penetrations of projectiles or fragments (Ghassemieh, 2011). Several research studies have been conducted to evaluate the relationship between weight reduction and crashworthiness. For an automotive body structure, the B-pillar is considered a critical major component for side impact and roof crush safety. A study by Yang et al. (2012) has evaluated the weight reduction and crashworthiness performance of B-pillars, which generally affect the crash performance of full vehicle side impact and door crush, which is directly related to the safety of the car occupants.

The current technique of automotive weight optimization is by using higher-strength steel for car bodies. High-strength steels of different strength levels have been developed to enable a wide selection of materials for different applications according to the required function. Compared with other lightweight materials, high-strength steels are advantageous in terms of costs and availability of production facilities. Thus, they are used as the principal material for the improvement of body weight reduction and collision safety. However, high-strength steels have limited formability which increases the risk of sheet breakage during sheet forming due to their low elongation properties, and their higher yield stresses would cause dimensional defects in the form of springback and wrinkles. To address these problems, forming dies have to be modified many times during the trial runs, which leads to increased die manufacturing costs and other technical problems.

In modern transportation engineering, the application of lightweight construction is challenging. Lightweight constructions are optimal if the materials are used only in areas where stresses appear and change near yield stresses (Kleiner et al., 2003). Lightweight component design can be categorized into two types. Firstly, lightweight construction deals with the use of light materials. In this approach, the use of light material keeps the same workpiece geometry but reduces the total component weight. Secondly, lightweight construction deals with different design strategies including the process of forming the component (Girubha and Vinodh, 2012). The formability of the material depends on many factors including the thickness, properties of the material and complexity of the formed part. This factor may initiate defects such as springback which lead to increased part rejection.

There are several methods that are being investigated to optimize the weight of vehicles, such as multi-material automotive bodies. The design of an automotive body with an optimized multi-material combination requires complete and accurate data about the performance of the constituent materials to form complex structural components. The concept of the multi-material is utilizing the right material types in the right location, suitable for its intended application. In the study by Cui et al. (2011), it was found that a 14% weight reduction from the original design can be obtained by applying the right material in the design at reduced costs, without neglecting the safety factors (Table 7.1).

Pan and Zhu (2010) stated that mass reduction can be achieved by using high-strength steel and other lightweight materials or utilizing optimization techniques to develop more robust and lightweight structures. However, the arising issue is that the lightweighting of the vehicle should improve or at least maintain the required performance, such as its stiffness performance, crashworthiness performance, noise,

**TABLE 7.1**

**Performance Comparison of Body Design with Equivalent Bending Stiffness (Cui et al., 2011)**

| Body Structures | Bending Stiffness[a] (mm) | Weight (kg) | Cost ($) | V Value ($) |
|---|---|---|---|---|
| Original design | 1.5744 | 236.3 | 191.2 | 545.6 |
| Optimal design | 1.5735 | 205.4 | 205.5 | 513.6 |
| Worst design | 1.5739 | 171.7 | 324.0 | 581.6 |

[a] The bending stiffness was evaluated as the deflection at the load application point under the fixed load.

vibration, harshness (NVH) performance as well as durability. Thus, it can be concluded that the goal of lightweight construction is to minimize the weight of auto body without compromising the performance of vehicle.

In recent years, the automotive industry is striving to reduce the weight of vehicles, in line with the demand of global development trend towards lighter, fuel-efficient cars with increased resistance to collisions. There are several ways of weight reduction techniques available for the automotive manufacturing industry. For example, the hot stamping of high-strength steels offers the possibility of obtaining significant weight reductions. Li et al. (2014) have discussed the hot stamping of door-impact beams made with high-strength steel. Hot stamping is a combination of three processing phases, namely, approach, forming and quenching (Abdulhay et al., 2011). Nowadays, hot stamping is gaining popularity for the production of numerous automotive parts where there are variations in thicknesses requiring special processes to form the shape. Currently, hot stamping processes are utilized for the fabrication of the A-pillar, B-pillar, roof, rear bumper beam and door beams. The manufacturing of such components can be carried out either by using variations of the conventional hot stamping process or by utilizing tailored semi-finished products (Figure 7.1).

There are two main variants of the hot stamping process, the direct and the indirect hot stamping methods. For the direct hot stamping process, the blank is heated in a furnace and then transferred to the press to be subsequently formed and quenched in a closed tool. On the other hand, the indirect hot stamping process utilizes a semi-complete cold pre-formed part which is subjected only to a quenching and calibration operation in the press after austenitization (Karbasian and Tekkaya, 2010).

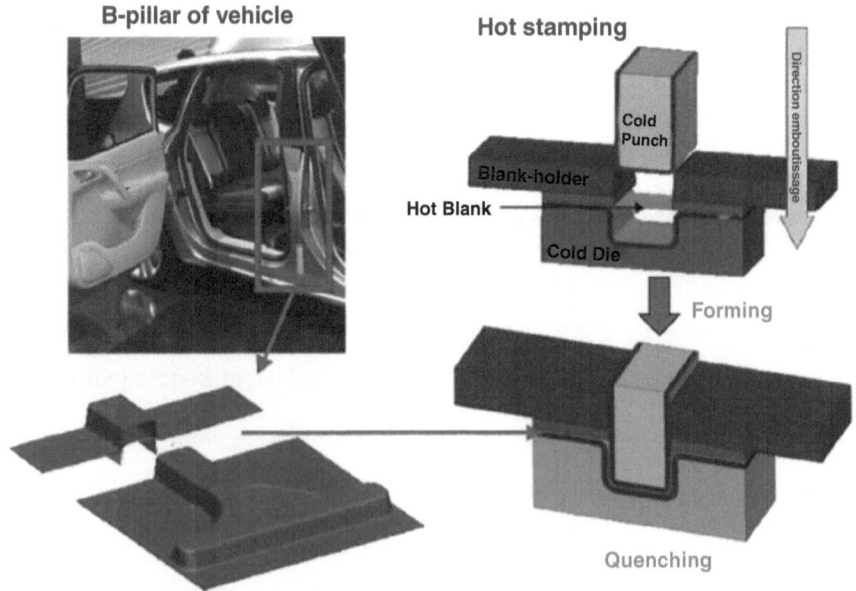

**FIGURE 7.1** Manufacturing of B-pillars by the hot stamping process. (With permission from Abdulhay et al., 2011.)

Many recent efforts to improve the sheet metal manufacturing process and forming capability for complex shapes have failed to produce defect-free components. To overcome these challenges, the use of semi-finished products, such as tailored blanks can be an attractive solution for the manufacturers. Merklein and Opel (2012) stated that tailored blanks can potentially reduce the vehicle weight by 20%–34%, resulting from the combinations of local variation of the thickness, sheet material, coating and material properties. In a subsequent study, Merklein et al. (2014) divided tailored blanks into four sub-groups, namely, tailor welded blanks, patchwork blanks, tailor-rolled blanks (TRBs) and tailor heat-treated blanks (THHBs), as shown in Figure 7.2.

Tailor welded blanks (TWBs) can be described as sheet metal blanks with similar or dissimilar thicknesses, materials or coating welded in a single plane prior to forming. The formability characteristics of TWBs are affected by weld conditions such as weld properties, weld orientation and weld location, thickness differences and variations in strength between the sheet (Veera Babu et al., 2010). TWBs are mostly used in complex designs to minimize weight and to optimize engineering properties. A patchwork blank is a process where welded blanks of one or more pieces of reinforcing sheet metals (patches) are lap-welded onto the main sheet (Kashani and Kah, 2015). The advantages of patchwork blanks are the higher fitting accuracy between the two sheet blanks and even small areas of the blanks can be reinforced (Merklein et al., 2014). TTHBs are sheets that exhibit locally different material properties. Local heat treatment is used to change the material properties, with the main objective being to improve the formability of the products. In some cases, the forming process of tailored blanks is even more difficult compared to conventional blanks because of the inhomogeneous thicknesses or strength distributions (Merklein et al., 2014). TRBs are sheets with a continuous thickness transition. Zhang (2012), in their research, prepared a TRB with the desired thickness in different rolling directions and roll gap adjustments. The TRB process is suitable for blanks requiring optimal

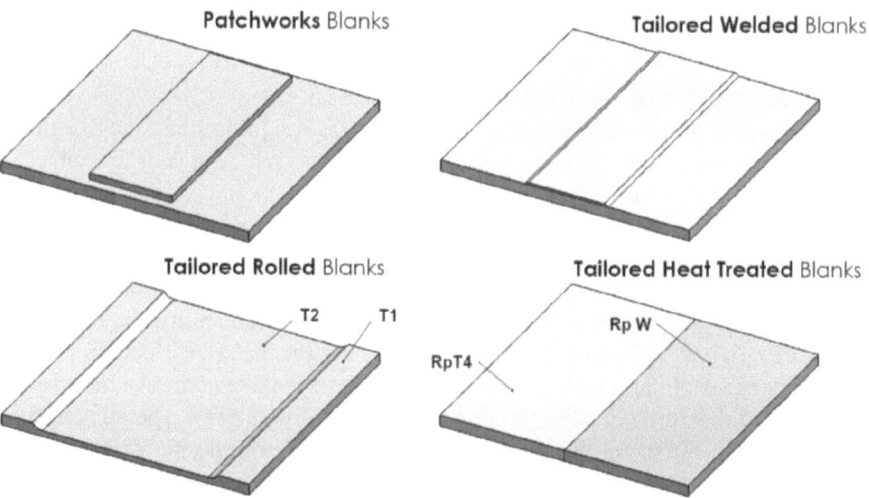

**FIGURE 7.2**    Classification of tailored blanks. (With permission from Merklein et al., 2014.)

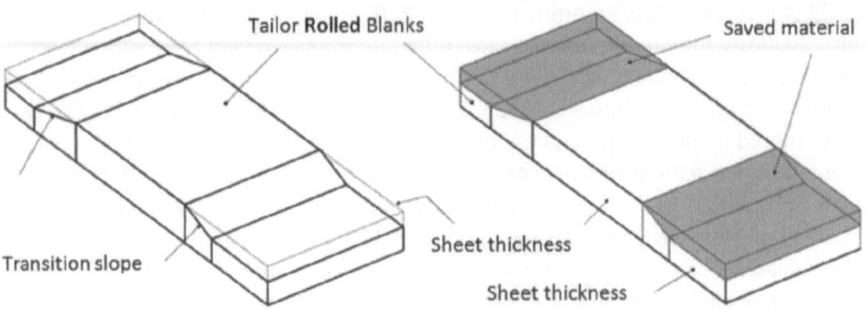

**FIGURE 7.3** Sheet with the constant thickness compared with the TRB with the optimized thickness distribution. (With permission from Meyer et al., 2008.)

thickness distribution, especially if several thickness areas are needed in segments requiring high strengths (Mori et al., 2011). Meyer et al. (2008) have found that TRB is capable of achieving maximum depth in the deep drawing process as compared to the blank having a constant thickness. Usually, the maximum thickness of the TRB is set to have equal strength to the constant thickness of the normal blank, as shown in Figure 7.3.

Yang et al. (2012) mentioned in their study of the lightweight design of the vehicle structure of the front longitudinal beam (FLB) part with TRB for crashworthiness, TRB allows continuous metal thickness transition, which leads to better formability and greater weight reductions. The two main challenges in the automotive industry concerning lightweight and crashworthiness performances often conflict with each other. Another study by Duan et al. (2015) has found that optimization of thickness using TRB has resulted in a 16.71% reduction of weight of the FLB inner, while the crashworthiness has improved as compared to the baseline design. Furthermore, TRBs have good forming characteristics as they have no welding seams and corresponding heat-affected zones. Thus, parts with potentially high-stress concentrations such as B-pillars and FLBs manufactured by the TRB process provide better crashworthiness results than other tailored blank processes.

Lightweight properties are crucial where the mass is significant for the functionality of the product, such as in aeronautical applications. Where masses are subject to acceleration, lightweight components have a direct impact on product performance. In automotive applications, lightweight components enable lighter crankshafts capable of producing higher revolutions. Benefits of lightweight components are increased driving comfort, safety and improved fuel consumption. However, detrimental effects of lightweight materials, such as reduced formability and ductility of high-strength materials should also be considered in the material selection.

Applications of parts with non-uniform thickness in the automotive industry are still limited due to the challenge of springback and formability. The applied force gives stress to the metal beyond its yield strength, causing plastic deformation to occur, but does not cause the failure of the material. Therefore, the sheet can be formed into a variety of complex shapes. However, due to the elastic response of the material, there will be changes to the sheet geometry after the applied force is removed. This response, termed springback, is not totally categorized as defects but causes difficulty in subsequent assembly processes. Springback causes geometrical non-compliance to the final desired shape and may not satisfy the design specifications. However, springback is a common phenomenon in sheet metal forming, mainly caused by the elastic recovery and redistribution of internal stresses during the unloading process (Burchitz, 2005).

During a forming operation, the applied load causes the sheet to deform and the contour of the sheet section takes on the shape of the die. However, as the applied load is released, the elastic deformation is removed due to the release of the elastic strain energy causing the internal contour of the deformed sheet to take a different shape from the desired shape of the die. The springback in sheet metal parts is the result of the sequence of deformation experienced in the metal forming process. As the sheet metal slides over a die shoulder, it undergoes bending–unbending deformation developing cyclic bending loads on the sheet sections, and as a result, the unbalanced stress distribution is developed over the thickness. To compensate for the issue of springback, the die designer would usually consider a springback factor so that the final shape after springback complies with the assembly requirements.

### 7.1.1  Problem Statement

Bending is a mechanical process in which the sheet metal experiences plastic deformation along a straight line, causing it to bend over a certain angle and form the anticipated shape. The springback effect is when the material undergoing bending has a tendency to partially return to its original shape due to the elastic recovery of the material. Springback is generally defined as the additional deformation of sheet metal parts after the loading is removed. It is influenced not only by the material and mechanical properties such as tensile and yield strengths but also by the thickness variations, bend radius, bend angle, rolling direction and other factors. Springback is a phenomenon of elastic nature determined by the distribution of stresses on the section of the formed part. In the manufacturing industry, springback needs to be anticipated, and designs of tools should include compensation for springback.

In recent years, automotive manufacturers have introduced parts having non-uniform thicknesses. This enables the fabrication of improved components with reduced weight and localized increase in strength. Springback still remains a major problem for complex shapes such as B-pillars, and predicting the behaviour and the effect of the manufacturing process on the springback pattern is still an ongoing effort.

### 7.1.2  Scope of Study

In this study, the parameters under consideration are thickness ratio, bend angle, annealing temperature, alignment and rolling direction. The material sample chosen is an aluminium AA6061 metal sheet with a maximum thickness of 3 mm. The specimen geometry is fixed at 50 mm length and 20 mm width, with thickness variation either in a curved or tapered shape. The bending test is based on the ASTM D790 Standard test method. A profile projector is used to measure the bending angle after unloading, and an ALICONA Infinite Focus Machine (IFM) is used to validate the measurements.

## 7.2 METHODOLOGY

The methodology in this study is divided into three main stages, namely, analysis of parameters, optimization of springback and preparation of specimen profile (non-uniform and experimental apparatus). All stages related to the V-bending test will be described here for a better understanding of the parameters that were considered in the experiment.

### 7.2.1 OPTIMIZATION OF SPRINGBACK PARAMETERS USING THE TAGUCHI METHOD

Taguchi methods are statistical methods, sometimes called robust design methods, which can reduce the large number of experiments required using traditional experimental design procedures. The reduction in the number of experimental runs would lessen the time and resources required, and the results obtained would give an indication of the effect of factors (variables) on the desired quality characteristics (Kamaruddin et al., 2010). Taguchi's design of experiment utilizes specially constructed tables known as an 'orthogonal array' (OA). Orthogonal arrays can be used with useful statistic information and reliable factorial effect (Kuo et al., 2007). This factorial effect is different from the full factorial experiment that requires a large number of experiments as the number of parameters increases. Orthogonal arrays make the design of experiment very easy and consistent, and most importantly, they require a relatively lesser number of experimental trials to study the entire experiment space.

#### 7.2.1.1 Springback Parameters

Non-uniform profiles can either be linear or parabolic profiles. In this study, two sets of experiments were conducted for (1) the tapered shape, representing the linear profile, and (2) the curved shape, representing the parabolic profile. Table 7.2 shows the six sets of specimens with different thickness ratios for the experiment on linear-shaped profiles. The thickness ratio is defined as the ratio of the minimum thickness to the maximum thickness values. The feasible space for the experiment parameters was defined by varying the thickness in the range of 2.45–3.00 mm, the alignment of the puncher in the range of 7–9 mm, and the punch stroke in the range of 3–7 mm. These ranges of values were selected based on past research works. Similarly, for the second set of experiments on the curve-shaped profile, the thickness values were in the range of 2.60–3.00 mm, the alignment of the puncher was in the range of 7–9 mm,

---

## TABLE 7.2
### Springback Parameter and Their Levels for Tapered Shape

| Factor | Parameters | Level 1 | Level 2 | Level 3 | Level 4 | Level 5 | Level 6 |
|--------|-----------|---------|---------|---------|---------|---------|---------|
| A | Thickness ratio | 1 | 0.923 | 0.902 | 0.881 | 0.859 | 0.838 |
| B | Stroke (bend angle) | 3 mm | 5 mm | 7 mm | - | - | - |
| C | Alignment | 7 | 8 | 9 | - | - | - |

**TABLE 7.3**

**Springback Parameter and Their Levels for Curved Shape**

| Factor | Parameters | Level 1 | Level 2 | Level 3 | Level 4 | Level 5 | Level 6 |
|---|---|---|---|---|---|---|---|
| A | Thickness ratio | 1 | 0.985 | 0.975 | 0.947 | 0.907 | 0.871 |
| B | Stroke (bend angle) | 3 mm | 5 mm | 7 mm | - | - | - |
| C | Annealing temperature | 24°C (RT) | 150°C | 300°C | - | - | - |

**TABLE 7.4**

**Chemical Composition of AA6061-T6 (See Mahdi et al., 2022)**

| Component | Al | Cr | Cu | Fe | Mg | Mn | Si | Ti | Zn | Other |
|---|---|---|---|---|---|---|---|---|---|---|
| Wt% | 95.8–98.6 | 0.04–0.35 | 0.15–0.4 | Max 0.7 | 0.8–1.2 | Max 0.15 | 0.4–0.8 | Max 0.15 | Max 0.25 | Max 0.15 |

**TABLE 7.5**

**Mechanical Properties of AA6061-T6 (See Chandu et al., 2013)**

| Properties | Metric |
|---|---|
| Density | 2.7 g/cc |
| Vickers hardness | 107 |
| Ultimate tensile strength | 310 GPa |
| Tensile yield strength | 276 MPa |
| Elongation at break | 12% |
| Modulus of elasticity | 68.9 GPa |
| Shear strength | 207 MPa |
| Shear modulus | 26 GPa |
| Melting point | 582°C–652°C |

the punch stroke was in the range of 3–7 mm, the annealing temperatures were in the range of 24°C, i.e., room temperature (no annealing) to 300°C and, finally, the rolling direction in the range of 0°–90°. These parameters are summarized in Table 7.3.

## 7.2.2 MATERIALS

AA6061-T6 aluminium alloy sheets of 3 mm thickness were used in this study. This alloy offers excellent joining characteristics and has a good acceptance of applied coatings. It has relatively high strength, good workability and high resistance to corrosion and is widely available. Applications of AA6061-T6 can be found in components such as aircraft fittings, magneto parts, brake pistons, hydraulic pistons, appliance fittings, bike frames, valves and valve parts. The chemical and mechanical properties of AA6061-T6 are shown in Tables 7.4 and 7.5, respectively.

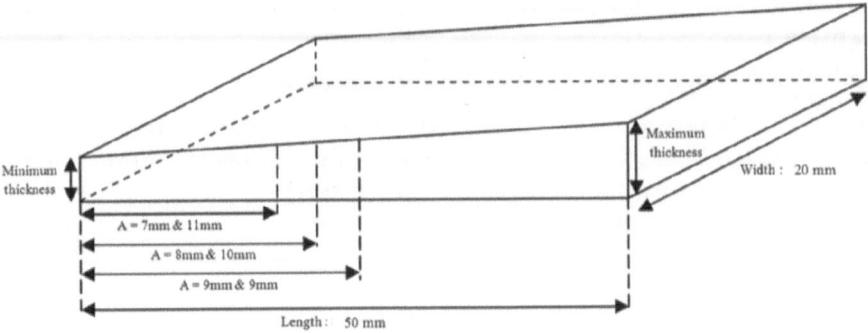

**FIGURE 7.4**   Tapered-shaped aluminium strip for a non-uniform profile.

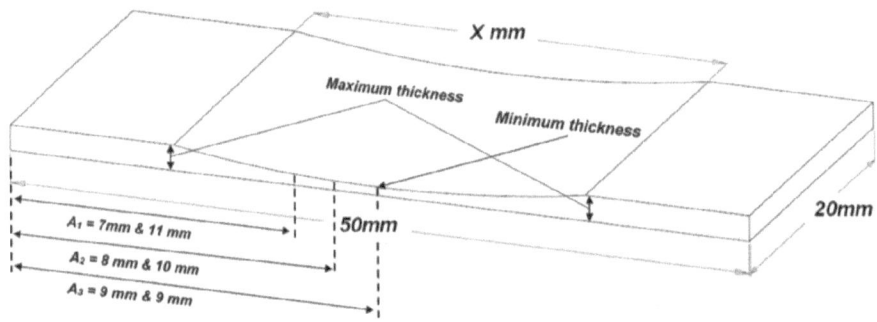

**FIGURE 7.5**   Curve-shaped aluminium strip for a non-uniform profile.

**TABLE 7.6**
**The Resulted Thickness Ratio**

| Set | Length (mm) | Wide (mm) | Tapered-Shaped | Curved-Shaped |
|---|---|---|---|---|
| 1. | 50 | 20 | 1.000 | 1.000 |
| 2. | 50 | 20 | 0.923 | 0.985 |
| 3. | 50 | 20 | 0.902 | 0.975 |
| 4. | 50 | 20 | 0.881 | 0.947 |
| 5. | 50 | 20 | 0.859 | 0.907 |
| 6. | 50 | 20 | 0.838 | 0.871 |

### 7.2.3   SPECIMEN PREPARATIONS

For this stage, the aluminium strip specimens are cold formed into non-uniform profiles (tapered shape and curved shape), as shown in Figures 7.4 and 7.5, following the thickness ratio listed in Table 7.6. For the curved specimen, the dimension of "X mm" indicates the length of the curved area. The specimens are then cut into the

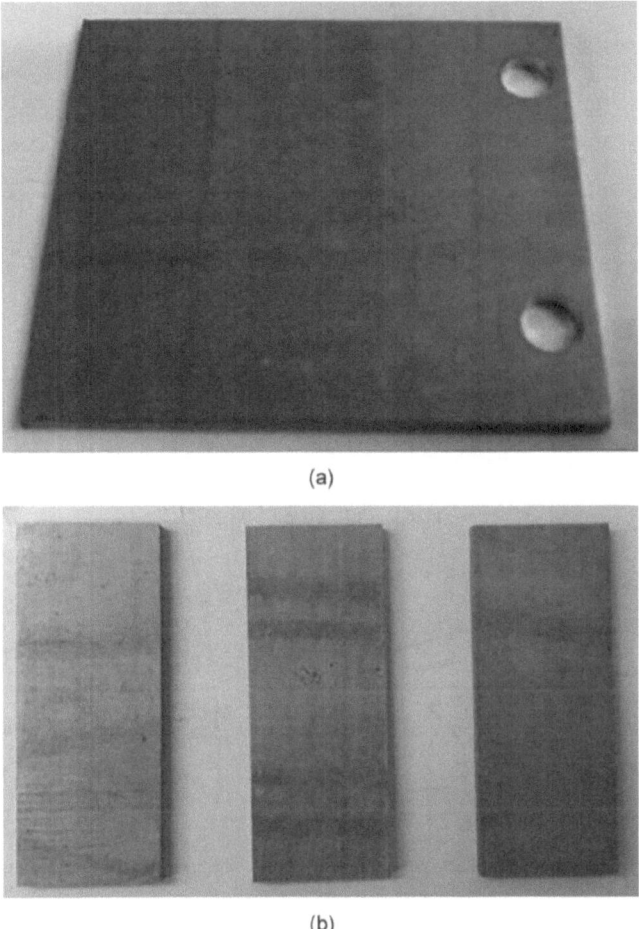

(a)

(b)

**FIGURE 7.6**    Specimen (a) before deformation and (b) after cutting with a dimension of 50 mm × 20 mm.

desired dimension of 50 mm (length) × 20 mm (width). A total of six specimen sets of different thickness ratios were prepared, as listed in Table 7.6. Figure 7.5 shows (a) the specimen before deformation and (b) the specimen after it was cut into 50 mm × 20 mm using an EDM wire cutter (Figure 7.6).

### 7.2.4  SAMPLE PREPARATION

The non-uniformity of the specimen is produced by a cold forming process using an appropriate tool and die profiles. Figure 7.7 shows the die and puncher sets used for producing the tapered and curved specimens. The die sets were designed in CAD and fabricated out of D2 tool steel material. A 100-ton mechanical press is used for the cold forming process.

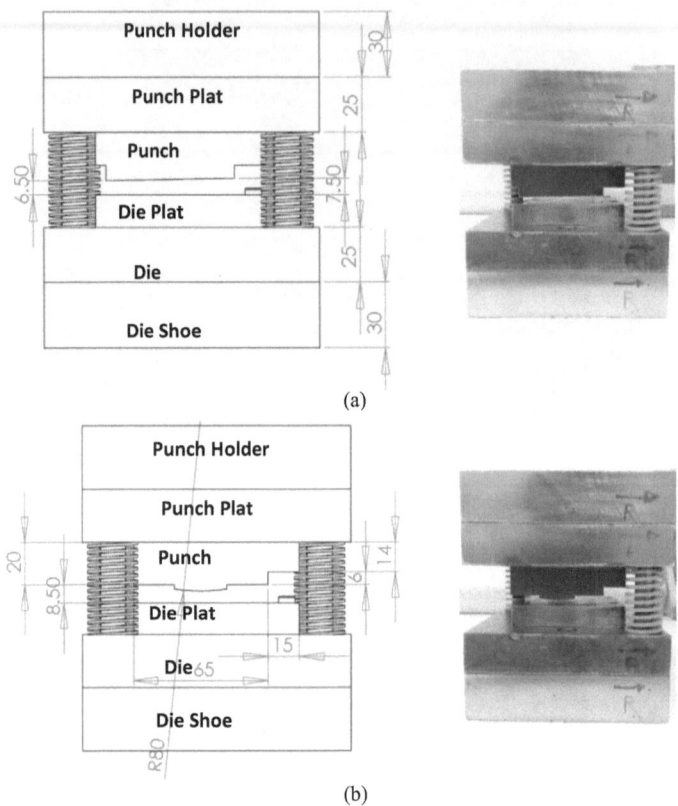

**FIGURE 7.7** Die and puncher for non-uniform profiles: (a) tapered-shaped and (b) curved-shaped.

### 7.2.5 Experimental Setup

#### 7.2.5.1 V-Bending Experiment

V-bending is a common method in a bending process in which the punch and die have matching 'V'-shaped profiles. Here, the sheet metal is placed between the matching punch and the die and is then compressed (Diegal, 2002). In the V-bending experiment, observations were made at different strokes. A schematic of the V-bending test and its related definition is elaborated in Figure 7.8a and b. In this study, the bend radius is set at 10°, 17° and 24° which corresponds to a stroke of 3, 5 and 7 mm, respectively.

The springback response of the specimen is measured based on the variation of the bend angle, thickness ratio and alignment of the workpiece during the experiment. Alignment can be described as the point where the punch is located on the specimen during the bending process. Figure 7.9 shows the illustration for the alignment of 9 mm where the orange line marked on the puncher is parallel to the orange line marked on the specimen.

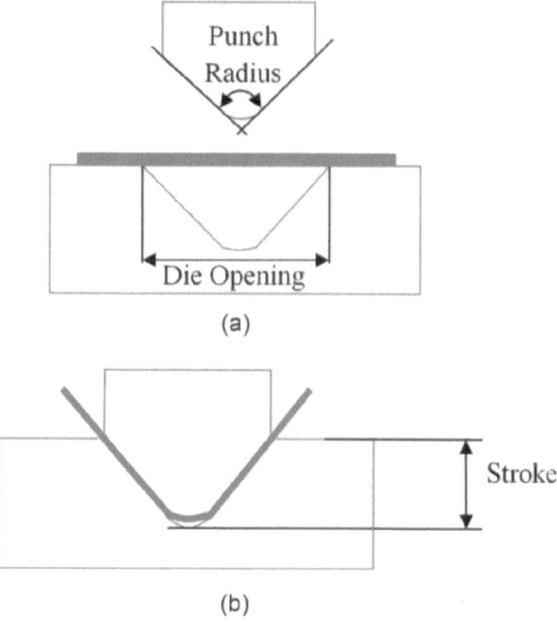

(a)

(b)

**FIGURE 7.8**    The definition of the V-bending process (a) initial and (b) end of the process.

**FIGURE 7.9**    Definition of alignment.

### 7.2.6   Springback Measurement

#### 7.2.6.1   After-Unloading Measurement

The amount of springback is defined as the difference between the angle after unloading and before unloading. In this case, the stroke values play an important role in the springback measurement. The measurements of the springback are made on a Rax Vision Mitutoyo Profile Projector (PC 3000) using the two lines technique to measure the bending angle. For the measurements of the angle after loading, the die valley radius, punch radius and the die opening were at 90°, 2 mm and 32 mm, respectively, as shown in Figure 7.10. The die radius and punch radius values are assumed to be equal in this study. Figure 7.11 shows the scanned image of the specimen after unloading using ALICONA Infinite Focus Machine (IFM) use for comparison and result validation.

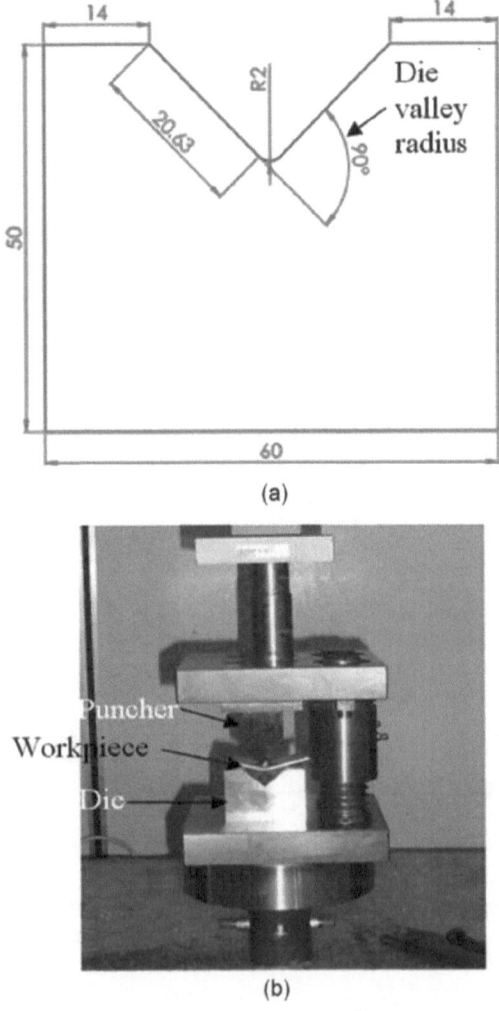

(a)

(b)

**FIGURE 7.10**   (a) V-bending bottom die and (b) V-bending experimental setup.

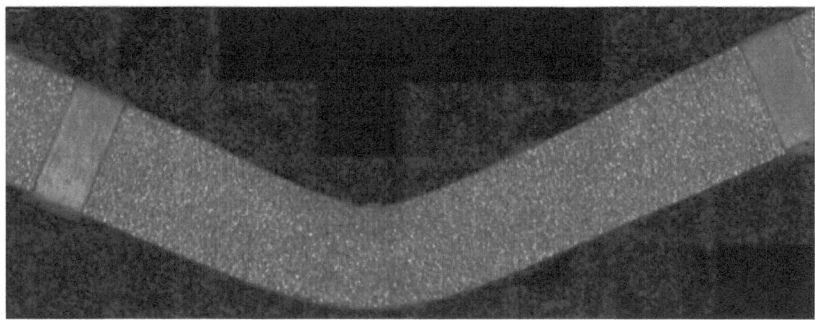

**FIGURE 7.11** Scanned image of the unloading specimen using ALICONA Infinite Focus Machine (IFM).

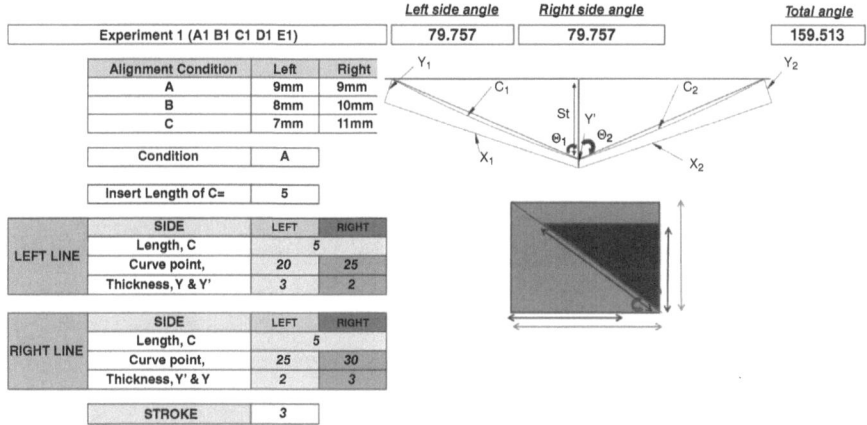

**FIGURE 7.12** Angle calculation window.

### 7.2.6.2 Before Unloading Measurement

#### 7.2.6.2.1 Linear Line Angle Calculation

I. Firstly, the angles of the linear line of the left side, $\theta_{L_1}$, and the right side, $\theta_{L_2}$, are determined. An example of the calculation of these angles is shown in Figure 7.12. Figure 7.13 list the calculated data for the left side of the specimen.

II. Figure 7.14 shows the data for the left side. The stroke values in the experiment are 3, 5, and 7 mm for full-length specimens. For the curved specimen, the length of $c$ covers only the curved region.

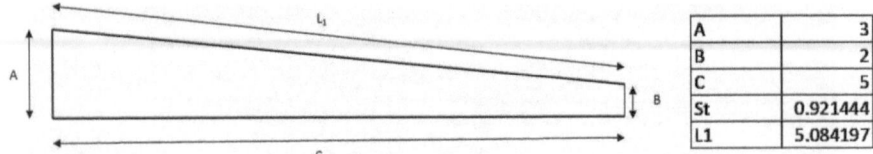

| A | 3 |
|---|---|
| B | 2 |
| C | 5 |
| St | 0.921444 |
| L1 | 5.084197 |

**FIGURE 7.13**　Data for the left side of a specimen (specimen front view).

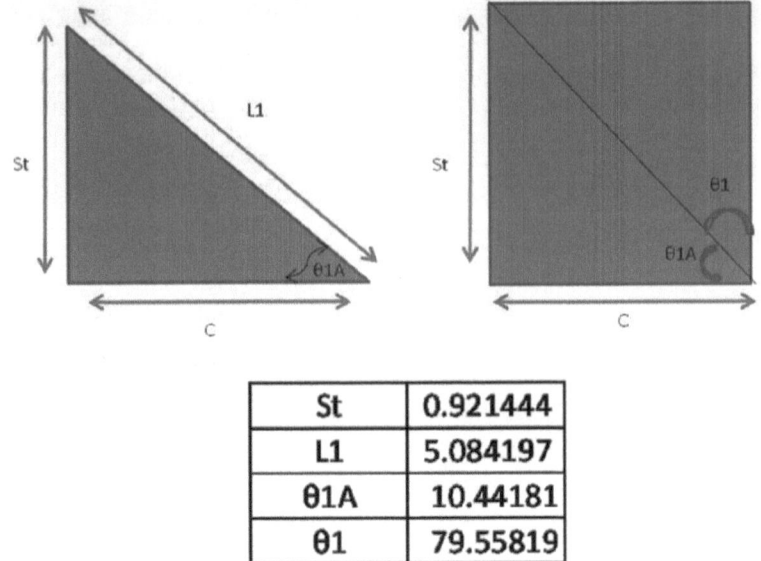

| St | 0.921444 |
|---|---|
| L1 | 5.084197 |
| θ1A | 10.44181 |
| θ1 | 79.55819 |

**FIGURE 7.14**　Data for the left side of a specimen (die opening view).

| T1 | 1 |
|---|---|
| L'1 | 5.181608 |

**FIGURE 7.15**　Tapered surface profile.

III. The value of $L_1$ can then be determined by trigonometrical calculations. The value of angle $\theta_{1A}$ is calculated by using the trigonometry formula without considering thickness reduction along the length of $L_1$.

IV. The value of angle $\theta_1$ is determined by using the value for $L_1'$ which considers the thickness reduction in the long length of $L_1$. The value of $L_1'$ is calculated by considering the tapered surface profile, as shown in Figure 7.15.

V. The value of $T_1$ is the difference between the maximum thickness and minimum thickness along the length of $L_1'$.

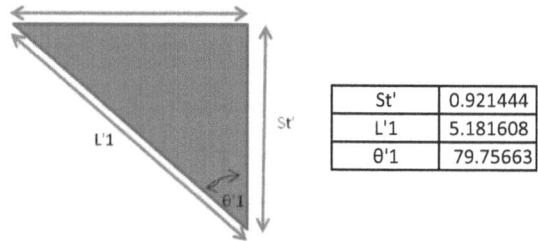

| St' | 0.921444 |
| L'1 | 5.181608 |
| θ'1 | 79.75663 |

**FIGURE 7.16**   The angle before unloading.

VI. The angle of $\theta_1'$ is calculated using trigonometric formulas by using the St' and $L_1'$ values. The result of the angle $\theta_1'$ is shown in Figure 7.16.

VII. The total angles before unloading is calculated by summing up the values of both the left side and right side ($\theta_1'$ and $\theta_2'$) angles.

## 7.3   RESULTS AND DISCUSSION

Analysis of the springback and the factors that influence it by using the Taguchi method is useful in springback prediction and optimization. In this chapter, the Taguchi experimental design model was developed and used to determine the level of influence of the parameters on the springback responses.

Analysis in this chapter also covers the effect of stress, strain, yield criteria and work hardening on the springback behaviour. Note that it is important to determine the material properties because it is closely linked to specimen characteristics and also the parameters set in the experimental works. In addition, the discussion will also include the external factors that might influence the results. It is expected that the results of this study can be used as a reference for future research and used as a platform to understand the behaviour of the springback and its influencing parameters.

### 7.3.1   V-Bending Experiment

The amount of springback can be determined by the difference between the angles of the specimen strip after loading and after unloading. In this case, stroke values play an important role in springback measurement after loading conditions. Many past studies have evaluated the effect of process parameters in the V-bending process, such as the bending angle, material thickness, punch radius, and punch height (Thipprakmas and Phanitwong, 2011). However, most of the studies have focused on specimens with uniform section thicknesses. The effect of process parameters on non-uniform thickness specimens is yet to be explored. In the V-bending experiment, observations were made for selected parameters of thickness ratio, bend angle, annealing temperature, alignment and rolling direction. The bend angle is set at 10, 17 and 24 degrees or at 3, 5 and 7 mm stroke, respectively. The maximum stress obtained is approximately 700 MPa. The results in terms of the average of the springback were obtained after conducting the V-bending experiment. The results of the experiments are summarized in Table 7.7 for tapered shape specimens and Table 7.8 for the curve-shaped specimens.

## 7.3.2 STRAIN HARDENING

Strain also plays a significant role in the springback behaviour in which the increase of strain hardening will theoretically increase the springback. Figure 7.17 shows the stress–strain behaviour of the material at a maximum stroke of 7 mm. Strain hardening (also called work hardening or cold working) is the process of making a metal harder and stronger through plastic deformation. This explains why the flow stress increases with the increase in the axial strain amount. Strain hardening will increase the hardness of the section with the minimum thickness. Figures 7.18 and 7.19 show the relationship between the thickness of the workpiece and Vickers hardness number for tapered and curve-shaped specimens, respectively. The average values of hardness were obtained from five readings around the same location. Figure 7.20 shows the relationship between the effect of annealing temperature on the specimen

**TABLE 7.7**

**Experimental Result of Tapered Shape for V-Bending**

| | Parameters | | | | | |
|---|---|---|---|---|---|---|
| | A | B | C | D | E | |
| | Thickness Ratio | Bend Angle | Annealing Temperature | Alignment | Rolling Direction | Average |
| Experiment Number | | (°) | (°C) | (mm) | | Springback (%) |
| 1 | 1.00 | 3 | 24(RT) | 7 | 0 | 0.02749 |
| 2 | 1.00 | 5 | 150 | 8 | 45 | 0.01122 |
| 3 | 1.00 | 7 | 300 | 9 | 90 | 0.00503 |
| 4 | 0.923 | 3 | 24(RT) | 8 | 45 | 0.02206 |
| 5 | 0.923 | 5 | 150 | 9 | 90 | −0.01272 |
| 6 | 0.923 | 7 | 300 | 7 | 0 | 0.00287 |
| 7 | 0.902 | 3 | 150 | 7 | 90 | 0.02150 |
| 8 | 0.902 | 5 | 300 | 8 | 0 | 0.01543 |
| 9 | 0.902 | 7 | 24(RT) | 9 | 45 | 0.00050 |
| 10 | 0.881 | 3 | 300 | 9 | 45 | 0.02143 |
| 11 | 0.881 | 5 | 24(RT) | 7 | 90 | 0.01842 |
| 12 | 0.881 | 7 | 150 | 8 | 0 | 0.00322 |
| 13 | 0.859 | 3 | 150 | 9 | 0 | 0.02906 |
| 14 | 0.859 | 5 | 300 | 7 | 45 | 0.00988 |
| 15 | 0.859 | 7 | 24(RT) | 8 | 90 | 0.01163 |
| 16 | 0.838 | 3 | 300 | 8 | 90 | 0.02592 |
| 17 | 0.838 | 5 | 24(RT) | 9 | 0 | 0.01247 |
| 18 | 0.838 | 7 | 150 | 7 | 45 | 0.01293 |

**TABLE 7.8**

**Experimental Result of Curved Shape for V-Bending**

| | Parameters | | | | | |
|---|---|---|---|---|---|---|
| | A | B | C | D | E | |
| | Thickness Ratio | Bend Angle | Annealing Temperature | Alignment | Rolling Direction | |
| Experiment Number | | (°) | (°C) | (mm) | | Average Springback (%) |
| 1 | 1.00 | 3 | 24(RT) | 7 | 0 | 0.01027 |
| 2 | 1.00 | 5 | 150 | 8 | 45 | 0.01943 |
| 3 | 1.00 | 7 | 300 | 9 | 90 | 0.03606 |
| 4 | 0.985 | 3 | 24(RT) | 8 | 45 | 0.01872 |
| 5 | 0.985 | 5 | 150 | 9 | 90 | 0.01077 |
| 6 | 0.985 | 7 | 300 | 7 | 0 | 0.01462 |
| 7 | 0.975 | 3 | 150 | 7 | 90 | 0.00077 |
| 8 | 0.975 | 5 | 300 | 8 | 0 | 0.01021 |
| 9 | 0.975 | 7 | 24(RT) | 9 | 45 | 0.01075 |
| 10 | 0.947 | 3 | 300 | 9 | 45 | 0.00020 |
| 11 | 0.947 | 5 | 24(RT) | 7 | 90 | 0.01917 |
| 12 | 0.947 | 7 | 150 | 8 | 0 | 0.02748 |
| 13 | 0.907 | 3 | 150 | 9 | 0 | 0.00372 |
| 14 | 0.907 | 5 | 300 | 7 | 45 | 0.00708 |
| 15 | 0.907 | 7 | 24(RT) | 8 | 90 | 0.02478 |
| 16 | 0.871 | 3 | 300 | 8 | 90 | 0.00534 |
| 17 | 0.871 | 5 | 24(RT) | 9 | 0 | 0.01291 |
| 18 | 0.871 | 7 | 150 | 7 | 45 | 0.04339 |

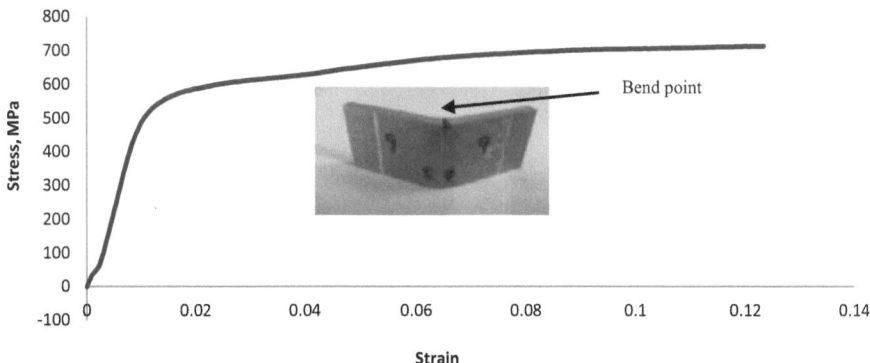

**FIGURE 7.17**   The stress–strain behaviour of the material.

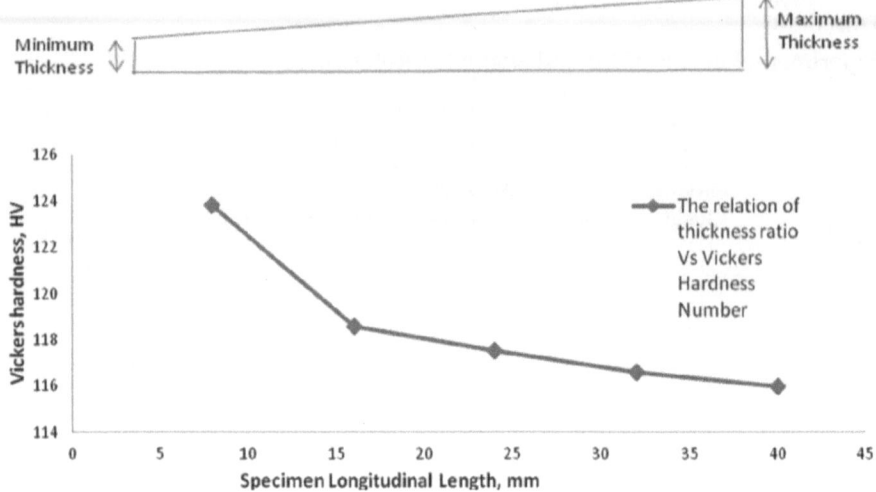

**FIGURE 7.18**   The relation of thickness vs Vickers hardness number (curved shape).

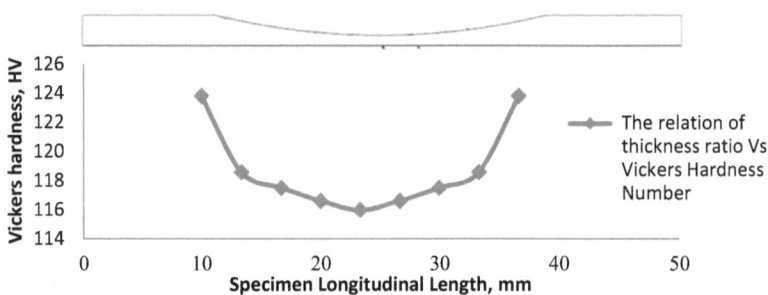

**FIGURE 7.19**   The relation of thickness vs Vickers hardness number (curved shape).

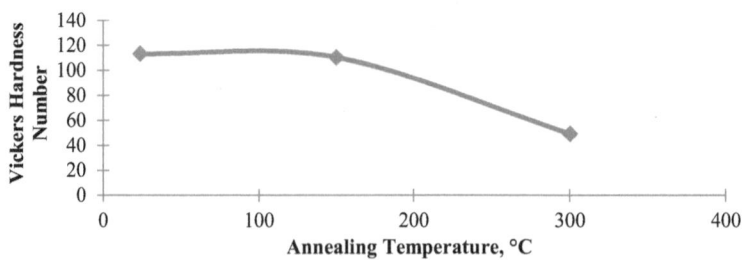

**FIGURE 7.20**   The relation of annealing temperature vs Vickers hardness number.

and Vickers hardness number. Strain-hardening constraint, which is known as the strain-hardening index, is usually represented by $n$. This constant is used to calculate stress–strain behaviour in work hardening. The work-hardening expression can be determined at higher strain values in the range of plasticity zone. By changing the value of true stress–strain on the curve to a logarithmic graph the value of the slope, the slope of $n$ (strain-hardening exponent) and an intercept of log $K$ ($K$ as strength coefficient value) can be obtained. Generally, if the $n$ value is 0, it is shown that the material is perfectly plastic solid and if the value is 1, it represents the material is totally elastic.

Strain also plays a significant role in the occurrence of springback; increased strain hardening increases the springback. The observation is then converted into a logarithm scale. The $n$ values are then obtained from the slope of the graph. The result shows that the required stress increases with the decrease of the $n$ value. Figure 7.21 shows the behaviour of stress and strain for different thicknesses by taking $n$ as the parameter. The result revealed that thickness also affects the value of stress and strain. Figure 7.22 shows values known as the stress and strain for different annealing temperatures by taking $n$ as the parameter. Based on the graph, specimens with high annealing temperatures are less elastic than specimens with low

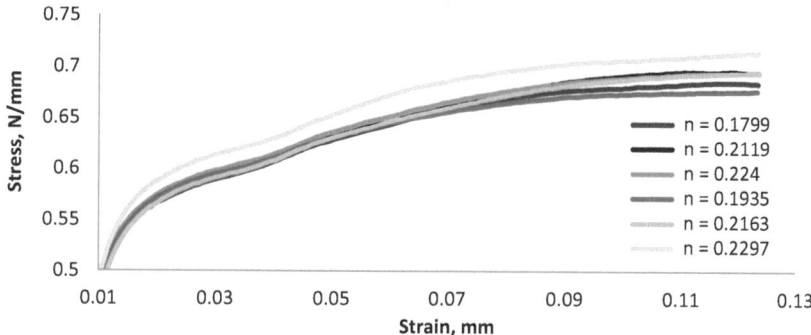

**FIGURE 7.21**    Stress vs strain for different thicknesses taking $n$ as the parameter.

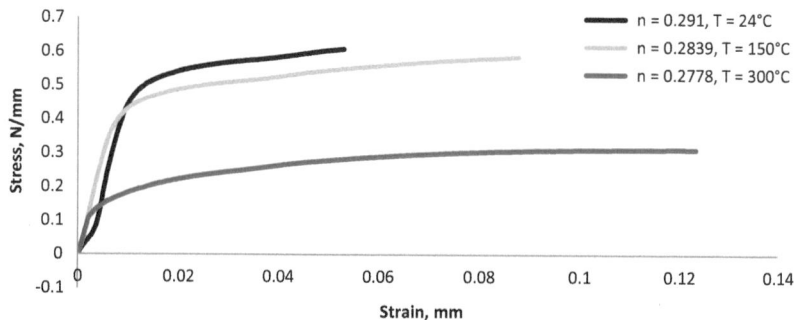

**FIGURE 7.22**    Stress vs strain for different annealing temperatures taking $n$ as the parameter.

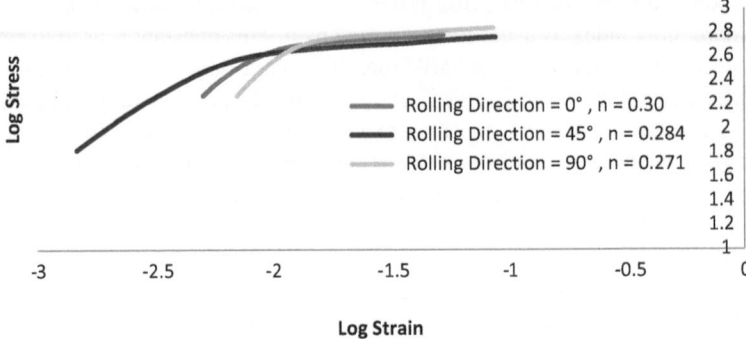

**FIGURE 7.23** Stress–strain in various rolling directions in the logarithmic diagram taking *n* as the parameter.

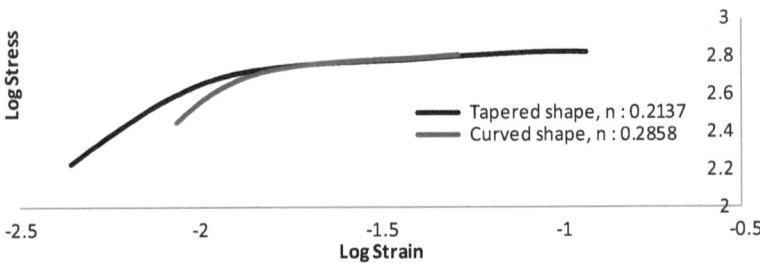

**FIGURE 7.24** Stress–strain of tapered and curved shapes in the logarithmic diagram taking *n* as the parameter.

annealing temperatures. Figure 7.23 shows the values known as the strain-hardening index for different rolling directions (0°, 45° and 90°) which are 0.30, 0.284 and 0.271, respectively. The results almost coincided with studies made by Dwivedi et al. (2002). Figure 7.24 shows the stress–strain behaviour of tapered and curved shapes in the logarithmic diagram taking *n* as the parameter. From the result, the curved shape is more elastic than the tapered shape.

## 7.4 CONCLUSIONS

In this study, the main aim is to investigate the effect of the thickness ratio, bending angle, annealing temperature, alignment and the rolling direction on the springback behaviour of AA6061 strip with non-uniform thickness distribution. The parameters and specimen profiles were selected taking into consideration the industrial requirement and from the previous studies.

It was found that for tapered-shaped specimens, increasing the bend angle will result in a decrease in the springback. However, for curve-shaped specimens, the springback value increases with increments of bend angle. For both cases, an increase in the thickness ratio will result in a decrease in the springback.

# REFERENCES

Abdulhay, B., Bourouga, B. and Dessain, C. Experimental and theoretical study of thermal aspects of the hot stamping process, *Appl. Therm. Eng.* 31 (2011) 674–685.

Air, C., Board, R. and Lutsey, N. Review of technical literature and trends related to automobile mass-reduction technology, California Air Resources Board (2010).

Burchitz, I. Springback: Improvement of its predictability: Literature study report (2005).

Chandu, K., Venkateswara, R.E., Rao, A., and Subrahmanyam. The Strength of Friction Stir Welded Aluminium Alloy 6061. International *J. Res. Mech. Eng. Tech.* 4(1) (2013) 119–122.

Cui, X., Zhang, H., Wang, S., Zhang, L. and Ko, J. (2011). Design of lightweight multi-material automotive bodies using new material performance indices of thin-walled beams for the material selection with crashworthiness consideration, Materials & Design, 32(2), 815-821,

Davis, U.C. Institute of Transportation Studies Review of technical literature and trends related to automobile mass-reduction technology, UCD-ITS-RR-10-10 (2010).

Diegel, O. *Bend Works: The Fine Art of Sheet Metal Bending.* Complete Design Service (2002).

Duan, L., Sun, G., Cui, J., Chen, T. and Li, G. Lightweight design of vehicle structure with tailor rolled blank under crashworthiness. *11th World Congress on Structural and Multidisciplinary Optimisation*, Sydney, Australia, 07th-12th, June 2015.

Dwivedi, J.P., Shah, S.K., Upadhyay, P.C. and Das Talukder, N.K. Springback analysis of thin rectangular bars of non-linear work-hardening materials under torsional loading, *Int. J. Mech. Sci.* 44 (2002) 1505–1519.

Ghassemieh, E. Materials in automotive application, state of the art and prospects. In: M. Chiaberge (Ed.), *New Trends and Developments in Automotive Industry* (pp. 347–394). In Tech: London (2010).

Ghassemieh, E. Materials in automotive application, state of the art and prospects. In: M. Chiaberge (Ed.), *New Trends and Developments in Automotive Industry* (pp. 365–390). InTech: London (2011).

Girubha, J. and Vinodh, R.S. Application of fuzzy VIKOR and environmental impact analysis for material selection of an automotive component, *Mater. Des.* 37 (2012) 478–486.

Joost, W.J. Reducing vehicle weight and improving U.S. energy efficiency using integrated computational materials engineering, *JOM* 64 (2012) 1032–1038.

Kamaruddin, S., Khan, Z.A. and Foong, S.H. Application of Taguchi method in the optimization of injection moulding parameters for manufacturing products from plastic blend, *Int. J. Eng. Technol.* 2 (2010) 574–580.

Karbasian, H. and Tekkaya, A.E. A review on hot stamping, *J. Mater. Process. Tech.* 210 (2010) 2103–2118.

Kashani, H.T. and Kah, P.J. Martikainen, laser overlap welding of zinc-coated steel on aluminum alloy, *Phys. Procedia.* 78 (2015) 265–271.

Kleiner, M., Geiger, M. and Klaus, A. Manufacturing of lightweight components by metal forming, *CIRP Ann. Manuf. Technol.* 52 (2003) 521–542.

Kuo, C.F.J., Su, T.L. and Tsai, C.P. Optimization of the needle punching process for the non-woven fabrics with multiple quality characteristics by grey-based Taguchi method, *Fibers Polym.* 8 (2007) 654–664.

Li, M.F., Chiang, T.S., Tseng, J.H. and Tsai, C.N. Hot stamping of door impact beam, *Procedia Eng.* 81 (2014) 1786–1791.

Mahdi, E., Eltai, E., Alabtah, F. and Eliyan, F. Mechanical characterization of AA 6061-T6 MIG welded aluminum alloys using a robotic arm. *Key Eng. Mat.* 913 (2022) 271–278. doi: 10.4028/p-rhrr3n.

Mayyas, A., Qattawi, A.M. and Shan, D. Design for sustainability in automotive industry: A comprehensive review, *Renew. Sustain. Energy Rev.* 16 (2012) 1845–1862.

Merklein, M. and Opel, S. Investigation of tailored blank production by the process class sheet bulk metal forming investigation of tailored blank production by the process class sheet-bulk metal forming, *API Conf. Proc.* 395 (2012) 1315.

Merklein, M., Johannes, M., Lechner, M.A. and Kuppert, A. A review on tailored blanks: Production, applications and evaluation, *J. Mater. Process. Tech.* 214 (2014) 151–164.

Meyer, A.Ã., Wietbrock, B. and Hirt, G. Increasing of the drawing depth using tailor rolled blanks: Numerical and experimental analysis, *Int. J. Mach. Tools Manuf.* 48 (2008) 522–531.

Mori, K., Abe, Y., Osakada, K. and Hiramatsu, S. Plate forging of tailored blanks having local thickening for deep drawing of square cups, *J. Mater. Process. Tech.* 211 (2011) 1569–1574.

Pan, F., Zhu, P., and Zhang, Y. Metamodel-based lightweight design of B-pillar with TWB structure via support vector regression. *Comp. Struct.* 88(1–2) (2010) 36–44.

Prawoto, Y., Djuansjah, J., Tawi, K. and Martin-Fanone, M. Tailoring microstructures: A technical note on an eco-friendly approach to weight reduction through heat treatment. *Mater. Des.* 50 (2013) 635–645. doi: 10.1016/j.matdes.2013.03.062.

Thipprakmas, S. and Phanitwong, W. Process parameter design of spring-back and spring-go in V-bending process using Taguchi technique, *Mater. Des.* 32 (2011) 4430–4436.

Veera Babu, K., Ganesh Narayanan, R. and Saravana Kumar, G. An expert system for predicting the deep drawing behavior of tailor welded blanks, *Expert Syst. Appl.* 37 (2010) 7802–7812.

Yang, Z., Peng, Q. and Yang, J. Lightweight design of B-pillar with TRB concept considering crashworthiness, *IEEE 2012 Third International Conference on Digital Manufacturing and Automation (ICDMA),* Guilin, China (2012) 510–513.

Zhang, H. Springback characteristics in U-channel forming of tailor rolled blank, *Acta Metall. Sin. (Engl. Lett.)* 25 (2012) 207–213.

# 8 Evaluation of Dissimilar Grade Aluminium Alloy-Tailored Blanks Fabricated by FSW

*AF Pauzi and AB Abdullah*
Universiti Sains Malaysia

*MF Jamaludin*
Universiti Malaya

## CONTENTS

## 8.1 INTRODUCTION

One of the recent trends in the automotive industry is moving toward the utilization of lightweight metals, such as aluminium, for the fabrication of vehicle body structures. Aluminium alloys have numerous desirable characteristics for engineering applications. For example, AA5052 is the strongest non-heat-treatable aluminium, usually used as an alternative to steel in aerospace, marine and automotive industries because of its lightweight, good formability, good strength and high corrosion resistance [1,2]. While AA6061 is suitable for the application that required good tensile strength at high temperatures and dimensional stability and even has good corrosion resistance to seawater [3]. AA6061 has very good weldability characteristics and is

DOI: 10.1201/9781003164241-8

typically used for heavy-duty structures in shipbuilding and rail coaches [4]. On the other hand, AA1100 is rarely used in sheet metal work that requires high strength and good machinability, although the alloy has excellent electrical conductivity and a reflective finish [5]. The automotive industry prefers to use the 6000 series aluminium alloys due to their excellent strength, surface quality and good weldability [6]. Using welding techniques such as FSW, there is an opportunity in joining different combinations of aluminium alloys for better blank performance and cost efficiency. However, the welding of dissimilar aluminium can be difficult to achieve due to their differences in material properties [7].

Being a solid-state joining method, FSW is a potential candidate for welding dissimilar aluminium alloys. Several studies have explored the combinations of different alloy grades and rotational speeds and their effect on hardness and tensile properties. For example, Balaji et al. [8] have found that the most suitable tool rotational speed is between 600 and 1200 RPM for the joining of AA2024 and AA7075. It was also found that the tensile strength of the material was reduced only by 10% as compared to the base materials. Similar findings were noted by Liu et al. [9] who found that the maximum tensile strength of dissimilar aluminium alloy joints is ~85% of the base Al alloy. Similarly, Hassan et al. [10] indicated that weld hardness drops at the heat-affected zone (HAZ) on the softer material side. In addition, the tool profile also influences the performance of the weld, based on the study by Krishna et al. [11]. Several works on the FSW of dissimilar aluminium alloys are summarized in Table 8.1.

Past studies have mostly focused on the pre-forming evaluation of the welded blank such as the strength and hardness of the weldments. Post-forming evaluations, such as the forming pattern or behaviour of the welded blank after forming, are still

## TABLE 8.1
## Recent Studies of Friction Stir Welding on Various Aluminium Alloys

| References | Combination | Effect on Hardness | Effect on Strength |
|---|---|---|---|
| Illangovan et al. [12] | AA6061 – AA5086 | Hardness in the stir zone higher than HAZ and TMAZ | Higher tensile strength for the threaded pin and lower tensile strength on the straight cylindrical pin due to the presence of defects |
| Piccini and Svoboda [13] | AA5052 – AA6063 | | |
| Barbini et al. [14] | AA2024 – AA7050 | Minimum hardness value at the AS | Yield strength and UTS increase with welding speed increase |
| Rodriguez et al. [15] | AA6061 – AA7050 | Hardness peak value at the SZ region adjacent to the AS | Tensile loading increases with the increase of the rotational speed |
| Koilraj et al. [16] | AA2219 – AA5083 | Lowest hardness accrued at the HAZ on AA5083 | Tensile test failed in the AA5083-side HAZ because of the strength of the aluminium alloy |
| Hasan et al. [17] | AA7075 – AA6061 | - | Threaded cone pin and single flat tool result in higher tensile value |

lacking. In this chapter, post-forming measurements focusing on springback patterns would be conducted. Various process parameters including bending location, bending direction, rolling direction, material combination and sides, either retreating or advancing, will be considered. The present work would form the basis for subsequent studies in the fabrication of aluminium-based TWBs using FSW technology.

## 8.2 METHODOLOGY

The main objective of this study is to evaluate the springback phenomenon that occurs on a tailored blank made from different combinations of aluminium alloy grades fabricated by FSW. Three aluminium alloy grades of AA6061, AA5052 and AA1100 were chosen and prepared in 200 mm×80mm size with a thickness of 3 mm. The material elemental compositions of AA6061-T6, AA5052-H32 and AA1100-H14 are shown in Table 8.2a–c, respectively. The specimen pair is arranged in a butt-weld configuration and clamped on a conventional milling machine, as shown in Figure 8.1. The optimum feed rates (welding speed), tool angle and spindle speed (rotational speed) were identified from a previous study. The clamp was fabricated based on the design recommendation by Kamble [18], as shown in Figure 8.2. It was designed to firmly hold the specimens during welding to avoid specimen tilting, to ensure the gap between the specimens is maintained, and to prevent the specimen from lifting or bending during welding. The cylindrical friction stir welding (FSW) tool, as shown in Figure 8.3, is made from hardened tool steel having 2 mm pin and 10 mm shoulder profiles, and both are unthreaded.

For the V-bending experiment, the die set is prepared as shown in Figure 8.4. The specimens were cut into 50 mm×20mm strips. The stroke depths of 3, 6 and 9 mm were selected for evaluation. In addition, the position of the weld line, as shown in Figure 8.5, is also observed. The bending location, as shown in Figure 8.6, is defined as the distance from the point of bending to the midpoint of the weld line. A

---

**TABLE 8.2**
**Material Composition of AA6061-T6, AA5052-H32 and AA1100-H14 (% wt.) [1,3,5]**

| % | Si | Fe | Cu | Mn | Mg | Cr | Zn | Ti | Al |
|---|----|----|----|----|----|----|----|----|----|
| **AA6061-T6** | | | | | | | | | |
| Min | 0.4 | | 0.15 | | 0.8 | 0.04 | | | Remainder |
| Max | 0.8 | 0.7 | 0.4 | 0.15 | 1.2 | 0.35 | 0.25 | 0.15 | |
| **AA5052-H32** | | | | | | | | | |
| Min | | | | | 2.2 | 0.15 | | | Remainder |
| Max | 0.25 | 0.4 | 0.1 | 0.10 | 2.8 | 0.35 | 0.1 | - | |
| **AA1100-H14** | | | | | | | | | |
| Min | | | 0.05 | | | | | | Remainder |
| Max | 0.95 | - | 0.20 | - | 0.05 | - | 0.1 | - | |

**FIGURE 8.1**   Experimental set-up on the conventional milling machine.

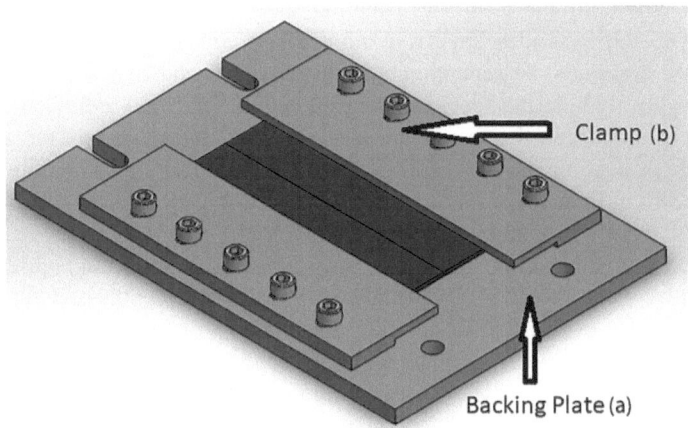

**FIGURE 8.2**   Design of the fixture.

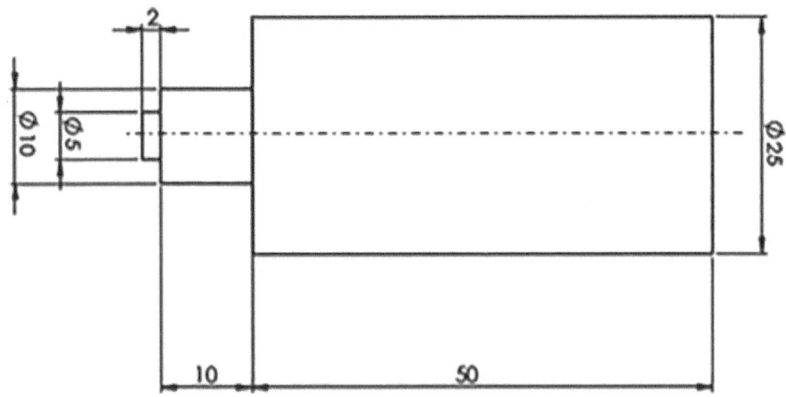

**FIGURE 8.3**   Schematic diagram of the tool (units in mm).

**FIGURE 8.4**    V-bending set-up on the hydraulic press machine.

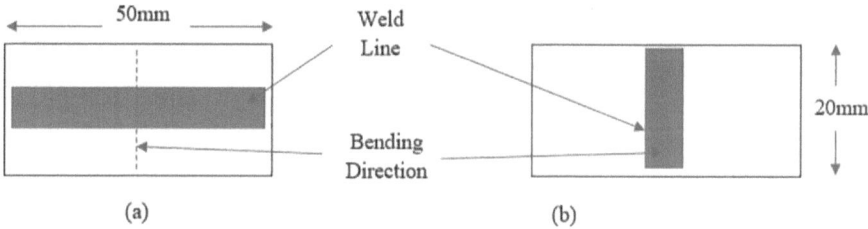

**FIGURE 8.5**    Bending direction: (a) perpendicular and (b) parallel.

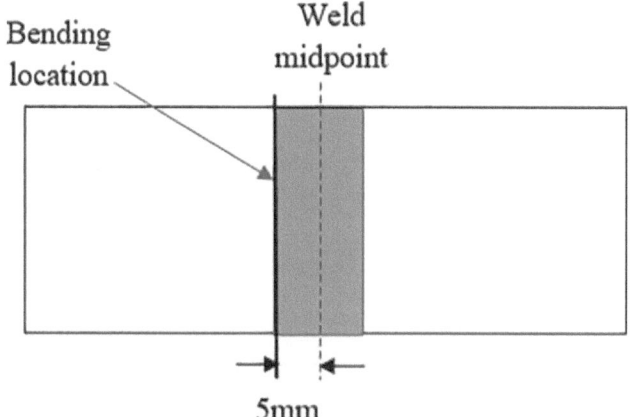

**FIGURE 8.6**    Bending location on similar grades 5 mm from the weld midpoint.

coordinate measuring machine (CMM) with a resolution of 0.0005 mm and a USB microscope (image sensor up to 3 megapixels with magnification ratio: 10× to 300×) were used for angular measurements related to springback.

## 8.3 RESULTS AND DISCUSSION

In this section, the effects of various process parameters such as bending direction and bending location, rolling direction, retreating and advancing side of the material on post-forming evaluation would be discussed, focusing on the pattern of springback.

### 8.3.1 LOAD PATTERN

It was found that the FSW specimen of similar alloy grades would experience a maximum bending load that is 50%–60% lower than the base materials. Furthermore, if the combination is between different Al grades, the bending load will be lower than the lowest bending loads of the parent materials. As can be seen from Table 8.3, the maximum bending load for AA6061-T6 – AA1100-H14 is lower than the lowest bending load for the AA1100-H14 parent material. This behaviour can be manipulated in the forming process as lower machine capacity is required for the lesser bending load.

### 8.3.2 SPRINGBACK PATTERN

In this section, the effect of various parameters on springback will be discussed for similar and dissimilar material combinations.

#### 8.3.2.1 Effect of Rolling Direction of Similar and Dissimilar Alloys

The results obtained after performing V-bending are shown in Figure 8.7a–e. In general, the springback increases as the stroke is increased for all combinations.

---

**TABLE 8.3**

**Maximum Stroke and Maximum Bending Recorded from the V-Bending Experiment**

| No. | Material | Maximum Stroke before the Elastic Limit (mm) | Maximum Bending Load (kN) |
|---|---|---|---|
| 1 | AA60661-T6 | 0.67 | 2555.5 |
| 2 | AA5052-H32 | 0.58 | 1964.1 |
| 3 | AA1100-H14 | 0.50 | 1034.5 |
| 4 | AA6061-T6 – AA6061-T6 | 0.33 | 1047.3 |
| 5 | AA5052-H32 – AA5052-H32 | 0.29 | 847.2 |
| 6 | AA1100-H14 – AA1100-H14 | 0.25 | 556.8 |
| 7 | AA6061-T6 – AA5052-H32 | 0.42 | 1082.4 |
| 8 | AA6061-T6 – AA1100-H14 | 0.33 | 560.0 |

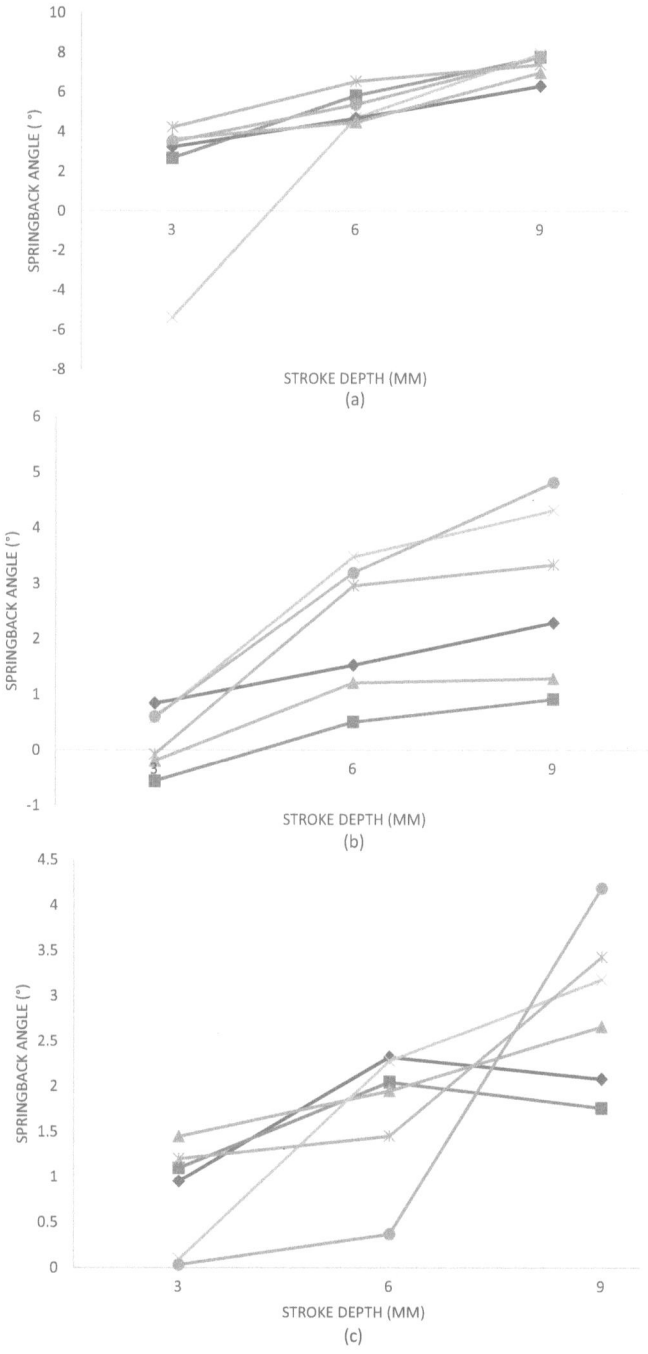

**FIGURE 8.7** The effect of rolling direction (RD) and bending direction (BD) to spring-back on the combination of AA6061-T6 – AA6061-T6 (a), AA5052-H32 – AA5052-H32 (b), AA1100-H14 – AA1100-H14 (c), AA6061-T6 – AA5052-H32 (d), and AA6061-T6 – AA1100-H14 (e).

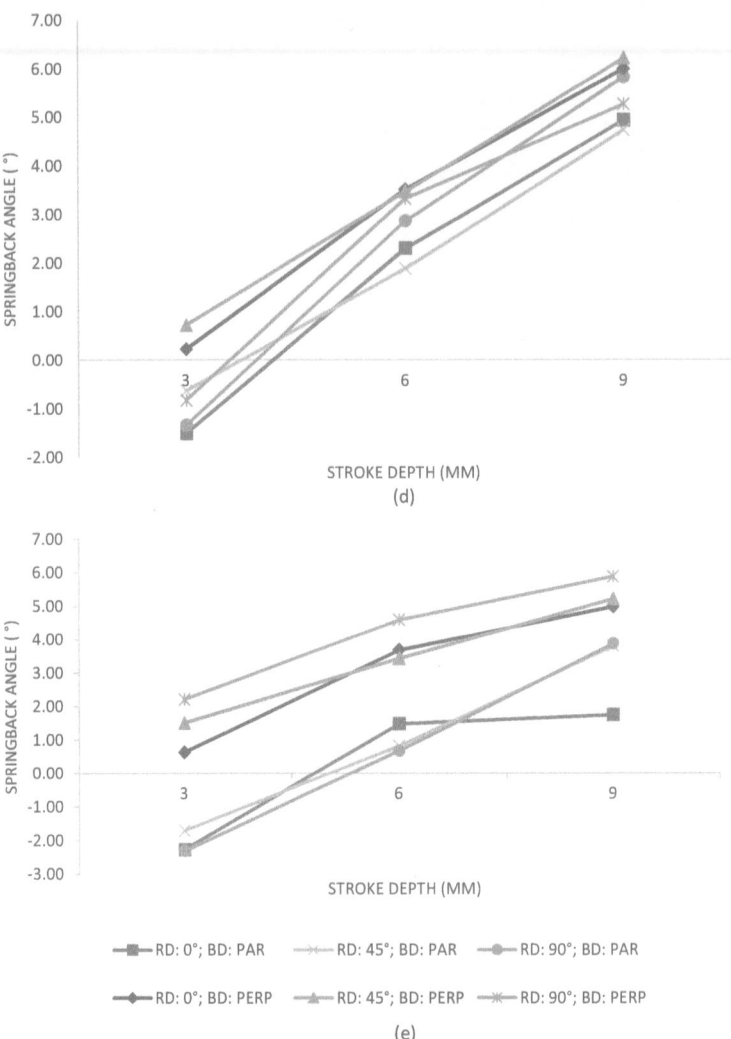

**FIGURE 8.7 *(Continued)*** The effect of rolling direction (RD) and bending direction (BD) to springback on the combination of AA6061-T6 – AA6061-T6 (a), AA5052-H32 – AA5052-H32 (b), AA1100-H14 – AA1100-H14 (c), AA6061-T6 – AA5052-H32 (d), and AA6061-T6 – AA1100-H14 (e).

For similar material combinations, the highest springback value was obtained for the AA6061-T6, as shown in Figure 8.7a–c. However, low springback values were obtained when AA6061-T6 was combined with other Al grades of alloys, as shown in Figure 8.7d and e. This could be caused by the reduction in dislocation density in the weld zone during the thermal cycle. Thus, combining dissimilar Al grades will tend to reduce the amount of average springback of the welded specimen. These results are in agreement with those of Durga et al. [19]. The highest springback angle of 7.91° was obtained for the AA6061-T6 – AA6061-T6 combination, while the lowest springback angle of 4.18° was obtained for the AA1100-H14 – AA1100-H14 welded specimen.

The results also show that the springback is lower for bending parallel to the weld line as compared to bending perpendicular to the weld line. This could be due to the surface contact areas of the weld with the bending tool. For bending parallel to the weld line, the bending tool will be in contact only at the weld area that has lower strength and hardness compared to the base materials, while perpendicular bending will involve the inclusion of the base materials in the bending. The rolling direction also influences the springback response. Specimens with a 90° rolling direction, shown in Figure 8.7a, have the highest springback values as compared to 0° and 45° rolling directions for both AA6061-T6 – AA6061-T6 and AA6061-T6 – AA1100-H14 specimens, shown in Figure 8.7c. Similar observations were noted for the combination of AA6061-T6 – AA5052-H32, where the highest springback value was obtained in the perpendicular direction, with a rolling direction of 45°, as shown in Figure 8.7b. These results agree well with the findings by Mansor [20] and Ling et al., [21].

### 8.3.2.2 Effect of Retreating and Advancing Sides of Dissimilar Alloy

In friction stir welding processes, the specimen to be joined can be referred to be located on the retreating (RS) or advancing side (AS), depending on the tool rotation and travel direction. This section will discuss the springback based on the retreating and advancing sides of the joined specimen. Figure 8.8 shows the springback responses for similar alloy welds of AA6061-T6, AA5052-H32 and AA1100-H14. On the AS, the highest springback was obtained for AA6061-T6 with a springback angle of nearly 4.5°, followed by the springback of AA5052-H32 with a springback angle of around 1.8°. In contrast, higher springback values were obtained for AA1100-H14 on the RS for all three rolling directions. Generally, the AA5052-welded specimens resulted in the lowest springback angle as compared to AA6061-T6 and AA1100-H14. The differences in springback angles of the AS and RS are likely due to the result of the material flow from RS to AS during FSW that will result in different hardness values, even though the material being welded is the same.

Figure 8.9a shows that for the AA6061-T6 – AA5052-H32-welded specimens, the springback was higher on the AA6061-T6 side as compared to the AA5052-H32 side. This is due to the higher strength or hardness of the material itself. Theoretically, the harder materials will have lower springback if compared to softer materials. Similarly, for the AA6061-T6 – AA1100-H14 weld, a higher springback was observed on the AA1100-H14 side, which corresponded to the lower hardness value of the material. In general, the higher springback values were obtained on the AS as compared to the RS, for all strokes and rolling directions evaluated. This can be attributed to the HAZ that would usually soften the material at the AS, as compared to the RS which is more solid.

### 8.3.2.3 Effect of Bending Location of Similar Alloys

The effect of bending location was evaluated by moving the point of bending away along a 5 mm distance from the centre of the weld line on both the RS and AS. The resultant springback angle values at 5 mm away from the centre of the weld lines are shown in Figure 8.10. The distance was chosen based on the tool size and the welded area. In general, it was seen that a higher springback is observed in the AS

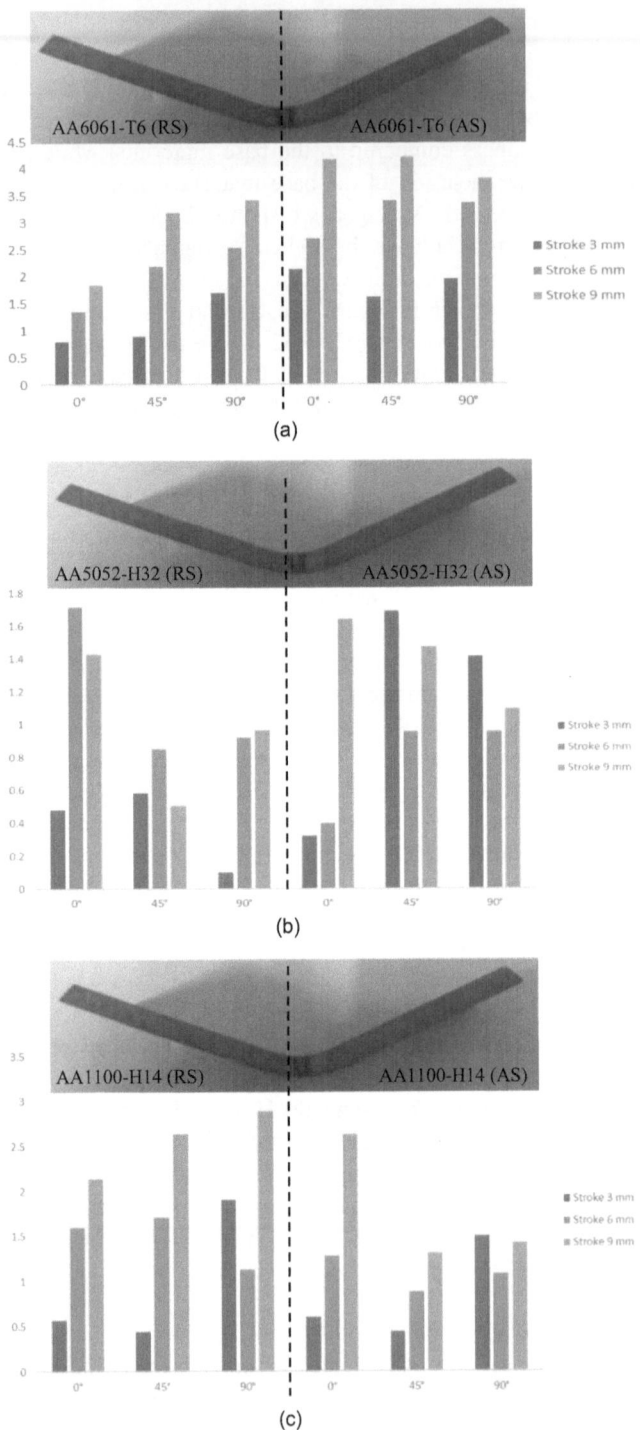

**FIGURE 8.8** Effect of springback on the centre of the weld line for all combinations and RD of AA6061-T6 – AA6061-T6 (a), AA5052-H32 – AA5052-H32 (b), and AA1100-H14 – AA1100-H14 (c).

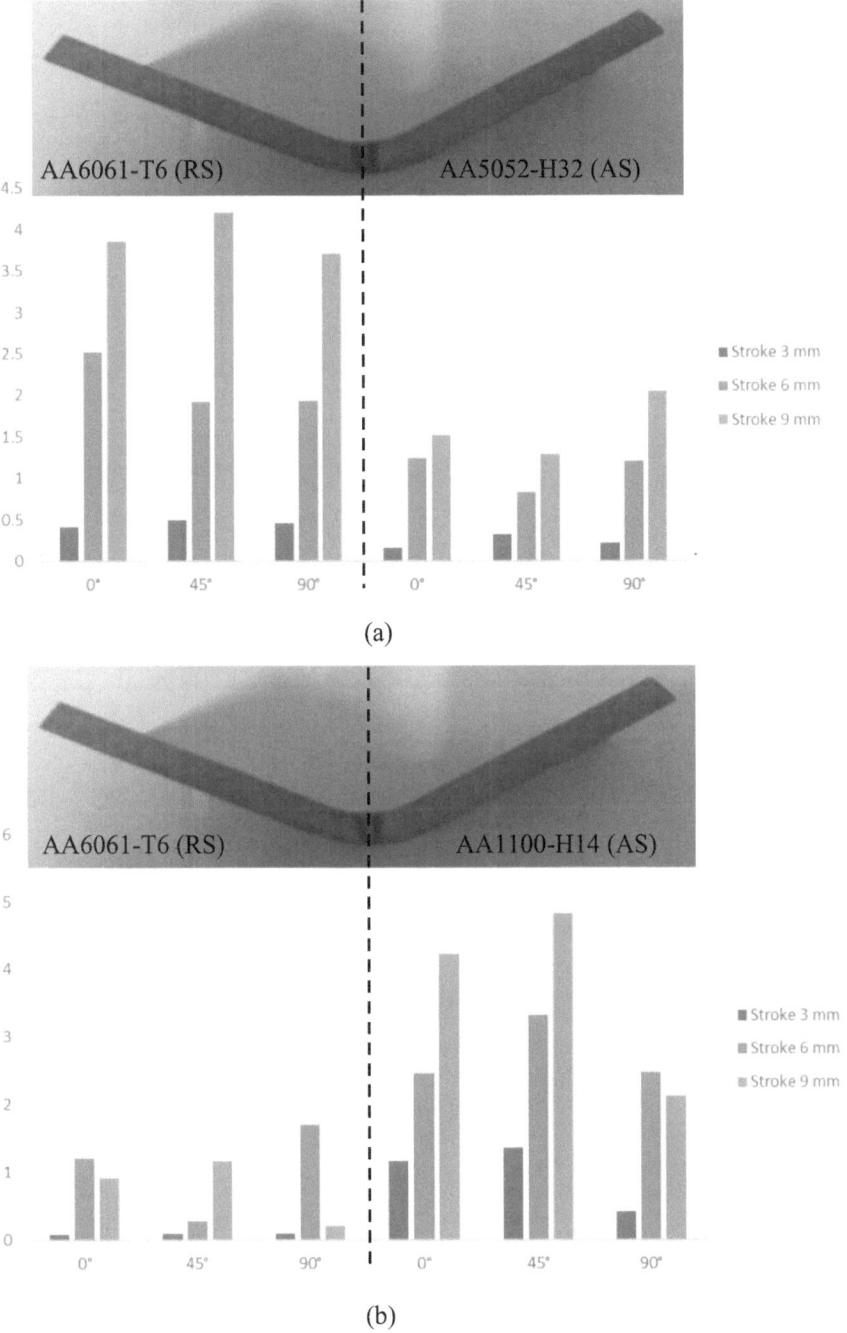

**FIGURE 8.9** Effect of springback on the centre of the weld line for all combinations and RD: (a) AA6061-T6 – AA5052-H32 and (b) AA6061-T6 – AA1100-H14.

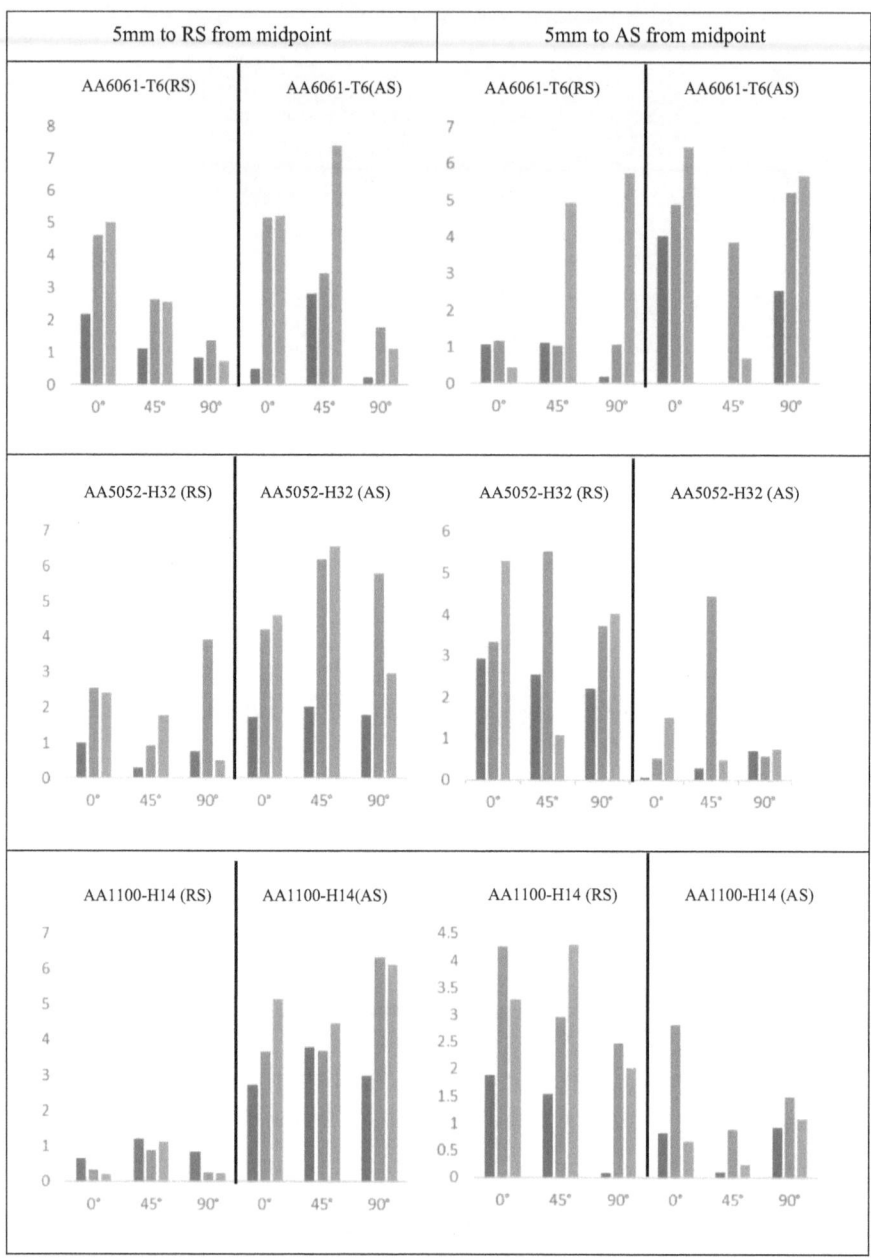

**FIGURE 8.10** Effect of springback on different bending locations of the same grades of aluminium alloy.

for the AA6061-T6-welded specimen, which is similar to the results of the bending at the weld line. This could be because the bending at 5 mm was still on the weldment area where both of the materials were mixed. However, for the AA5052-H32 and AA1100-H14 welds, a higher springback value was observed on the material side of the bending location.

## 8.4 CONCLUSIONS AND FUTURE WORK

In this chapter, the springback patterns of FSR blanks of aluminium alloys of similar and dissimilar grades were evaluated. The effect of rolling directions, strokes, bending location and bending direction with respect to the weld line was determined experimentally. The following conclusions from the study can be drawn:

i. FSW soften the joined materials, therefore reducing the springback. The combinations of similar AA6061-T6 resulted in the highest springback value, while the combination of AA6061-T6 – AA5052-H32 gave the lowest springback.
ii. The rolling direction has an effect on the springback. High springback values were obtained for both similar (AA6061-T6 – AA6061-T6) and dissimilar (AA6061-T6 – AA1100-H14) welds at the 90° rolling direction. However, for the combination AA6061-T6 – AA5051-H32, the highest springback was obtained for the 45° rolling direction.
iii. The location of bending in the AS resulted in a higher springback as compared to the bending in the RS, as the FSW tends to soften the material in the AS.
iv. Bending parallel to the weld line resulted in lower springback as compared to bending perpendicular to the weld line.

In the study, it was noted that the joining of AA5052-H32 – AA1100-H14 was not successful due to surface tunnel defects, as shown in Figure 8.11. This type of defect results in weak joints, which may be related to the mismatch in the welding speed, tool rotation, and tool pin size. Surface tunnel defects can also be caused by insufficient tool shoulder contacts with the base material. Further study can be conducted to determine the most suitable processing parameters to join these two Al grades together.

Future studies can also explore the FSW of non-uniform thickness combination, which can further improve weight reduction and optimized component properties. Subsequently, the determination of springback properties of non-uniform TWB produced by FSW can be explored.

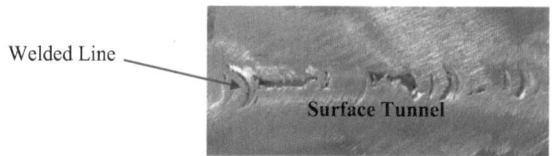

**FIGURE 8.11** One of the failure patterns found on welded AA5052-H32 – AA1100-H14 combination blanks.

## ACKNOWLEDGEMENT

This work was partly supported by the Universiti Sains Malaysia RUI Grant (Grant No. 1001/PMEKANIK/8014031).

## REFERENCES

1. S. Shanavas and J. E. Raja Dhas, "Weldability of AA 5052 H32 aluminium alloy by TIG welding and FSW process: A comparative study," *IOP Conf. Ser. Mater. Sci. Eng.*, vol. 247, no. 1, 2017.
2. S. Shanavas, J. Edwin Raja Dhas, and N. Murugan, "Weldability of marine grade AA 5052 aluminum alloy by underwater friction stir welding," *Int. J. Adv. Manuf. Technol.*, vol. 95, no. 9–12, pp. 4535–4546, 2018.
3. T. V. Christy, N. Murugan, and S. Kumar, "A comparative study on the microstructures and mechanical properties of Al 6061 alloy and the MMC Al 6061/TiB$_2$/12$_p$," *J. Miner. Mater. Charact. Eng.*, vol. 9, no. 1, pp. 57–65, 2010.
4. K. S. Kumar, "Tensile strength and hardness test on friction stir welded aluminium 6061-T6 and 5083-H111-O alloys," *Int. J. Sci. Dev. Res. (IJSDR)*, vol. 2, no. 1, pp. 88–93, 2017.
5. R. Kumar, S. Varghese, and M. Sivapragash, "A comparative study of the mechanical properties of single and double sided friction stir welded aluminium joints," *Procedia Eng.*, vol. 38, pp. 3951–3961, 2012.
6. J. Hirsch, "Aluminium in innovative light-weight car design," *Mater. Trans.*, vol. 52, no. 5, pp. 818–824, 2011.
7. H. Lin, Y. Wu, S. Liu, and X. Zhou, "Effect of cooling conditions on microstructure and mechanical properties of friction stir welded 7055 aluminium alloy joints," *Mater. Charact.*, vol. 141, pp. 74–85, 2018.
8. N. Balaji, S. Kannan, and S. Arun, "Performance analysis of friction stir welding on aluminium Aa7075 and Aa2024 alloy material," *Int. J. Eng. Res. Adv. Technol.*, vol. 3, no. 4, pp. 10–16, 2017.
9. X. Liu, S. Lan, and J. Ni, "Analysis of process parameters effects on friction stir welding of dissimilar aluminum alloy to advanced high strength steel," *Mater. Des.*, vol. 59, pp. 50–62, 2014.
10. R. Hassan, G. Hassan, and B. Sudarsanam. "Effects of heat affected zone softening extent on strength of advanced high strength steels resistance spot weld," *10th International Conference on Trends in Welding Research*, Tokyo, 2016.
11. K. G. Krishna, A. Devaraju, and B. Manichandra, "Study on mechanical propreties of friction stir welded dissimilar AA2024 and AA7075 aluminum alloy joints," *Int. J. Nanotechnol. Appl.*, vol. 11, no. 3, pp. 285–291, 2017.
12. M. Ilangovan, S. Rajendra Boopathy, and V. Balasubramanian, "Effect of tool pin profile on microstructure and tensile properties of friction stir welded dissimilar AA 6061–AA 5086 aluminium alloy joints," *Def. Technol.*, vol. 11, no. 2, pp. 174–184, 2015.
13. J. M. Piccini and H. G. Svoboda, "Effect of the Tool Penetration Depth in Friction Stir Spot Welding (FSSW) of Dissimilar Aluminum Alloys," *Procedia Mater. Sci.*, vol. 8, pp. 868–877, 2015.
14. A. Barbini, J. Carstensen, and J. F. dos Santos, "Influence of a non-rotating shoulder on heat generation, microstructure and mechanical properties of dissimilar AA2024/AA7050 FSW joints," *J. Mater. Sci. Technol.*, vol. 34, no. 1, pp. 119–127, 2018.
15. R. I. Rodriguez, J. B. Jordon, P. G. Allison, T. Rushing, and L. Garcia, "Microstructure and mechanical properties of dissimilar friction stir welding of 6061-to-7050 aluminum alloys," *Mater. Des.*, vol. 83, pp. 60–65, 2015.

16. M. Koilraj, V. Sundareswaran, S. Vijayan, and S. R. Koteswara Rao, "Friction stir welding of dissimilar aluminum alloys AA2219 to AA5083 - Optimization of process parameters using Taguchi technique," *Mater. Des.*, vol. 42, pp. 1–7, 2012.

17. M. M. Hasan, M. Ishak, and M. R. M. Rejab, "Influence of machine variables and tool profile on the tensile strength of dissimilar AA7075-AA6061 friction stir welds," *Int. J. Adv. Manuf. Technol.*, vol. 90, no. 9–12, pp. 2605–2615, 2017.

18. L. V. Kamble, S. N. Soman, and P. K. Brahmankar, "Understanding the fixture design for friction stir welding research experiments," *Mater. Today Proc.*, vol. 4, no. 2, pp. 1277–1284, 2017.

19. B. Durga Rao and R. Ganesh Narayanan, "Springback of friction stir welded sheets made of aluminium grades during V-bending: An experimental study," *ISRN Mech. Eng.*, vol. 2014, pp. 1–15, 2014.

20. K. K. Mansor, "Study the optimization parameters for spring back phenomena in U-die bending," *Eng. Tech. J.*, vol. 31, no. 9, pp. 1705–1718, 2013.

21. J. S. Ling, A. B. Abdullah, and Z. Samad, "Application of Taguchi method for predicting springback in V-bending of aluminum alloy AA5052 strip," *J. Sci. Res. Dev.*, vol. 3, no. 7, pp. 91–97, 2016.

# 9 Springback Simulation of Tailor Welded Blank and The Compensation by Using Displacement Adjustment and Spring Forward Methods

*AD Anggono and WA Siswanto*
Muhammadiyah University of Surakarta

*B Omar*
Universiti Tun Hussein Onn Malaysia

*RI Riza*
Muhammadiyah University of Surakarta

## CONTENTS

DOI: 10.1201/9781003164241-9

## 9.1 INTRODUCTION

The weights of cars have risen significantly in recent decades due to the increased additions of electronics, safety, and comfort standards. An increase in weight results in added fuel consumption. To compensate for the increased weights of the extra components, the structure of the car, especially the body-in-white, must be made lighter. One such solution is to utilize bespoke blanks to strengthen the body of the car in locations where more strength or stiffness is required.

Tailor welded blanks (TWBs) are semi-finished sheet items that are made by welding metal sheets of identical or varying thicknesses to form a larger blank. A single piece of a TWB can be made up of different plate thicknesses, sheet material, coating, or substance qualities. Due to these alterations, the updated blanks are now more equipped for the subsequent forming process or final use. As technology progresses, thinner patches are joined by welding or adhesively bonded on top of the main sheet to provide local reinforcement. These patchwork blanks result in process integration because the blank and reinforcing patch is made in the same tool. In addition, rolling techniques have also been used to generate metal sheets with gradual changes in thickness. The fundamental benefit of a TWB is that it allows certain qualities at specific portions of the blank, which saves weight and money. The use of low-density or high-strength materials, such as aluminium and high-strength steel can also help conserve weight, although they are restricted in terms of formability. Nevertheless, the forming behaviour of these materials can be improved with local heat treatment, allowing them to be used in the production of the body-in-white in the form of tailor heat-treated blanks. Although most of the applications of TWBs are in the automotive sector, they can also be utilized in other domains where weight optimization is important, such as in the aerospace industry.

However, there are some drawbacks to TWBs. For example, the extra step of welding multiple flat sheets in TWB's production process might compromise the blank's formability by generating martensitic structures. This research will focus on the drawing ability of TWB before proceeding to the deep-drawing approach. The drawing ability of sheet metal can be utilized to assess its stamping process flexibility.

## 9.2 TAILOR WELDED BLANKS

TWBs are semi-finished products comprising at least two single panels welded to one another before being shaped. Weld joints are divided into two types: linear and nonlinear joints. For engineered blanks, the weld seams are typically not linear. These sheets can come in a variety of mechanical grades, thicknesses, and coatings. Usually, laser welding or mash seam welding is used to join these blanks, although technologies such as high-frequency, friction stir, electron beam, or induction welding are also theoretically possible, and are currently being explored.

The mash seam welding process requires less blank cutting accuracy and can utilize high weld velocities (Miyazaki et al., 2007). On the other hand, laser welding produces a narrow weld seam with a restricted heat-affected zone (HAZ), and weight savings, due to the absence of material overlap. Laser welding also allows nonlinear paths, such as curved welding. New welding techniques, such as friction stir welding, allow the joining of 'unweldable' materials such as aluminium because the parts do not melt and harden during the welding process (Buffa et al., 2007).

Using customized welded blanks enables the adaptation to regionally diverse loading circumstances or other engineering needs of the component (Kinsey et al., 2000). A continuous weld line, rather than discrete weld sites, increases structural rigidity and crash performance. Sheet joining before forming has several advantages, including a reduction in the number of necessary forming tools, increased forming precision, and greater material utilization, all of which result in lower manufacturing costs. The weight reduction of items created from custom-welded blanks is the most significant gain as compared to regular products. However, custom-welded blanks will have different properties of formability, failure, and mechanical and microstructural characteristics as compared to their base metal counterparts. Past studies have evaluated these differences such as the evaluation of tensile samples with decreasing gauge range by Zadpoor et al. (2007), or the testing of tensile specimens that spanned the HAZ by Lechler et al. (2010).

Mechanical parameters such as strength and ductility, the breadth of the HAZ, and formability, are all influenced by the sample, welding process, weld line direction, thickness proportion of the blanks, and percentage of the weld material in the cross-section of the sample. For example, in the welding of aluminium, the alteration of the alloy composition reduces the strength of the material. The temperature contrast created by the welding process would substitute the wrought structure with a cast system with large equiaxed grains at the weld line and columnar grains in the nearby areas. Although the chemical structure of the alloying elements is preserved, the alloying elements may be distributed more irregularly than in a monolithic sheet. Moreover, the welding process alters the form and size of the second particles, resulting in the appearance of pores in the material. Thus, due to all these factors, a TWB may have variations in the mechanical values as compared to the monolithic phase. For aluminium, the yield strength of non-heat-treatable aluminium is only slightly reduced owing to healing effects, but the yield strength of heat-treatable alloys can be considerably reduced due to averaging. Metals, especially welded blanks, would harden as the carbon content of the material increases. The particle sizes in the weld are smaller than those in the monolithic sheet, although the welding process produces a cast structure. As an outcome, the strength of interstitial-free steel weld lines increases threefold. The hardness of the HAZ is comparable to that of the initial sheet and weld line. Yield strength and ultimate tensile strength both increase when total elongation is lowered.

The formability of welded sheets can be evaluated using stretch flanging experiments, deep-drawing experiments, and spherical punch stretch forming investigations with a strain state that approximates plane strains. Blending the characteristics of the welded joint and base materials can result in a solitary forming limit curve. Nevertheless, for more precision, it is desirable to depict them as separate curves. The rule of mixing can also be used to determine weld seam characteristics. TWBs are less formable than normal blanks if there is a large number of weld reinforcements. The most critical criteria in the formability of customized welded blanks are the thickness proportionality and the orientation of the weld line in relation to the loading direction (Khan et al., 2014).

Kusuda et al. (1997) estimated the loss in formability due to the thickness ratio and weld line orientation for low-carbon steel. In a spherical punch stretch forming experiment, it was found that a 30% drop in forming height was obtained when a sample made from the same thickness blanks was stretched along the weld line. This number did not vary much regardless of the thickness ratio. However, when loaded perpendicular to the weld line, the reduction in forming height was only 10% for similar blank thickness, which dramatically lowered when the thickness ratio increased. Optimum formability was obtained when the weld line was at 45° to the stretching direction. In stretch flanging experiments, Kusuda et al. (1997) found a 20% reduction in forming height, with the reduction being substantially larger when the thickness percentage was <1.

In most circumstances, the weld seam has less of an influence on formability than the thickness gradient's morphological discontinuities. Since non-uniform displacement rises as the thickness ratio grows, the formability would decrease. Deviations and porosity in the weld line produce non-uniform strain gradients. Imperfections and permeability in the weld line contribute to non-uniform strain gradients. The thicker and thinner blanks will be subjected to the same main strain if the major strain treatment is applied to the weld line. As an outcome, the blanks have the same stress levels, causing greater power to be applied to the thicker sheet. The thick and thin blanks, and the joint line, are all exposed to the same force once the major strain is parallel to the weld line. Weld line migration is considerable because of irregular strain transfer, higher stressing of the thinner material and severe weld line migration. The formability of the dissimilar sheet in deep drawing resembles that of the monolithic sheet in deep drawing when cold-rolled steel with a small carbon concentration is employed and two blanks of comparable thickness are welded around each other. The restricted drawing proportion is the difference between both the thin and thick blank quantities if the thickness ratio is not 1.

The weaker material changes shape more than the stronger material in the punch face, enabling the weld seam to travel towards the thicker blank, resulting in ripping. The inferior material in the side wall and flange is prone to wrinkling because of compressive stress. The weld seam would be displaced from the weaker/thinner section to the side having greater strength. This issue impacts any blank with a non-homogeneous thickness or material qualities. As a result, the weld seam should be placed in less deformed areas. In addition, draw beads or customized blank holders could be used to regulate the flow of material. As shown in Figure 9.1, adaptation can be accomplished by segmenting the welded blank depending on thickness differences or by applying variable blank holding stresses locally.

The earliest investigations on regulating material flow using different blank holder forces were undertaken by Ahmetoglu et al. (1995). Kinsey and Cao (2003) established a method for predicting the required forces in the segments of a blank holder by using quantitative analyses of weld line movement in a deep-drawing operation. The formability of customized welded blanks manufactured from aluminium alloy 5182-H00 was also tested using this method. A 22% improvement in drawing depth was obtained using a fragmented die and flexible controllers. Deep drawing using draw beads and varied blank thickness proportions was examined in simulated and experimental studies by Kim et al. (2000) and Heo et al. (2001).

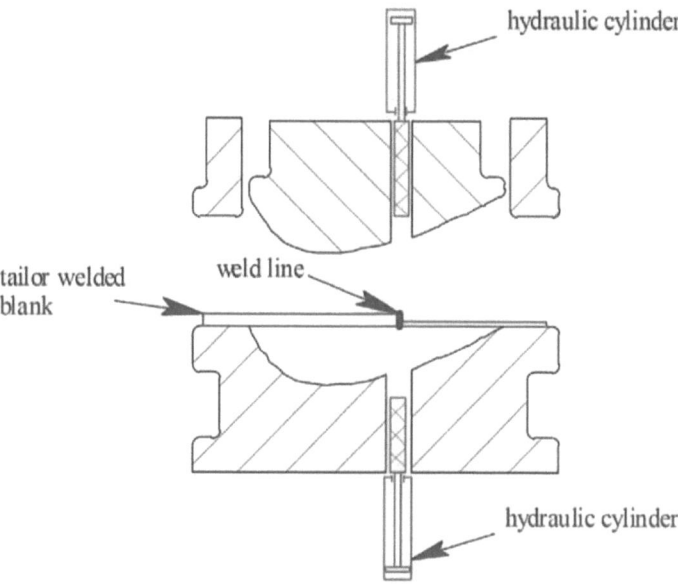

**FIGURE 9.1**    Segmented die with adaptable controllers (see Kinsey et al., 2001).

There are two failure situations as an outcome of the changed blank's weld line orientation (Zadpoor et al., 2007). When the loading direction is parallel to the weld line, the fracture starts at the weld line and spreads to other locations. This is due to the weld line's poorer formability. Failure occurs in the weaker (i.e., thinner) element when straining occurs parallel to the weld line (Azuma et al., 1988). As a result, the strain is concentrated in the thinner material.

Panda and Kumar (2008) investigated custom-welded interstitial clean steel blanks of various thicknesses and finishes (galvanized vs. ungalvanized). Evaluations were made on the tensile qualities measured longitudinally and transversely to the weld line, as well as the weld quality and formability, in a biaxial stretch forming investigation. As a result of bainite formation and internal stresses, the hardness of the weld region increases. The transverse welded samples fracture in the thinner, uncoated material, while the longitudinally welded samples were more durable than base metal samples and have less elongation. The strain is transferred to the adjacent regions because the weld line is significantly less ductile than the parent metal. Consequently, the breakdown occurred underneath the weld area in the thinner plate. Formability can be improved by providing counter pressure and lubricant on the stretched region.

Improved formability can also be obtained by increasing the welding speed used to connect the blanks. Cheng et al. (2007a) and S. M. Chan et al. (2003) evaluated the weld line orientation and effects of thickness ratio on the formability of customized welded blanks made of AISI 304 and SPCC steel. It was found that the formability of TWBs of similar thickness was similar to the formability of the parent materials. However, the formability starts to decrease as the combinations of blanks increase.

Bayraktar et al. (2008) evaluated the macro- and microstructural characteristics of laser-welded steel blanks. Impact testing was utilized to estimate the temperature for the brittle-ductile transition process at various temperatures. Gaied et al. (2009) evaluated the formability of custom-welded blanks using both experimental and computational approaches. Hill's anisotropic yield criteria were used to characterize the structural response, and the failure criterion was based on plastic instability (necking) and ultimate fracture.

Narayanan and Naik (2010) examined four distinct failure criteria and their variations in their evaluation of projected forming boundaries to experimental results for unwelded and tailor welded blanks. Chan et al. (2005) established an anisotropic damage-coupled model and a damage criterion for steel TWBs. Formation limiting stresses could be predicted, and a stamping process could be efficiently modelled using the damage model and failure parameter created in LS-DYNA.

Zadpoor et al. (2008) investigated the mechanical properties and granularity of friction stir welded customized blanks of varying thicknesses and materials. It was found that the crack is partly brittle and partly ductile, and the hardness and mechanical properties of the fracture would decrease with increasing sheet thickness, either with or without mechanical post-weld treatment. Saunders and Wagoner (1996) also evaluated the forming ability and mechanical parameters of laser-welded and mash seam welded blanks and subsequently created a model to forecast the deformation pattern and the motion of the weld lines.

Anand et al. (2006) examined the fatigue strength and fracture processes of $CO_2$-laser-welded steel tailored blanks using a tension-tension fatigue cycle. The morphology of the weld line was examined to determine the quality of the weld as a material's fatigue strength is influenced by its surface features. It was found that the weld line has an increased fatigue strength than the base material; thus, the fracture tends to start in the thinner section of the blank, which is further amplified by the presence of porosities and inclusions.

Customized blanks usually have stronger notch effects due to the mixture of mash seam welding and high thickness ratio percentage. As a result, the fatigue strength is reduced, and the fracture tends to be nearer to the weld line. Clapham et al. (2004) evaluated residual stresses in laser-welded modified blanks using neutron diffraction. After analysing the samples in their as-welded form and applying uniaxial force, it was discovered that residual stresses remained constant or even decreases with further distortion. Thus, residual tensions in laser-welded blanks during forming are not a cause for concern. Weld quality can be determined using magnetic flux leakage investigation, as used by O'Connor et al. (2002), in identifying holes and concavities measuring as small as 0.34 mm in length. Thus, this testing method can be used for checking the weld lines of custom-welded blanks used in the automotive industry.

## 9.3   ALUMINIUM TWBs

Customized tailor welded aluminium alloy blanks could further decrease body weights in automobile applications. However, aluminium is hard to weld due to its high reflectivity, low molten viscosity, and the existence of oxide layers. Welding causes the loss of alloying components, porosity, hot cracking in the contact region, and a decrease in strength. Shakeri et al. (2002) investigated the deterioration of non-vacuum electron beam welded and Nd: YAG laser-welded blanks. Both welding processes caused transverse loading failure in the weld line and the formability decreased as the thickness ratio increased.

Friedman and Kridli (2000) investigated the microstructure and mechanical properties of blanks with varying thickness ratios. Since the cross-section at the weld site was lowered, blanks made from the same thickness panels crumpled at the weld line. In tailored blanks of various thicknesses, a transition zone is created between the two sheets, allowing them to withstand higher tensile distortion.

In another study, Buffa et al. (2007) investigated the mechanical properties and formability of aluminium TWBs produced by laser welding and friction stir welding. Raw items and the tailored blank were subjected to tensile and cup deep-drawing tests with the rolling direction parallel to the joint line. While friction stir welding for welding aluminium blanks outperforms other procedures by a slight margin, it still has limited formability. As a result, it is recommended that the customized welded blank be laser heat processed to improve formability during the forming process. Bayley and Pilkey (2005) looked at the effect of weld flaws on the formability of aluminium TWBs, with a focus on ductility and shear localized fracture.

The direction of the weld line to the path of considerable strain affects the forming and collapsing characteristics of custom-welded steel blanks. When the weld line is perpendicular to the loading direction, weld quality, weld strength, and sheet thickness ratio affect the forming reaction and malfunction. However, when a significant strain is applied to a weld line, the ductility is determined by flaws at the interface as well as faults inside the weld line.

Topography studies (external faults) and X-ray micro-focus radiography can be used to uncover these issues (internal defects). Internal weld faults, such as pores, were discovered to have a far higher influence on formability than surface flaws. The findings of the trials were incorporated into a microscopic failure model that predicted when localization would begin. Davies et al. (2000) investigated the ductility of weld materials in customized welded blanks made of 5000 series aluminium alloys using tensile testing and subsequently proposed the limit diagrams for formability. Subsequently, Davies et al. (2002) studied the mechanical properties of aluminium TWBs at superplastic temperatures of 500°C–550°C and found that the weld material exhibited a higher strain rate sensitivity, higher flow stress, and decreased ductility.

Chien et al. (2003) investigated the fracture of laser-welded aluminium blanks using tensile and equibiaxial tests. The forming limit was determined using the Marciniak–Kuczynski defect technique. The strain was identified utilizing the weld profile's small notches. It was found that the forming limits of the plane strain and biaxial tension were lower than those of uniaxial tension.

## 9.4   TWB NUMERICAL SIMULATION

Lamprecht and Merklein (2004) investigated the mechanical properties and width of the welded joint in laser TWBs using tensile tests and microhardness examinations. In a custom-welded blank model, the dimensions and mechanical characteristics of the weld zone were also utilized as input factors. It was shown that simply modelling the weld seam was sufficient to be consistent with the experimental bulge testing and that no additional modelling of the HAZs outside of the weld seam was necessary. To simulate customized welded blank fracture, theoretical failure criteria or empirical studies from forming limit curve evaluations can be used (Zadpoor et al., 2007).

A finite element approach was utilized by Meinders et al. (2000) to describe a deep-drawing process for customized welded blanks. When an appropriate blank holder explanation is supplied, the findings correspond well with trials. Baptista et al. (2007) investigated the effects of sheet anisotropy caused by the rolling process on custom-welded blanks. A square cup deep-drawing method was recreated using the code DD3IMP and sheets with customizable rolling directions in the modified blank. It was suggested that an ideal combination of rolling orientations might improve the formability of TWBs since higher thinning was noticed along the cup boundaries for isotropic mild steel.

Zhao et al. (2001) built six finite element models for customized welded blank simulation using data from three-point bending tests, limiting dome height tests, and formability tests. A model with 3D shell components and disregard for the heat-impacted zone produced the best results in terms of adequate accuracy and affordable processing costs. The failure mode, buckling, and springback were all predicted accurately. In terms of consistency and practicality, Zadpoor et al. (2009) have conducted a similar study on the numerical implementation of weld specific for friction stir welded customized blanks. It was remarked that for the simulation study of friction stir welded blanks and modified blanks made of aluminium, the mechanical features of the weld seam cannot be overlooked. In friction stir welding, the weld seam is wide, and the weld zone in aluminium TWBs is weaker than the base materials. Ignoring the weld zone in restricting dome height testing could lead to an overestimation of the permitted dome heights.

In the prediction of failure and strain fields, the mechanical characteristics of the weld nugget must be addressed, which includes the attributes of the HAZ. Fratini et al. (2007) conducted computational and experimental studies to improve the mechanical quality of friction stir welded aluminium joints. In their study, a welding method was proposed for various sheet thickness ratios. Procedures, such as positioning the tool entry on the thinner blank, and setting suitable pin shoulder angles, can result in improved joint properties. The ultimate tensile strength of the welded blanks was 80% of the initial value and appeared to be unaffected by the thickness ratio. By modelling tensile testing, plane strain tests and restricted dome height tests, Raymond et al. (2004) have shown that weld properties and geometry need only to be considered if nonlinear weld seams are used or the weld lines are placed in high strain zones.

Lee et al. (1998) have explored weld line movements during the forming process of cylindrical and square cup designs. A multistep inverse finite element approach was developed to determine the initial weld line position in the blank based on the required weld line location after forming. Zhang (2007) developed a technique for improving contact forces in the production of custom-welded blanks. For the trials, limit diagram models and a response surface technique were used coupled with geographical localization. Jiang et al. (2004) investigated multi-stage techniques experimentally and numerically, with a focus on weld properties and a segmental blank holder, whereas Qiu and Chen (2007) assessed multi-stage processes experimentally and numerically, with a focus on material flow and production flaws.

TWBs are used in a variety of applications in the automotive industry, including side panels, doors, wheel arches, side members, and flooring. Inner panels are also made from customized welded blanks, resulting in better resource use and component integration. When used for exterior panels, the purpose is to increase assembly accuracy as well as the overall aesthetic. Members made from specially welded blanks increase the body's crash protection. Weight reductions would also be obtained because the number of fortifications may be kept low. Tailored tubes can also be created as a pre-product using TWBs.

Valente and colleagues (2008) have shown how bespoke welded blanks could also be utilized in hot stamping procedures. In the process, the hot-stamped object is exposed to regulated energy absorption, which prevents overstretching. This benefit is particularly advantageous in the manufacture of B-pillars. On the other hand, a material that is laser-welded to the boron-manganese steel must not harden (e.g., martensitic transition) during the hot stamping process, and its yield and ultimate tensile strengths both must be between 350 and 500 MPa. Variable cooling rates should not affect mechanical properties, and the overall elongation should be >10%.

According to tests, micro-alloyed steel (HX340 LAD + AS150) meets these requirements. The mechanical and forming qualities of the weld seam of customized welded blanks formed from this mixture are the same as the component materials. Consequently, hot-stamped B-pillars from custom-welded blanks inevitably fail at crack-sensitive locations, independent of the weld line position (e.g., flange and bending radii).

In recent years, custom-welded aluminium blanks are being considered in the automotive industry. For example, custom-welded blanks consisting of 5000- and 6000-grade Al alloys are used in Lamborghini Gallardo. The aerospace sector may also benefit from the use of tailored blanks, although welding procedures are not commonly used in high-performance aluminium alloys due to the loss of the carefully controlled microstructure. In these cases, adhesive bonding and (chemical) milling of a blank to varied thicknesses can be used instead. Custom-welded blanks offer added advantages in weight savings, even if the part is not designed for weight reduction. Pallett and Lark (2001) have also described the uses of custom-welded blanks in the construction and transportation sectors, including the use of sandwich panels in frameless building systems.

## 9.5   PATCHWORK BLANKS

Patchwork blanks are designed to partially strengthen a primary sheet by using one or more blanks (patch). In most cases, the patched blank is smaller than the original sheet and is joined together in an overlap joint configuration. Unlike normal reinforcement processes used in the automotive industry, the blank and the reinforcing patch are joined before forming. Therefore, only one shaping tool is needed, which reduces production costs. The additional benefit of patchwork blanks over normal reinforcement is the enhanced fitting precision between the two sheets because of the connection. Furthermore, even the tiniest sections of the blank may be strengthened.

Patchwork metal blanks can be joined using spot welding, laser welding and bonding methods. It is also possible to employ a mix of joining techniques. Since there is no surface linkage between the blanks with welded patchwork blanks, this might result in inhomogeneous flow behaviour. Crevice corrosion of the surface can develop when only a spot connection has been established. A bond connection, on either hand, closes the space between the blank and the patch with an adhesive bond, resulting in exceptional corrosion protection.

The bonding process between the blank has a noise and vibration dampening effect, as well as the possibility to mix various materials (e.g. steel and aluminium) without compromising mechanical or material attributes. However, surface maintenance is needed for adhesive bonding. On the other hand, laser or resistance spot welding is also used in producing patchwork blanks due to their flexibility and high-quality weld seam, with the processes able to be automated.

For the mass production of patchwork weld using resistance spot welding, the common approach is to initially weld a few spots to locate the patch on the blank, then the blank is formed, and finally, more weld spots are added to guarantee the joint's strength. In some cases, the patch blank is welded fully before forming to remove the extra post-forming welding step. Resistance spot-welded patchwork blanks are already in use for production.

Lamprecht and Geiger (2005) investigated the formability of laser-welded patchwork blanks using computational and empirical approaches. The goal was to provide a modelling approach for mending blanks in finite element analysis. A range of feed rates was selected to evaluate the effect of heat input on the plastic behaviour of the material. A $CO_2$ laser was used to create an autogenous continual fillet weld without a wire filler. Microhardness tests were used to assess the lateral diameters of the HAZ, and the plastic behaviour was investigated using tensile experiments with longitudinally welded samples reduced to the size of the HAZ's width. Low energy input into the substance is advantageous because a low laser speed results in superior weld seam strength and less rupture strain. When compared to experimental results from hydraulic bulging testing, it was determined that a numerical model that takes into consideration the various material behaviours of the weld seam produced the best outcomes.

Patchwork blank failure can be related to the blank layout and the tool geometry. The most common reason for failure is the abrupt shift in the part stiffness, which starts in the patch's interface region. Forming processes in these regions could result in localized stress concentration. Reducing the stress concentration and producing a more uniform stress distribution can be introduced by increasing the weld line geometry, reducing the patch's sheet thickness, and inserting holes or slots in the patch's border regions.

Patchwork blanks require fewer cutting procedures as an overlap joint is used as compared to custom-made butt-welded blanks. No edge preparation is required, reducing the need for alignment accuracy. This result in lower scrap and manufacturing costs. The disadvantage of patchwork blanks is that the spot-welded connections may be subjected to extremely high stresses, causing them to shatter. Additionally, the forming procedure is substantially more complicated than with homogeneous blanks because of the local variable thickness distributions.

## 9.6   TAILOR-ROLLED BLANKS

Sheets with a variety of predetermined thickness distributions can be produced by rolling. Unlike the abrupt dimensional changes in tailor welded and patchwork blanks, tailor-rolled blanks can be produced with a consistent transition from the thick and thin sections. Thus, tailor-rolled blanks have excellent formability as no stress peaks are produced due to abrupt thickness changes. Custom-rolled blanks can be made in any thickness transition, allowing for exact adaptability to the load in the application. Moreover, tailor-rolled blanks have greater surface quality since they do not have a weld seam.

A flexible rolling method may be used to achieve different thicknesses in the longitudinal direction. The roll gap is changed throughout sheet rolling to achieve the correct thickness distribution. After the rolling operation, the roll gap is altered electrically by measuring the sheet thickness. After the rolling operation, the roll gap is altered electrically by measuring the sheet thickness. The production process of flexible rolling is complex due to the online calibration of the roll gap. In addition, due to the strain hardening effect caused by the rolling process, the sheet may have regionally different properties, which can result in inhomogeneous springback behaviour in the materials.

The most cost-effective transition slope for a rolled blank has a thickness difference of 1 mm over a length of 100 mm. After rolling, custom-rolled sheets might be flexible rolled, heat-treated, coated, straightened and blanked, and stamped. The heat treatment of the tailored sheets gives them different material properties in different thickness areas. Kopp et al. (2005) employed deep-drawing tests to study the forming behaviour of tailor-rolled blanks made from steel DC04. A redesigned tool with an adaptable blank holder with a layered design was created to compensate for the height changes in the tailor-rolled blank. It was discovered that longer transition zones resulted in less wrinkling. The enhanced contact between the blank and the blank holder also reduced the wrinkles. Using the new adaptable blank holder, failure is not threatened by cracks or necking. It was also postulated that the transition's distance, direction and placement did not affect the constraining drawing proportion.

To compensate for the different springback of the thick and thin part sections, the forming die segments were locally shifted towards the punch. The number of movements is controlled by material qualities, which are approximated using flow curves for each resource and thickness value by Hirt and Dávalos-Julca (2012). A roll modification was made to maintain uniform bending radii across the length of the profile. The thick and thin portions were modified via experimentation, while the transition zone was adjusted using a nonlinear roll adjustment.

The strip profile rolling technique was invented by Kopp et al. (2005) to produce sheets having thickness variations in the longitudinal direction This method employs a unique roll mechanism to force the material to flow in a latitudinal orientation (similar to flow turning). By modifying process settings, it is possible to eliminate material movement along the longitudinal axis, which would result in flatness flaws. The material, the pass lowering, and the total decrease, as well as the strip and roll dimensions, all have an impact on material flow. The height decrease on every run is limited because the circulation stresses transverse to the rolling direction must be only elastic to avoid bulging.

The use of tiny disc shape rollers combined with minor height reductions per pass produces a restricted functional plastic zone when compared to the whole cross-section of the strip. A considerable number of rolling passes are required to prevent fractures, buckles, or bulges during the creation of wide grooves. In tensile testing, the customized rolled strips have restricted formability and inhomogeneous material flow, culminating in sample failure in the centre. Following strip profile rolling, the annealing method can be used to recover the strip's natural formability. Hirt and Dávalos-Julca (2012) demonstrated the fabrication of a customized tube with varying thicknesses using tailor-rolled strips and roll bending. A numerical model was used to compute the rolling variables required to achieve a certain role profile with the fewest roll passes possible. By symmetrically widening the groove rather than rolling it from one side to the other, the number of roll passes was minimized. To obtain the correct thickness distribution, 29 roll rounds were necessary. The concluding runs were flattening sweeps to prevent material from piling up beneath the groove. The flattening passes rolled up to expose a reverse version of the intended structure. The roll stands and strips again for the profile strip rolling approach must be exactly matched.

Before being welded and calibrated, the tailor-rolled strip is bent into a tube using several shaping techniques. Specific profiled rolls are used to increase the contact between the profiled strip and the forming rollers. The best tube design is established from a balance between optimum bending stiffness and minimal tube weight. The bending stiffness of the tube is raised by 12% by strengthening it locally. With roll forming, the modified tubes deviate from roundness twice as much as standard tubes of the same thickness. Custom-made tubes can be used as semi-finished components in the hydroforming process.

Chuang et al. (2008) combined the usage of custom-rolled blanks with an innovative multifunctional optimization technique for vehicle product creation. A multifunctional optimization examination is employed to assess the thickness and distribution of car parts.

The subsequent forming procedure for customized blanks may be more challenging than for normal blanks due to inhomogeneous thicknesses or strength distributions. Furthermore, lightweight materials such as high-strength steels and aluminium alloys have limited formability, thus it is difficult to create components with deep-drawing depth, sharp edges and small radii. Nevertheless, the forming limit and overall manufacturing process resistance can be improved by additional post-processing of the blanks. For example, the application of heat treatments on the blanks can be considered to increase the formability properties. Tailored heat-treated blanks (THTBs) on both sides to increase the formability of the items.

The review has outlined the research work undertaken on the behaviour of TWB, particularly on the qualities of various weld kinds of blanks. The analysis indicated that weld characteristics do not have a substantial impact on formability. However, different characteristics of two sheets can cause moulding problems such as poor mouldability (Zadpoor et al., 2007).

## 9.7  DISPLACEMENT ADJUSTMENT

The algorithm for the displacement adjustment (DA) approach is based on a real-world springback experiment. The idea behind this compensating method is to move the nodes in the opposite direction as the springback (Gan and Wagoner, 2004).

The blank is initially pressed with the original dies. After extracting the punch, springback occurs. To calculate the deviation, a comparison between the created portion and the reference part is conducted. The error is then added to the existing die tools for die compensation in the following step. A new set of dies will be created. A blank sheet gets deformed to the new dies in the next iteration. If the error is outside the tolerance range, another iteration will be carried out until the error is inside the tolerance range (Lingbeek et al., 2005).

## 9.8  SPRING-FORWARD METHOD

Springback is determined using the residual stress of the loaded portion multiplied by a negative factor in the spring-forward (SF) approach. To achieve the compensated shape, this inverse stress is applied to the distorted shape (Karafillis and Boyce, 1992). The SF algorithm is based on the idea that the inverse of stress corresponds to a forward deformation reaction rather than a spring backward response (Lingbeek et al., 2005).

Due to the complexity of stress distribution in the loaded shape, the SF method does not hold in the reality of springback analysis in the experiment. On the other hand, the SF approach is capable of compensating in any direction, as well as providing local compensation and has the possibility for further investigation (Cheng et al., 2007b).

## 9.9  METHODOLOGY

The material characteristics and thickness of the weld blank influence the formability of TWBs. Past studies have evaluated the effects of welding settings on the tensile and forming behaviour of TWBs based on the stress–strain curves, deep pull ability, yield points and weld lines (Bhagwan et al., 2003).

In this study, a uniaxial tensile test was selected to evaluate the deformation behaviour of TWBs with variations in the welding direction. The tensile specimen was prepared under DIN 50125H 12.5×50, as shown in Figure 9.2, obtained from

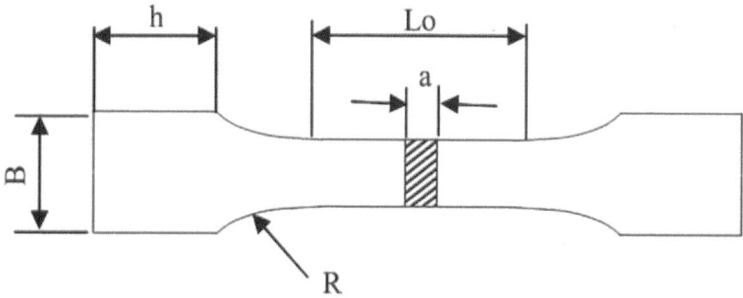

**FIGURE 9.2**  Specimen dimensions.

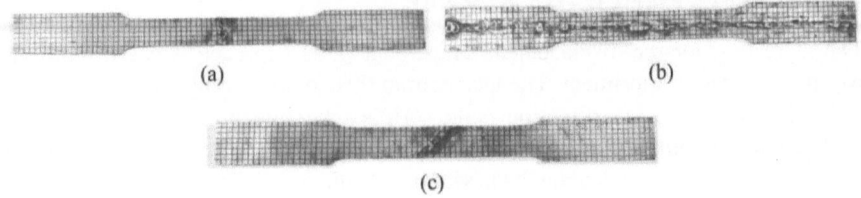

(a)                                                    (b)

(c)

**FIGURE 9.3**   Specimen with weld orientations of (a) 0°, (b) 90°, and (c) 45°.

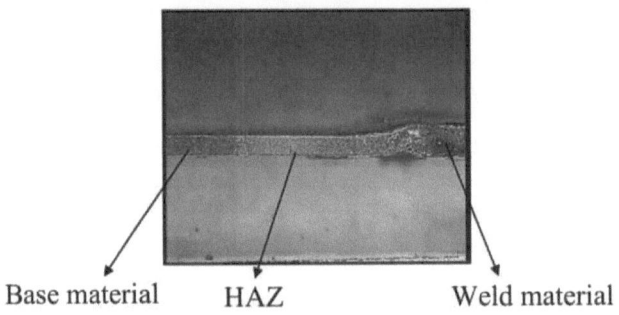

Base material        HAZ                    Weld material

**FIGURE 9.4**   Cross sectional area for the hardness test.

tailored blanks made of the same material. The dimensions of the specimens are $h = 50\,\text{mm}$, $B = 20\,\text{mm}$, $L_o = 50\,\text{mm}$, $R = 20\,\text{mm}$ and a thickness of 1 mm. Tensile tests were performed on samples with a variation in the weld line orientation, a, at 0°, 90°, and 45°, as indicated in Figure 9.3.

The weld line direction with respect to the transition region determines the shape-ability of TWBs (such as the weld and HAZs). The length and position of the transitional phase vary according to the welding process. For TWBs, the thickness differentiation and the presence of the weld existence of the weld metallic purpose local pressure and pressure distribution on the transition zone. Grids are marked on the surface of the specimen to aid in visualizing the pressure distribution during the test. Vickers hardness tests were conducted on the regions of the specimens, identified as the base material, weld zone, and the HAZ, as shown in Figure 9.4.

## 9.10   FINITE ELEMENT (FE) ANALYSIS OF TAILORED BLANK

FE simulations of the sheet metal forming process allow for the calculation of complicated material parts and the optimization of the sensitive parameter (Anggono and Siswanto, 2014). This is particularly essential in FE analyses for multi-material TWBs.

In this study, the TWB is represented as a two-dimensional surface, as shown in Figure 9.5, with denser mesh densities at the welded junctions. The core material was HX260LAD steel of 1.0 mm thickness. The input for Young's modulus, yield strength, and Poisson's ratio are 210, 176 and 0.3 MPa, respectively.

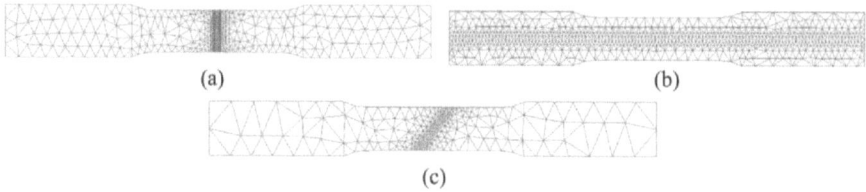

**FIGURE 9.5** Meshing of the welding zone using weld orientations of (a) 0°, (b) 90°, and (c) 45°.

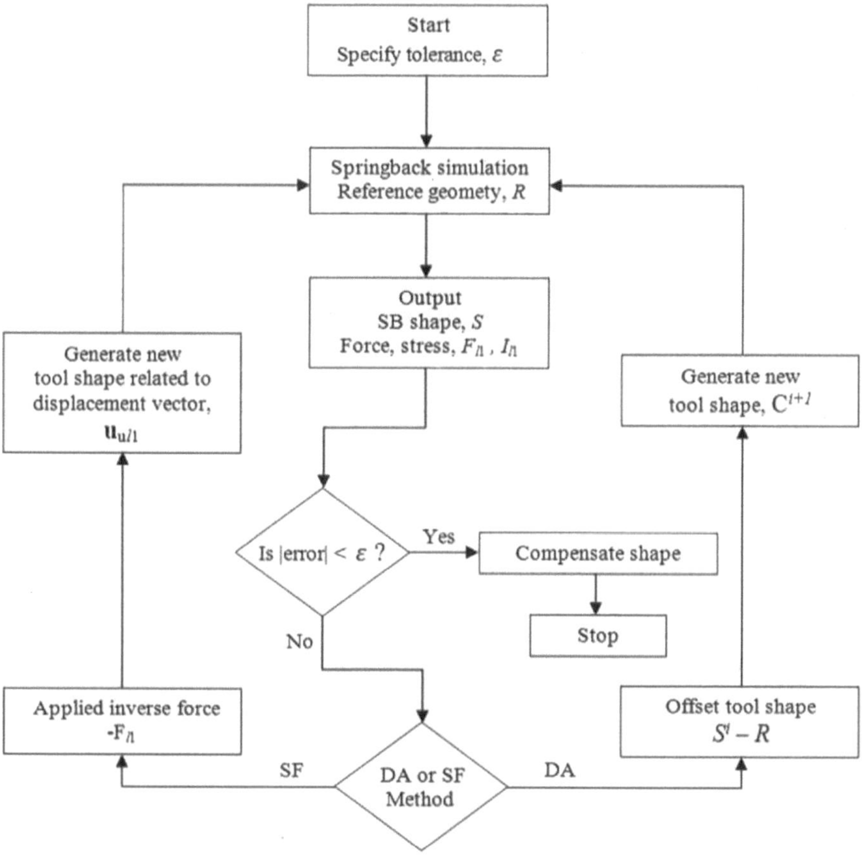

Start
Specify tolerance, $\varepsilon$

Springback simulation
Reference geometry, $R$

Output
SB shape, $S$
Force, stress, $F_{\bar\Lambda}$, $I_{\bar\Lambda}$

Generate new
tool shape related to
displacement vector,
$\mathbf{u}_{u/1}$

Generate new
tool shape, $C^{l+1}$

Is |error| $< \varepsilon$ ?    Yes → Compensate shape

No

Stop

Applied inverse force
$-F_{\bar\Lambda}$

Offset tool shape
$S^l - R$

SF      DA or SF      DA
Method

**FIGURE 9.6** Springback compensation by using DA and SF.

## 9.11 SPRINGBACK SIMULATION

The DA and SF methods are combined in this methodology as shown in Figure 9.6. The DA approach could converge quickly, but it could not adjust for the dies' parallel-to-the-punch-direction wall area. While the SF approach can adjust for springback in any direction of the die shape, due to the presence of residual stress, convergence

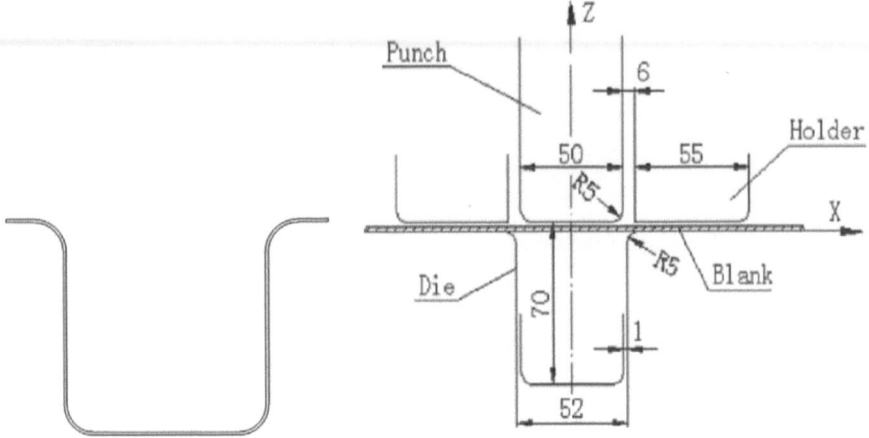

**FIGURE 9.7**    U-Bending for springback compensation test, unit in mm.

is slower and solutions are more complex. Figure 9.7 depicts the 2D simple analytical models that explain bending under tension and allow for the calculation of stress resultant and dimension variation during unloading.

Mild steel DC04 with Young's modulus of 210 GPa and Poisson's ratio of 0.3 was used as the material. An initial yield stress of 167.9 MPa, a reference stress of 550 MPa, and a work hardening exponent of 0.223 were used. The Ramberg–Osgood relationship between actual stress and logarithmic strain is believed to apply to the material.

The dividing and combining element is used to do adaptive discretization or refining. Mesh refinement ranged from coarse, standard to fine. The simulation is run in five stages all around the world. To establish contact, the blank holder is pressed onto the blank with a predefined displacement. The boundary condition is removed in the second phase and replaced by a 1kN applied force (BHF) on the binder. The punch has then moved a total of 70 mm toward the blank in the third stage. The model's nodes are all fixed in their current places in the fourth phase, and the contact pairs are eliminated. The standard set of boundary conditions is reintroduced in the final phase, and the springback is permitted to occur (Figure 9.8).

## 9.12    RESULTS AND DISCUSSION

Comparisons between the fracture test specimens of the tension test conducted experimentally and produced by FE analysis are shown in Figures 9.9 and 9.10, respectively. The data used to define the material properties for the FE were based on the results obtained from the tensile testing. The FE analysis shows similar results to the experimentation for the 0°, 90°, and 45° weld orientations. In general, fractures occur in the base materials. This is due to the high hardness of the welded zone as compared to the base material, which corresponds to higher strength and less ductility.

The orientation of the weld in relation to the direction of strain has a big impact on the strain distribution. The specimen with a longitudinal weld has the lowest strains.

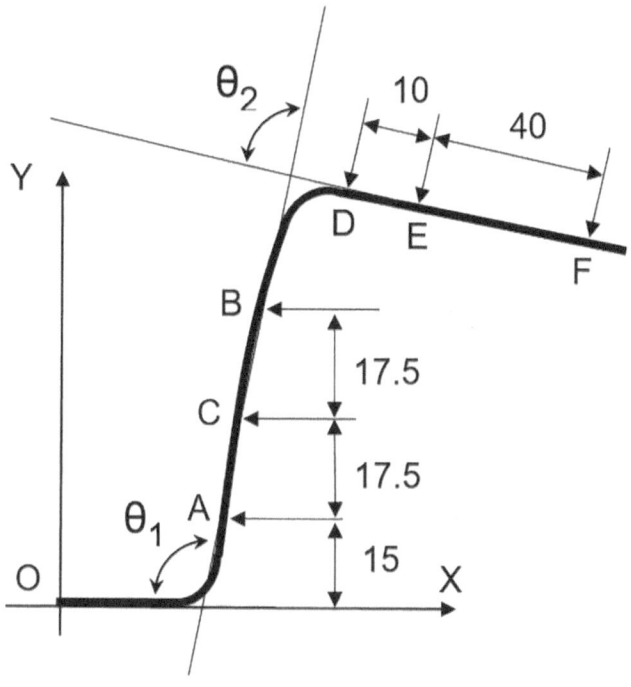

**FIGURE 9.8**   Measurement of the springback angle.

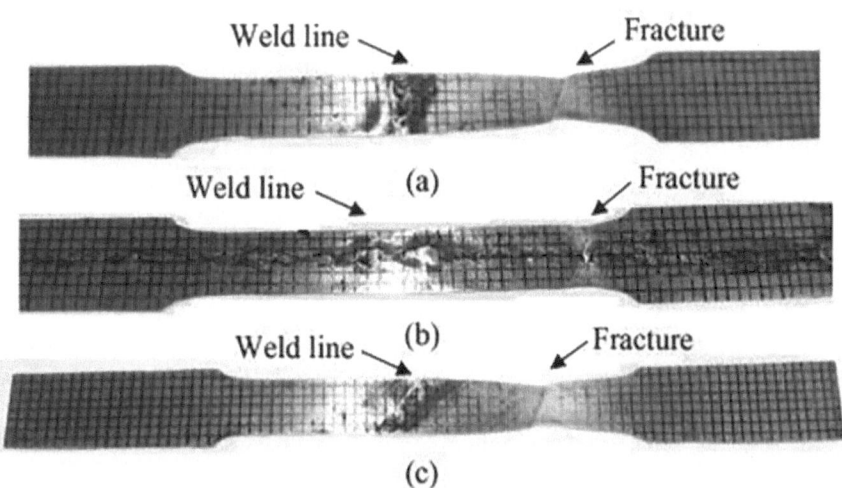

**FIGURE 9.9**   TWB experiment results on the tension test using weld orientations of (a) 0°, (b) 90°, and (c) 45°.

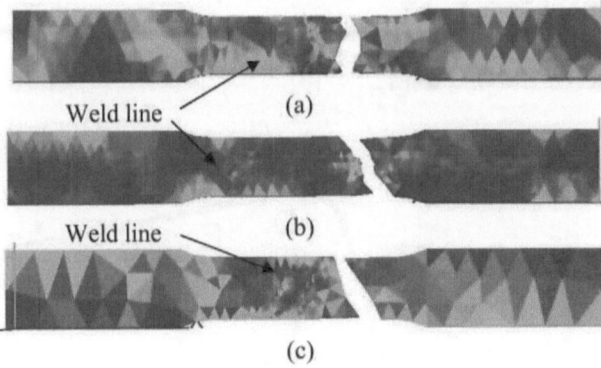

**FIGURE 9.10** TWB FE results on the tension test using weld orientation of (a) 0°, (b) 45°, and (c) 90°.

**TABLE 9.1**
**TWBs Properties**

|  | 0° | | 45° | | 90° | | Average | |
|---|---|---|---|---|---|---|---|---|
|  | **Exp.** | **FEA** | **Exp.** | **FEA** | **Exp** | **FEA** | **Exp** | **FEA** |
| UTS (MPa) | 159.8 | 168.6 | 170 | 166.5 | 208 | 210 | 179.27 | 181.7 |
| Yield stress (MPa) | 105.8 | 118.7 | 113 | 115.4 | 167.61 | 151.6 | 109.4 | 128.7 |
| Total strain | 0.56 | 0.64 | 0.43 | 0.56 | 0.33 | 0.48 | 0.44 | 0.56 |

Table 9.1 shows the results for Ultimate Tensile Strength (UTS), yield stress and total strain (elongation). UTS is the maximum stress that a material can withstand while being stretched or pulled before breaking.

Hardness can be used as an indirect method of assessing a specimen's ductility. In this study, the measured Vicker Hardness Number (VHN) values for the weld zone, HAZ, and the base material were 233.8, 119.9, and 87.76, respectively. The welded section has a higher hardness than the HAZ and the base material. A high hardness zone in the weld suggests that it is more brittle and warps less easily than other regions of the material. Zadpoor et al. (2007) remarked that the welding process would change the hardness of the weld area considerably, with increasing value in the weld zone.

Figure 9.11 shows the simulations and experimental results of the tensile tests for welding directions of 0°, 45°, and 90°. The tensile strength of the weld produced from the simulation is nearly identical to that obtained in the experiment. The results show that the strength factor increases as the welding angle increases.

Figure 9.12 shows the U-bending simulation results of the forming limit diagram (FLD) under this BHF 10 kN. Deformation redundancy is a criterion for determining whether or not cracking occurred during the forming process. The difference between deformation strain l and limiting strain k in the FLD defines it. Figure 9.12a

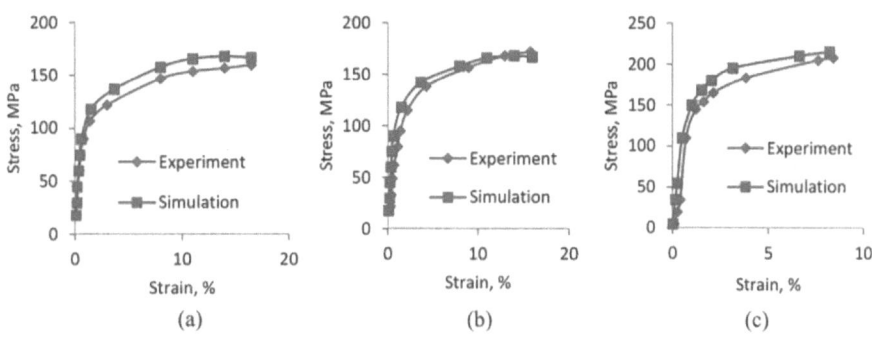

**FIGURE 9.11**   Specimen stress–strain curve with weld orientation of (a) 0°, (b) 45°, and (c) 90°.

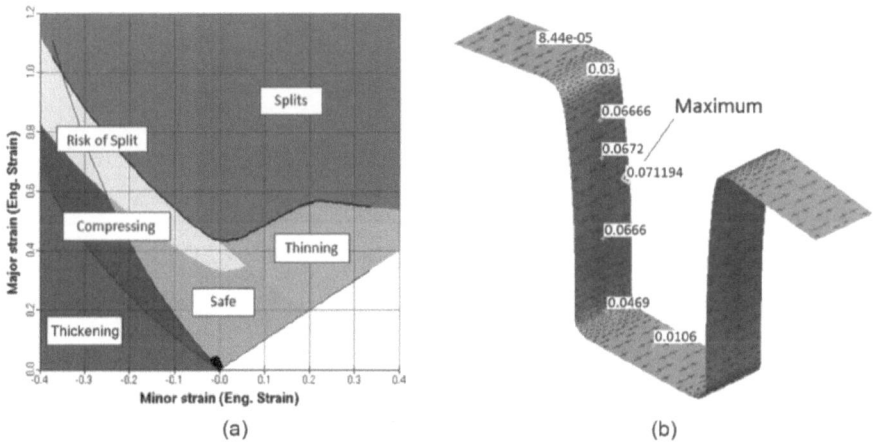

**FIGURE 9.12**   FLD (a) and major strain distribution (b).

shows that the deformation redundancy is extremely low. As a result, the procedure is in a safe zone. The plastic strain on the side wall is not uniform, as shown in Figure 9.12b, and the largest strain is in this location. The term "thinning" refers to the thickness of the sheet changing during simulation.

The effects of the BHF on springback deviation were studied using the FE technique. Table 9.2 shows the springback results, with the lowest deviation being 1.32mm under the greatest BHF of 25 kN and fine mesh type. Low springback will come from a larger BHF. Chan et al. (2003) produced a similar conclusion to determine the minimum BHF for various steels. Quality defects of products created by stamping procedures, such as fractures, wrinkles and surface distortion, can be suppressed by changing the BHF value during operation, according to Gan and Wagoner (2004). When using a higher BHF, the lowest springback was achieved.

The BHF impact was investigated further by raising the value until the forming simulation reached failure. In terms of determining the best BHF, the values

**TABLE 9.2**
**The Results of the Springback Value**

| BHF (kN) | $\theta_1$ (°) | $\theta_2$ (°) | Max. Springback (mm) |
|---|---|---|---|
| 1 | 96.84 | 82.12 | 7.27 |
| 5 | 95.98 | 81.38 | 7.23 |
| 10 | 95.67 | 81.82 | 4.75 |
| 15 | 95.7 | 81.88 | 3.81 |
| 20 | 95.7 | 81.88 | 1.98 |
| 25 | 96.07 | 82.05 | 1.32 |

**TABLE 9.3**
**The Result of the BHF Effect in Springback Analysis**

| BHF (kN) | Springback (mm) | Thinning (%) | Plastic Strain | Failure |
|---|---|---|---|---|
| 1 | 7.27 | 0.02 | 0.07 | 0.08 |
| 5 | 7.23 | 0.02 | 0.07 | 0.07 |
| 10 | 4.75 | 0.04 | 0.08 | 0.13 |
| 15 | 3.81 | 0.04 | 0.09 | 0.19 |
| 20 | 1.98 | 0.06 | 0.12 | 0.28 |
| 25 | 1.32 | 0.08 | 0.16 | 0.4 |
| 26 | 1.31 | 0.08 | 0.20 | 0.44 |
| 27 | 1.31 | 0.11 | 0.30 | 0.6 |
| 27.2 | 1.31 | 0.14 | 0.40 | 0.7 |
| 27.5 | 1.36 | 0.20 | 0.60 | 1.0 |

were 26, 27.2, and 27.5 kN. The high springback values of 1, 5, 10, and 15 kN were given by BHF, as shown in Tables 9.2 and 9.3. BHF of 27 and 27.2 kN were used to deliver the danger of failure, thinning, and high strain. Based on the FLD presented in Figure 9.13, the simulation failed under a BHF of 27.5 kN due to a high danger of splitting and failure. The ratio between the maximum major strain of an element and the major strain on the forming limit curve (FLC) for the same minor strain is described as the result of variable maximum failure. As a result, for analysis of safety and optimal BHF, 25–26 kN BHF is recommended.

The adaptive mesh was used to mesh the blank in this study. For the quick modelling of deep-drawing processes of huge and complex sheet-forming sections, spatial discretization with FEs of different sizes is critical. Small elements are used skillfully in zones with a sharply bent geometry. In terms of accuracy, the simulation allowed users to choose between rough, standard, and fine mesh. The springback outcomes are influenced by the adaptive refinement mesh type. In the FE simulation under a BHF of 10 kN, Figure 9.14 shows how element quality affects the springback outcome. Fine mesh type had the smallest springback of 4.75, 5.95 and 6.85 mm were used for rough and regular mesh, respectively.

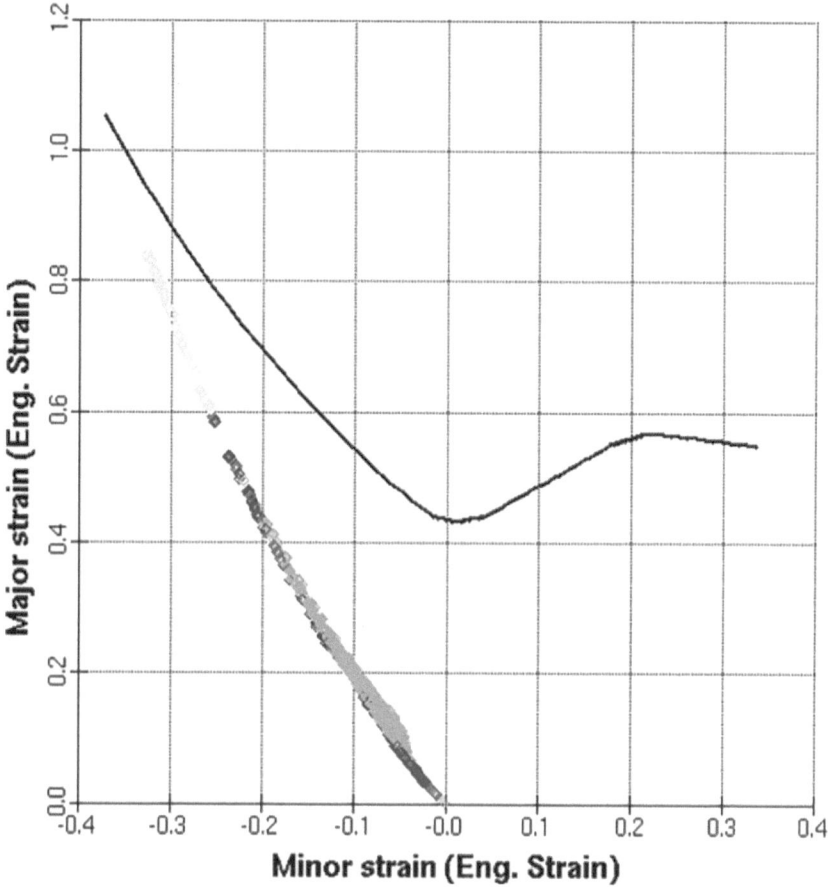

**FIGURE 9.13**    Forming simulation resulted in a BHF of 27.5 kN.

The springback findings utilized in this correction were taken from a simulation using a BHF of 20 kN, a friction coefficient of 0.3 and a finer mesh quality. Each region has its own sort of springback. The surfaces are compensated in the exact opposite direction of the springback simulation results when using the direct compensation type. The length of the transformation is determined by the compensation factor's value. The compensating factor must be bigger than zero to get the opposite direction of the springback.

The entire region as a rigid body is compensated relative to the averaged vector using rigid body motion. These surfaces maintain their original geometry while using the fixed draft type. Figure 9.15 depicts all compensation kinds with a compensation factor of 1.0, including the target, springback and compensated shapes.

Only the compensation element affects the form of direct movement. This adjustment performs translation in all directions ($x$, $y$ and $z$) using the springback shape as a reference. As illustrated in Figure 9.15a, the specified shape is compensated with the inverse vector of the springback simulation. The parameters of incremental

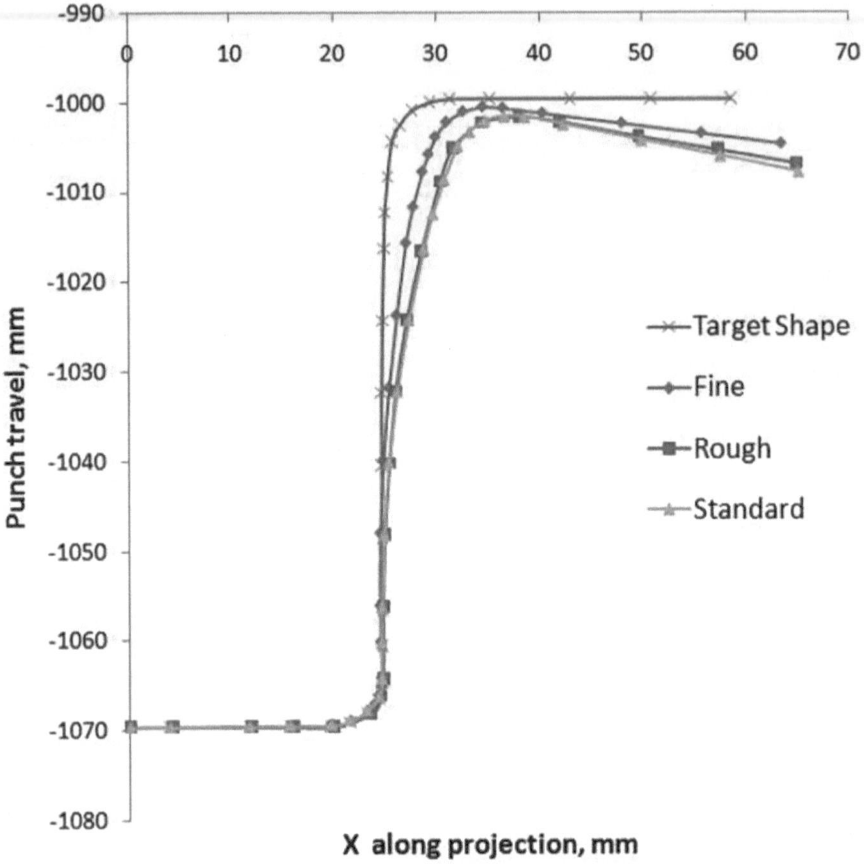

**FIGURE 9.14**   Mesh refinement's effect on the springback error under BHF 10 kN.

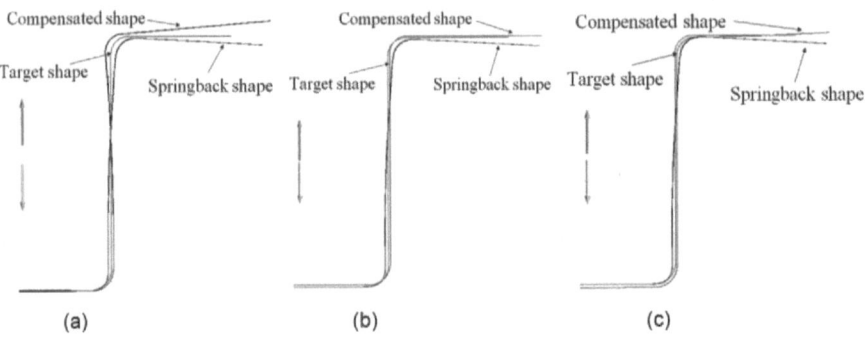

**FIGURE 9.15**   Compensation of the die surface when the compensation factor is 1.0, direct type (a), rigid body type (b), and fixed draft compensation type (c).

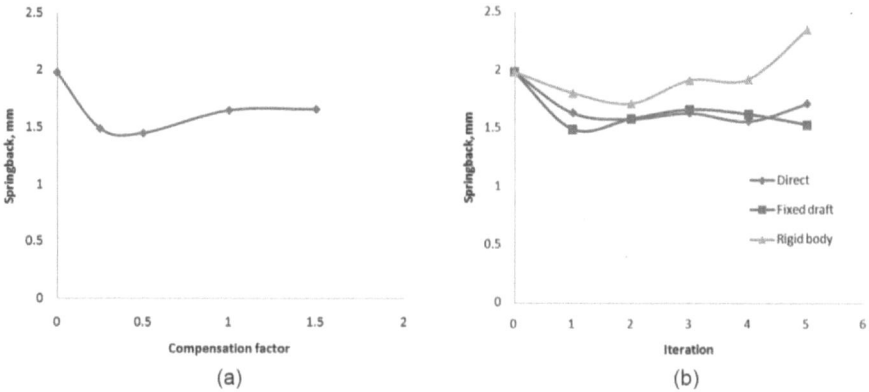

**FIGURE 9.16**   Springback results in every compensation factor (a) and iterations (b).

transformation related to the local coordinate system in the z-axis, move value 1.0, rotate value 5.0, and total transformation relative to the local coordinate system z-axis with no rotation, are described in Figure 9.15b. Figure 9.15c depicts the z-axis compensation result of a fixed draft type with compensation factor 1.0, automatic reference curve, and working orientation.

Figure 9.16a depicts the entire relationship between springback error and compensation factor. To get the lowest deviation, a modest compensation factor of 0.25 was used to correct the surface. The iteration was completed in five cycles of iterations using this value to see the trend of springback error. The iteration results for all types of compensation are shown in Figure 9.16b. The springback error decreased in the second iteration of each type but increased in the third and fifth iterations. The fixed draft type had lesser inaccuracy than the others, and it was even the lowest in the second iteration, with a maximum springback of 1.49 mm.

## 9.13   CONCLUSIONS

FEM software can be used to simulate the sheet metal forming of TWBs. In this study, the comparison between the results of simulation and experimentations was in close agreement with one another. It was seen that increasing the welding angle would increase the strength of the joint. In general, it was found that the 90° welding direction provides better strength than the 0° welding direction.

By minimizing modelling error, it is possible to get an accurate forecast of the springback calculation in sheet metal forming using FEs. One way to increase accuracy in spring back analysis is to utilize the appropriate material model. The springback error was reduced by using a larger blank holder force in sheet metal forming, however, thinning and wrinkling were more common due to the higher strain rate during forming. The magnitude of springback is determined by the bending moment, which is determined by the stress distribution through the sheet thickness. The springback was satisfactorily compensated using the DA and SF compensation algorithms. After five iterations, the springback error was shown to have decreased.

## REFERENCES

Ahmetoglu, M. A., Brouwers, D., Shulkin, L., Taupin, L., Kinzel, G. L., & Altan, T. (1995). Deep drawing of round cups from tailor-welded blanks. *Journal of Materials Processing Technology, 53*(3–4), 684–694. doi: 10.1016/0924-0136(94)01767-U.

Anand, D., Chen, D., Bhole, S. D., Andreychuk, P., & Boudreau, G. (2006). Fatigue behavior of tailor (laser)-welded blanks for automotive applications. *Materials Science and Engineering: A, 420,* 199–207. doi: 10.1016/j.msea.2006.01.075.

Anggono, A. D., & Siswanto, W. A. (2014b). Simulation of ironing process for earring reduction in sheet metal forming. *Applied Mechanics and Materials, 465–466,* 91–95. doi: 10.4028/www.scientific.net/AMM.465-466.91.

Azuma, K., Ikemoto, K., Arima, K., Sugiura, H., & Takasago, T. (1988). Sheet metals in forming processes. *Proceedings of 16th Biennial Congress,* Copenhagen, pp. 205–215.

Baptista, A. J., Oliveira, M. C., & Menezes, L. F. (2007). Effect of anisotropy on the deep-drawing of mild steel and dual-phase steel tailor-welded blanks. *Journal of Materials Processing Technology, 184*(1–3), 288–293. doi: 10.1016/j.jmatprotec.2006.11.051.

Bayley, C. J., & Pilkey, A. K. (2005). Influence of welding defects on the localization behaviour of an aluminum alloy tailor-welded blank. *Materials Science and Engineering: A, 403*(1–2), 1–10. doi: 10.1016/J.MSEA.2005.03.110.

Bayraktar, E., Kaplan, D., & Yilbas, B. S. (2008). Comparative study: Mechanical and metallurgical aspects of tailored welded blanks (TWBs). *Journal of Materials Processing Technology, 204*(1–3), 440–450. doi: 10.1016/j.jmatprotec.2007.11.088.

Bhagwan, A., Kridli, G., & Friedman, P. A. (2003). Formability improvement in aluminium tailor welded blanks via material combinations. *Proceedings of NAMRC,* Ontario.

Buffa, G., Fratini, L., Merklein, M., & Staud, D. (2007). Investigations on the mechanical properties and formability of friction stir welded tailored blanks. *Key Engineering Materials, 344,* 143–150. doi: 10.4028/www.scientific.net/KEM.344.143.

Chan, S. M., Chan, L. C., & Lee, T. C. (2003). Tailor-welded blanks of different thickness ratios effects on forming limit diagrams. *Journal of Materials Processing Technology, 132*(1–3), 95–101.

Chan, L. C., Cheng, C. H., Jie, M., & Chow, C. L. (2005). Damage-based formability analysis for TWBs. *International Journal of Damage Mechanics, 14*(1), 83–96. doi: 10.1177/1056789505045929.

Cheng, C., Chan, L., & Chow, C. (2007a). Weldment properties evaluation and formability study of tailor-welded blanks of different thickness combinations and welding orientations. *Journal of Materials Science, 42,* 5982–5990. doi: 10.1007/s10853-006-1126-0.

Cheng, H. S., Cao, J., & Xia, Z. C. (2007b). An accelerated springback compensation method. *International Journal of Mechanical Sciences, 49*(3), 267–279. doi: 10.1016/j.ijmecsci.2006.09.008.

Chien, W. Y., Pan, J., & Friedman, P. A. (2003). Failure prediction of aluminum laser-welded blanks. *International Journal of Damage Mechanics, 12*(3), 193–223. doi: 10.1177/105678950301200302.

Chuang, C. H., Yang, R. J., Li, G., Mallela, K., & Pothuraju, P. (2008). Multidisciplinary design optimization on vehicle tailor rolled blank design. *Structural and Multidisciplinary Optimization, 35*(6), 551–560. doi: 10.1007/s00158-007-0152-0.

Clapham, L., Abdullah, K., Jeswiet, J., Wild, P., & Rogge, R. (2004). Neutron diffraction residual stress mapping in same gauge and differential gauge tailor-welded blanks. *Journal of Materials Processing Technology, 148,* 177–185. doi: 10.1016/S0924-0136(03)00681-2.

Davies, R., Grant, G., Khaleel, M. A., Smith, M., & Oliver, H. (2001). Forming-limit diagrams of aluminum tailor-welded blank weld material. *Metallurgical and Materials Transactions A: Physical Metallurgy and Materials Science, 32,* 275–283. doi: 10.1007/s11661-001-0259-7.

Davies, R., Vetrano, J., Smith, M., & Pitman, S. (2002). Mechanical properties of aluminum tailor welded blanks at superplastic temperatures. *Journal of Materials Processing Technology, 128*, 38–47. doi: 10.1016/S0924-0136(02)00162-0.

Fratini, L., Buffa, G., & Shivpuri, R. (2007). Improving friction stir welding of blanks of different thicknesses. *Materials Science and Engineering: A, 459*, 209–215. doi: 10.1016/j.msea.2007.01.041.

Friedman, P. A., & Kridli, G. T. (2000). Microstructural and mechanical investigation of aluminum tailor-welded blanks. *Journal of Materials Engineering and Performance, 9*, 541–551.

Gaied, S., Roelandt, J.-M., Pinard, F., Schmit, F., & Balabane, M. (2009). Experimental and numerical assessment of Tailor-Welded Blanks formability. *Journal of Materials Processing Technology, 209*, 387–395. doi: 10.1016/j.jmatprotec.2008.02.031.

Gan, W., & Wagoner, R. H. (2004). Die design method for sheet springback. *International Journal of Mechanical Sciences, 46*(7), 1097–1113. doi: 10.1016/j.ijmecsci.2004.06.006.

Heo, Y., Choi, Y., Kim, H. Y., & Seo, D. (2001). Characteristics of weld line movements for the deep drawing with drawbeads of tailor-welded blanks. *Journal of Materials Processing Technolgy, 111*(1–3), 164–169.

Hirt, G., & Dávalos-Julca, D. H. (2012). Tailored profiles made of tailor rolled strips by roll forming: Part 1 of 2. *Steel Research International, 83*(1), 100–105. doi: 10.1002/srin.201100269.

Jiang, H., Li, S., Wu, H., & Chen, X. (2004). Numerical simulation and experimental verification in the use of tailor-welded blanks in the multi-stage stamping process. *Journal of Materials Processing Technology, 151*, 316–320. doi: 10.1016/j.jmatprotec.2004.04.294.

Karafillis, A. P., & Boyce, M. C. (1992). Tooling design in sheet metal forming using springback calculations. *International Journal Mechanical Science, 34*, 113–131.

Khan, A., Suresh, V. V. N. S., & Regalla, S. P. (2014). Effect of thickness ratio on weld line displacement in deep drawing of aluminium steel tailor welded blanks. *Procedia Materials Science, 6*(ICMPC), 401–408. doi: 10.1016/j.mspro.2014.07.051.

Kim, H., Heo, Y., Kim, N., Kim, H., & Seo, D. (2000). Forming and drawing characteristics of tailor welded sheets in a circular drawbead. *Journal of Materials Processing Technology, 105*, 294–301. doi: 10.1016/S0924-0136(00)00647-6.

Kinsey, B. L., & Cao, J. (2003). An analytical model for tailor welded blank forming. *Journal of Manufacturing Science and Engineering, 125*(2), 344–351. doi: 10.1115/1.1537261.

Kinsey, B., Liu, Z., & Cao, J. (2001). A novel forming technology for tailor-welded blanks. *Journal of Materials Processing Technology, 99*(1–3), 145–153. doi: 10.1016/S0924-0136(99)00412-4.

Kopp, R., Wiedner, C., & Meyer, A. (2005). Flexibly rolled sheet metal and its use in sheet metal forming. *Advanced Materials Research, 6–8*, 81–92. doi: 10.4028/www.scientific.net/AMR.6-8.81.

Kusuda, H., Takasago, T., & Natsumi, F. (1997). Formability of tailored blanks. *Journal of Materials Processing Technology, 71*(1), 134–140. doi: 10.1016/S0924-0136(97)00159-3.

Lamprecht, K., & Geiger, M. (2005). Experimental and numerical investigation of the formability of laser welded patchwork blanks. *Advanced Materials Research, 6–8*, 689–696. doi: 10.4028/www.scientific.net/AMR.6-8.689.

Lamprecht, K., & Merklein, M. (2004). Characterisation of mechanical properties of laser welded tailored and patchwork blanks. In: Geiger, M.; Otto, A. (eds): *Proceedings of the 4th International Conference on Laser Assisted Net Shape Engineering (LANE 2004)*, 21.–24.09.2004, Erlangen, pp. 349–358.

Lechler, J., Stoehr, T., Kuppert, A., & Merklein, M. (2010). Basic investigations on hot stamping of tailor welded blanks regarding the manufacturing of lightweight components with functionally optimized mechanical properties. *Transactions of the North American Manufacturing Research Institution of SME, 38*, 593–600.

Lee, C. H., Huh, H., Han, S. S., & Kwon, O. (1998). Optimum design of tailor welded blanks in sheet metal forming processes by inverse finite element analysis. *Metals and Materials International, 4*(3), 458–463. doi: 10.1007/BF03187809.

Lingbeek, R., Huétink, J., Ohnimus, S., Petzoldt, M., & Weiher, J. (2005). The development of a finite elements based springback compensation tool for sheet metal products. *Journal of Materials Processing Technology, 169*(1), 115–125. doi: 10.1016/j.jmatprotec.2005.04.027.

Meinders, T., Van Den Berg, A., & Huétink, J. (2000). Deep drawing simulations of tailored blanks and experimental verification. *Journal of Materials Processing Technology, 103*(1), 65–73. doi: 10.1016/S0924-0136(00)00420-9.

Miyazaki, Y., Sakiyama, T., & Kodama, S. (2007). Welding techniques for tailored blanks. *Nippon Steel Technical Report, 95*, 46–52.

Narayanan, R. G., & Naik, B. S. (2010). Assessing the validity of original and modified failure criteria to predict the forming limit of unwelded and tailor welded blanks with longitudinal weld. *Materials and Manufacturing Processes, 25*(11), 1351–1358. doi: 10.1080/10426914.2010.529588.

O'Connor, S., Clapham, L., & Wild, P. (2002). Magnetic flux leakage inspection of tailor-welded blanks. *Measurement Science and Technology, 13*(2), 157–162. doi: 10.1088/0957-0233/13/2/303.

Pallett, R. J., & Lark, R. (2001). The use of tailored blanks in the manufacture of construction components. *Journal of Materials Processing Technology, 117*, 249–254. doi: 10.1016/S0924-0136(01)01124-4.

Panda, S., & Kumar, D. (2008). Improvement in formability of tailor welded blanks by application of counter pressure in biaxial stretch forming. *Journal of Materials Processing Technology, 204*, 70–79. doi: 10.1016/j.jmatprotec.2007.10.076.

Qiu, X. G., & Chen, W. L. (2007). The study on numerical simulation of the laser tailor welded blanks stamping. *Journal of Materials Processing Technolgy, 187–188*, 128–131. doi: 10.1016/j.jmatprotec.2006.11.128.

Raymond, S., Wild, P., & Bayley, C. (2004). On modeling of the weld line in finite element analyses of tailor-welded blank forming operations. *Journal of Materials Processing Technology, 147*, 28–37. doi: 10.1016/j.jmatprotec.2003.09.005.

Shakeri, H. R., Buste, A., Worswick, M. J., Clarke, J. A., Feng, F., Jain, M., & Finn, M. (2002). Study of damage initiation and fracture in aluminum tailor welded blanks made via different welding techniques. *Journal of Light Metals, 2*, 95–110. doi: 10.1016/S1471-5317(02)00028-7.

Saunders, F.I., & Wagoner, R.H. (1996). Forming of tailor-welded blanks. *Metallurgical and Materials Transactions A, 27*, 2605–2616. https://doi.org/10.1007/BF02652354

Valente, R., Natal Jorge, R., Roque, A., Parente, M., & Fernandes, A. (2008). Simulation of dissimilar tailor-welded tubular hydroforming processes using EAS-based solid finite elements. *International Journal of Advanced Manufacturing Technology, 37*, 670–689. doi: 10.1007/s00170-007-1015-y.

Zadpoor, A., Sinke, J., & Benedictus, R. (2007). Mechanics of tailor welded blanks: An overview. *Key Engineering Materials, 344*, 373–382. doi: 10.4028/www.scientific.net/KEM.344.373.

Zadpoor, A. A., Sinke, J., Benedictus, R., & Pieters, R. (2008). Mechanical properties and microstructure of friction stir welded tailor-made blanks. *Materials Science & Engineering A, 494*(1–2), 281–290. doi: 10.1016/j.msea.2008.04.042.

Zadpoor, A. A., Sinke, J., & Benedictus, R. (2009). Finite element modeling and failure prediction of friction stir welded blanks. *Materials & Design, 30*(5), 1423–1434. doi: 10.1016/j.matdes.2008.08.018.

Zhang, J. (2007). Optimization of contact forces in tailor-welded blanks forming process. *The International Journal of Advanced Manufacturing Technology, 33*, 460–468. doi: 10.1007/s00170-006-0491-9.

Zhao, K. M., Chun, B. K., & Lee, J. K. (2001). Finite element analysis of tailor-welded blanks. *Finite Elements in Analysis and Design, 37*, 117–130. doi: 10.1016/S0168-874X(00)00026-3.

# 10 Springback Study of Aluminium and Steel Joint TWBs by FE Analysis

*MZ Rizlan, AB Abdullah, and Z Hussain*
Universiti Sains Malaysia

## CONTENTS

## 10.1 INTRODUCTION

Simulation of springback can be performed by using finite element (FE) code in ABAQUS software (Bakhshi-Jooybari et al., 2009). Compared to forming simulations, FE simulation of springback is much more sensitive to numerical tolerances and material models (Wagoner et al., 2013). Simulation of springback in the unconstrained bending test was done by Park et al. (2008) using the ABAQUS/ STANDARD implicit code with user-defined subroutine UMAT. Katre et al. (2014) have conducted FE simulation using commercially available elasto-plastic FE code that uses Lagrangian and explicit time integration techniques.

DOI: 10.1201/9781003164241-10

## 10.2 SIMULATION MODEL

Critical numerical procedures for springback simulation include the spatial integration scheme, element type and time integration schemes such as implicit/implicit, explicit/implicit, explicit/explicit, and one-step approaches (Wagoner et al., 2013). Significant material representations in springback simulation include the unloading scheme, strain-hardening rule, evolution of plastic properties, plastic anisotropy, Bauschinger effect and anticlastic curvature (Wagoner et al., 2013).

Park et al. (2008) studied the formability and springback of AA5052-H32 sheets made from surface friction stir methods, using FE prediction of springback from tests such as unconstrained cylindrical bending, 2D draw bending and draw-bend tests of friction stir welded sheets using combined isotropic-kinematic hardening law base on modified Chaboche model and Yld2000-2d yield function. For the simulation, the tools were represented by a four-node three-dimensional rigid body element (R3D4), while for the blank, the reduced four-node shell element (S4R) with nine integration points through thickness was used. The mesh size used was approximately $1.0 \times 1.0\,mm^2$, while the friction coefficient for no lubrication condition was assumed to be 0.17 and insignificant to the springback simulation (Park et al., 2008).

Katre et al. (2014) studied the springback of friction stir welded sheets of Al 5052-H32 and Al 6061-T6 using experiments and simulation. FE simulations were performed for the V-bending process using code that uses Lagrangian and explicit time integration techniques. Adaptive meshing automatically refines the mesh, and meshing was done with quadrilateral shell elements of Belytschko–Tsay formulation. The average mesh sizes for sheet specimens and tools were 2 mm, with the tools modelled as rigid bodies. To describe the stress–strain behaviour of the base material and weld zone, Hollomon's strain-hardening law was used. The plasticity model used Hill's 1990 yield criterion. To minimize the error compared to experimental results, the 'm' value in the yield criterion was optimized. Sheets subjected to explicit analysis and springback angle were evaluated after bending.

Bakhshi-Jooybari et al. (2009) studied the springback of CK67 steel sheets in V-die and U-die bending processes. Material properties from tensile tests were used in the simulation. To model the sheet anisotropy, Hill's anisotropy parameters were introduced into the FE code. For anisotropic materials, Hill's yield function is widely used in FE simulation due to its ease of formulation. The punch, die and blank holders were assumed as rigid bodies, while the blank was assumed as deformable. Quadrangle 4-node S4R shell elements were used for sheet modelling. For the V-bending simulation, the number of elements in the model was 80, consisting of 5 elements across the width and 16 elements across the length of the blank. While for the u-bending simulation, the number of elements in the model was 40, consisting of 2 elements across the width and 20 elements across the length of the blank. For springback measurement, one of the nodes on the blank was selected, and the position history was extracted from the loading and unloading steps. The springback was then calculated from the difference between blank bend angles during loading and after unloading. Figure 10.1 shows the number of elements in the modelling of V-die and U-die bending, while Figure 10.2 shows typical simulation results of springback in V-die and U-die bending.

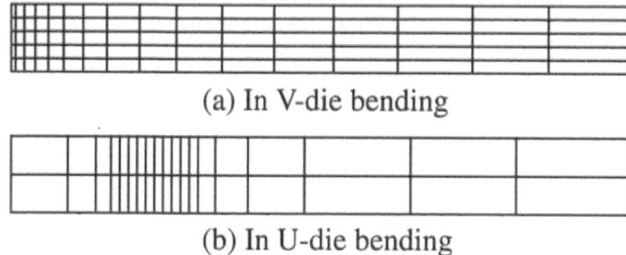

(a) In V-die bending

(b) In U-die bending

**FIGURE 10.1**   The number of elements in the modelling of V-die (a) and U-die bending (b). (With permission from Bakhshi-Jooybari et al., 2009.)

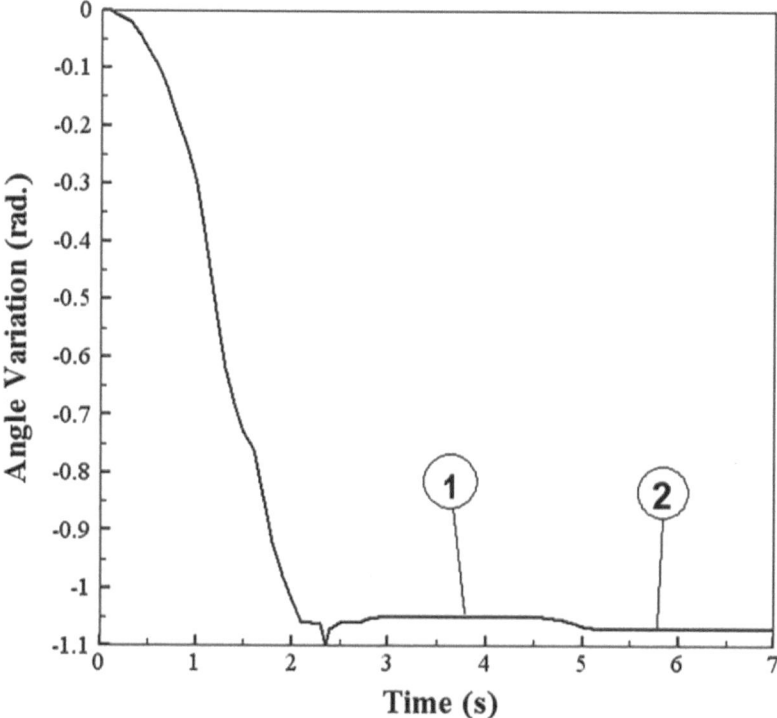

**FIGURE 10.2**   Typical simulation results of springback in V-die and U-die bending; part 1, the SET location before unloading; part 2, location after unloading. (With permission from Bakhshi-Jooybari et al., 2009.)

Gautam and Kumar (2018) investigated the springback in V-bending of interstitial free (IF) steel using experimental and numerical approaches. Simulation of the loading steps in the bending process utilized the ABAQUS-explicit solution procedure as the problems of nonlinear complicated contact can be handled with ease and shorter time. The completed bending simulation was imported from Explicit to ABAQUS-Implicit for the springback simulation. As the punch and die constraints were removed from the model in unloading, lower nonlinearity was offered by the springback simulation.

The bending process is usually treated as a nonlinear problem. The three nonlinearities in the bending problem are boundary nonlinearity, material nonlinearity and geometric nonlinearity. Material nonlinearity is caused by the plastic behaviour of anisotropic sheet metals in compliance with the power law of strain hardening in the true stress and true strain area. Boundary nonlinearity happens when the sheet metal comes in contact with the die shape during the bending operation. When the contact between the sheet and die occurs in the simulation, there is a large and instantaneous change in the model response, which results in nonlinearity caused by contact boundary conditions. Geometric nonlinearity is a result of geometric changes in the model in the analysis. The metal undergoes deflection in the bending process, and the bending load is not maintained perpendicular to the bent sheet. The bending load can be resolved into vertical and horizontal components, causing an alteration in model stiffness.

For conventional and TWB specimens, thickness integration by Simpson's rule with five-point integration can be adopted. Springback prediction relies on the integration scheme and integration points through the sheet thickness. As the springback amount depends on the bending moment which depends on stress distribution in sheet thickness, numerical integration of stress and strain through the thickness is required by the shell elements to determine the bending moment and force.

The punch and die sets are modelled as rigid surfaces since stress variation in the two components is not the focus of the simulation. For dies, the clearances are modelled to be equal to sheet thickness to prevent localized compression. The punch and dies are modelled as an analytical rigid shell, while the blank is modelled as a deformable shell planar with S4R shell elements. The S4R shell element is a four-node thin shell element with reduced integration, hourglass control and finite member strains. Point mass is assigned to the reference point on the punch and dies to compute the dynamic response. Figure 10.3 shows the tools assembly for bending simulation.

Hill's plasticity model or Hill's yield potential, an extension of Mises yield function for anisotropic materials, is used to incorporate sheets anisotropy in FE modelling. Springback in V-bending simulation for conventional blank and TWB utilizes

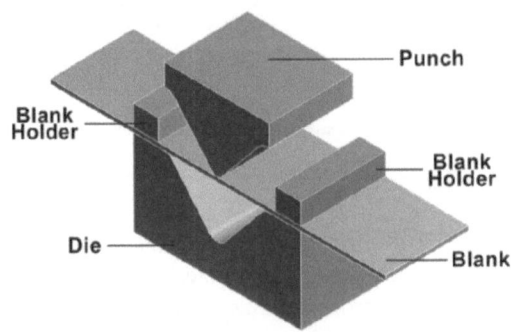

**FIGURE 10.3**   Tool assembly for bending simulation of (a) conventional blank and (b) TWB (see Gautam and Kumar, 2018).

the Newton–Raphson method by removing constraints such as die and punch so that the simulation adopted static-general procedure (ABAQUS standard) with nonlinearity due to geometry only. By using this method, the simulation is divided into a number of load increments and the approximate equilibrium configuration is found at the end of each load increment.

After all constraints are removed from the model, the blank is assigned the initial state of the bent data file containing a history of loading. To ensure that the central node remained stationary, it is assigned as zero velocity. To determine the springback, node coordinates from the load and unloaded frame are captured and plotted using the CAE interface in ABAQUS. The difference in node coordinates would give the change in the included angle after unloading. Figure 10.4 shows the initial and final stages of V-bending FE simulations, and Figure 10.5 shows the springback from the FE simulation of TWBs in V-bending.

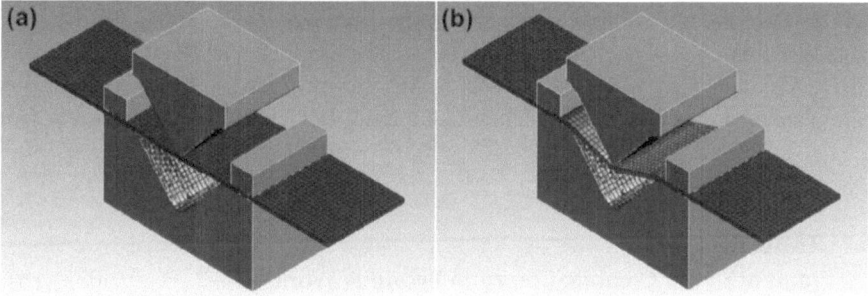

**FIGURE 10.4**   Initial and final stages of V-bending FE simulations: (a and b) unwelded sheet and (c and d) longitudinally welded sheet (see Gautam and Kumar, 2018).

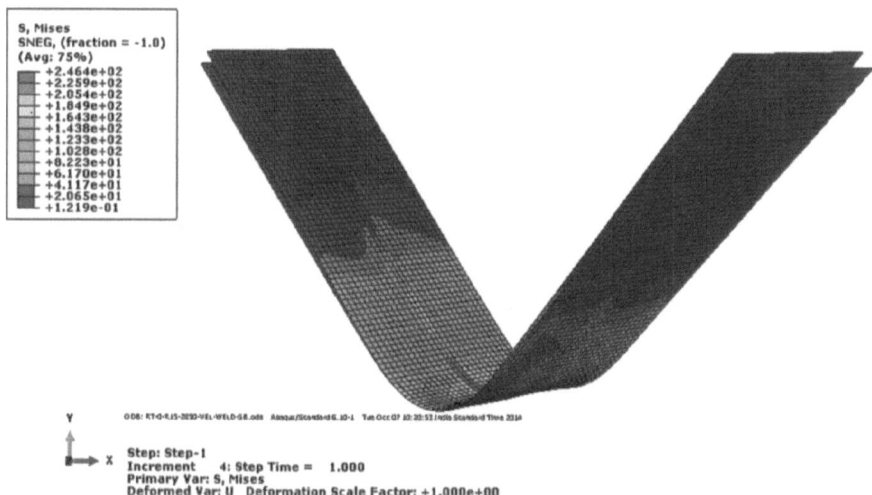

**FIGURE 10.5**   Springback in TWBs shown by the FE simulation of V-bending (see Gautam and Kumar, 2018).

## 10.3 SPRINGBACK SIMULATION USING ANSYS

Simulations for springback study were conducted using ANSYS software, utilizing SolidWorks software for modelling parts such as the die, puncher and plates. The materials used were selected based on past studies. Table 10.1 shows the materials selected for the experiments, categorized into thickness and material pair. The springback results obtained from the simulations were then compared with the experimental springback results from previous works.

The modelling for the V-bending apparatus comprising the die, puncher, blank holders and plates was done using SolidWorks software, as shown in Figure 10.6. There is a clearance of 0.05 mm between the puncher and the plates to avoid localized compression. The two plates, although separated, are combined as one part in the form of a conforming mesh.

The springback simulation in ANSYS uses static structural analysis. The stiffness behaviour for die, blank holders and puncher are classified as rigid since they are not the focus of the simulation. On the other hand, the plates are classified as flexible. The analysis is conducted in two stages. The original position of the puncher is defined as step 0. In step 1, the puncher moves downward in the -y direction towards the plate up to a predefined punch stroke length. In step 2, the puncher returns to

## TABLE 10.1
## Joint Material Combination from Previous Works

| Material Thickness (mm) | Material 1 | Material 2 | References |
|---|---|---|---|
| 1.5 | AA1050 | AA1050 | Simoncini et al. (2014) |
| | AA5052 | AA5052 | Park et al. (2008) |
| 2.1 | AA6061 | AA5052 | Katre et al. (2014) |
| | AA6061 | AA6061 | Rao and Narayanan (2014) |

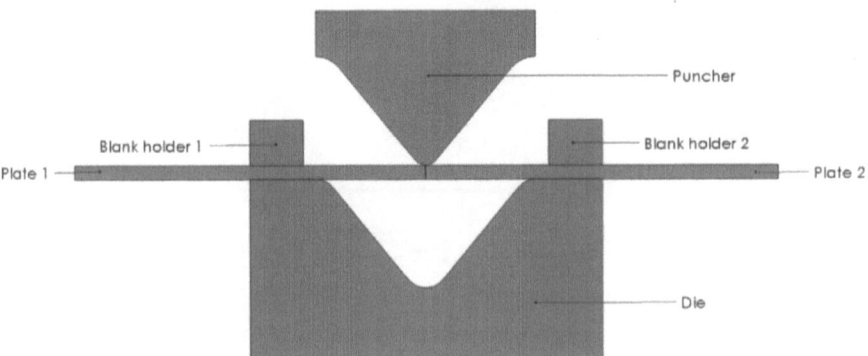

**FIGURE 10.6** SolidWorks modelling of the V-bending setup.

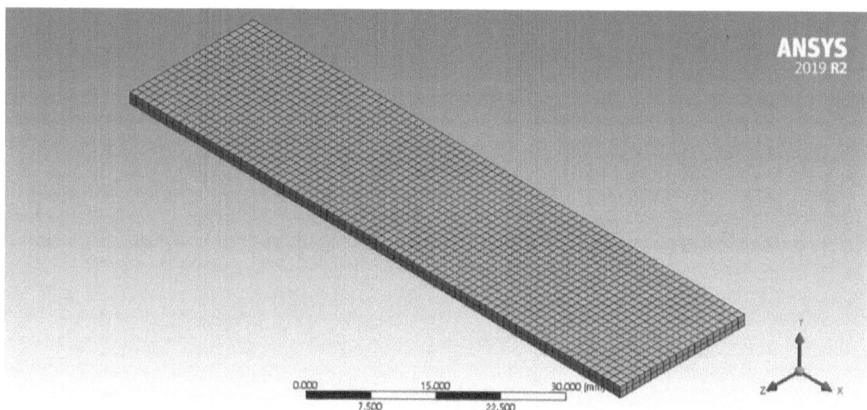

**FIGURE 10.7**   Mesh for the plates of the AA1050-AA1050 joint.

its initial position. The contact between the die and blank holders with the plates is assumed to be frictionless. On the other hand, the contact between the puncher and the plates is defined as frictional with a friction coefficient of 0.1 and asymmetric behaviour.

In terms of meshing, the mesh size for the plates is set to 1.0 mm with hard behaviour. For the puncher, die and blank holders, the mesh size is also set to be 1.0 mm but with soft behaviour. Figure 10.7 shows an example of the mesh done for the plates of AA1050-AA1050 joint. For the nonlinear springback simulation, the analysis uses a multilinear isotropic hardening model for its plasticity properties. The material properties and tensile data were extracted from the results in previous studies. The number of nodes and elements for the V-bending simulation are 25145 and 5324, respectively.

### 10.3.1   SPRINGBACK SIMULATION FOR JOINTS WITH A THICKNESS OF 1.5 MM

Two material combinations of AA1050-AA1050 and AA5052-AA5052 were considered for joints with thicknesses of 1.5 mm. Figure 10.8 shows the springback modelling and simulation for AA1050-AA1050 and AA5052-AA5052 joints.

The simulations show the results in the form of plate position relative to the starting point. Once the plate is bent according to the predefined punch stroke length, the puncher will return to its initial position. The plate will then attempt to return to its initial position as the external load from the puncher is removed, resulting in a difference in the final plate position. The measured bend angle and dimensions are shown in Figure 10.9, which are used in the calculation of the springback.

Since the thickness is constant, the following equation can be used to calculate the springback.

$$\tan\theta = \frac{\frac{1}{2}\,\text{Die opening}}{\text{Punch stroke}} \qquad (10.1)$$

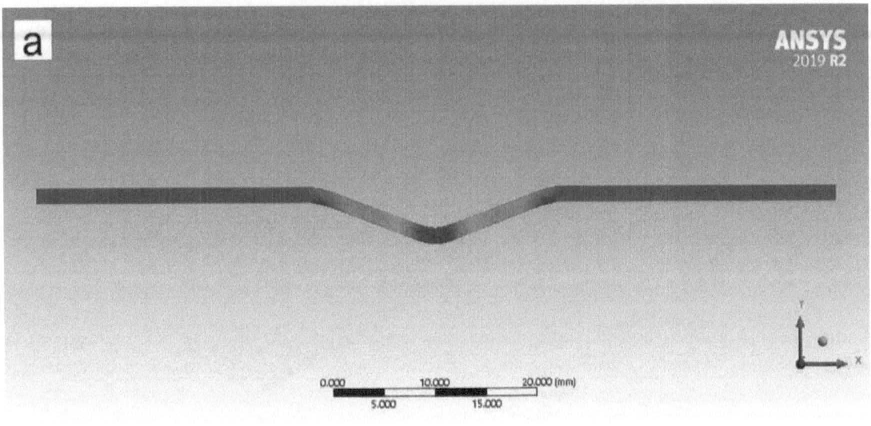

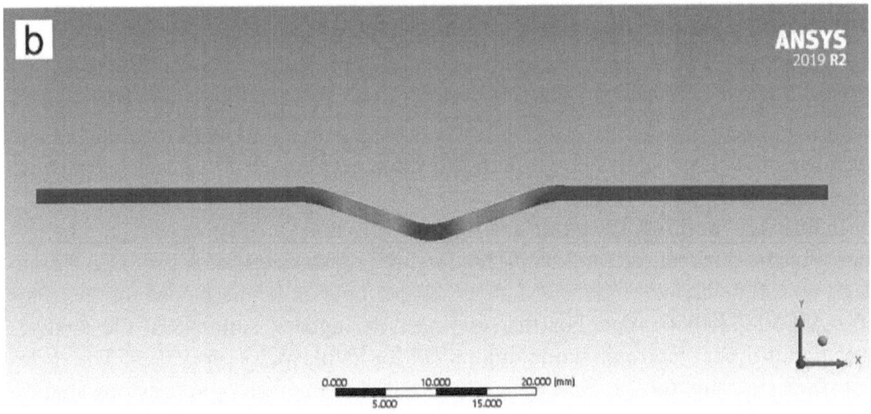

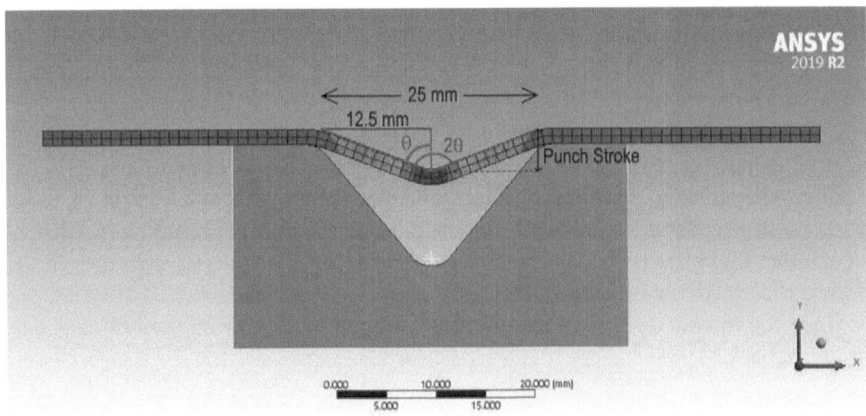

**FIGURE 10.8** Joint with a 1.5 mm thickness of (a) AA1050-AA1050 and (b) AA5052-AA5052.

**FIGURE 10.9** Bend angle and dimensions for springback calculation.

**TABLE 10.2**

**Springback Simulation Results for Joints with a 1.5 mm Thickness Compared with Previous Works**

| Material Thickness (mm) | Material 1 | Material 2 | References | Springback |
|---|---|---|---|---|
| 1.5 | AA1050 | AA1050 | ANSYS | 0.41° |
| | | | Simoncini et al. (2014) | 4.32°, 4.80°, 5.10° |
| | AA5052 | AA5052 | ANSYS | 2.78° |
| | | | Park et al. (2008) | 3.19°, 0.01°, 1.71°, 0.22° |

For the AA1050-AA1050 joint, from ANSYS results, punch stroke $= 4.1424$ mm

$$\tan\theta = \frac{12.5}{\text{Punch stroke}}$$

$$2\theta = 1\left[\tan^{-1}\left(\frac{12.5}{4.1424}\right)\right]$$

$$2\theta = 143.33°$$

Final deformation after springback $= 4.0936$ mm

$$2\theta_f = 2\left[\tan^{-1}\left(\frac{12.5}{4.0936}\right)\right]$$

$$= 143.74°$$

$$\text{Springback} = 143.73 - 143.33$$

$$= 0.41°$$

The springback angles from the results of V-bending simulations of AA1050-AA1050 and AA5052-AA5052 are calculated using Eq. (10.1), as shown in Table 10.2.

### 10.3.2 SPRINGBACK SIMULATION FOR JOINTS WITH A THICKNESS OF 2.1 MM

Two material combinations of AA6061-AA5052 and AA6061-AA6061 were considered for joints with thicknesses of 2.1 mm. Figure 10.10 shows the modelling and simulation for the springback of AA6061-AA5052 and AA6061-AA6061 joints.

Using Eq. (10.1), the results for the springback simulation of AA6061-AA5052 and AA6061-AA6061 are shown in Table 10.3.

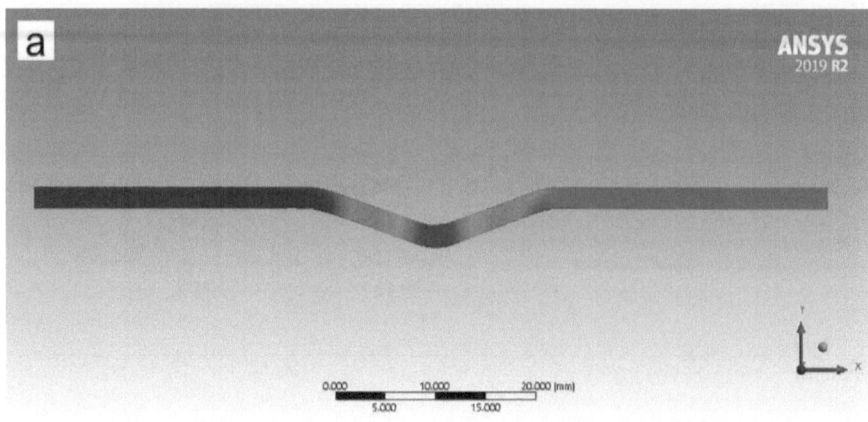

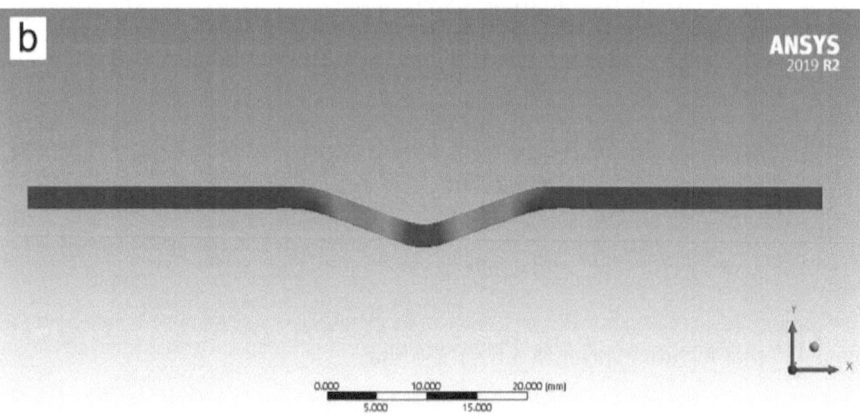

**FIGURE 10.10** Joint with 2.1 mm thickness of (a) AA6061-AA5052 and (b) AA6061-AA6061.

**TABLE 10.3**

**Springback Simulation Results for Joints with a 2.1 mm Thickness Compared with Previous Works**

| Material Thickness (mm) | Material 1 | Material 2 | Reference | Springback |
|---|---|---|---|---|
| 2.1 | AA6061 | AA5052 | **ANSYS** | **2.26°** |
| | | | Katre et al. (2014) | 3.66°, 0.23°, 1.26°, 3.90°, 3.12°, 1.85°, 2.32°, 1.44°, 0.75° |
| | AA6061 | AA6061 | **ANSYS** | **2.78°** |
| | | | Rao and Narayanan (2014) | 3.0°, 4.8°, 5.8°, 7.2°, 7.8°, 8.8°, 9.8°, 10,2° 11.8°, 12.4° |

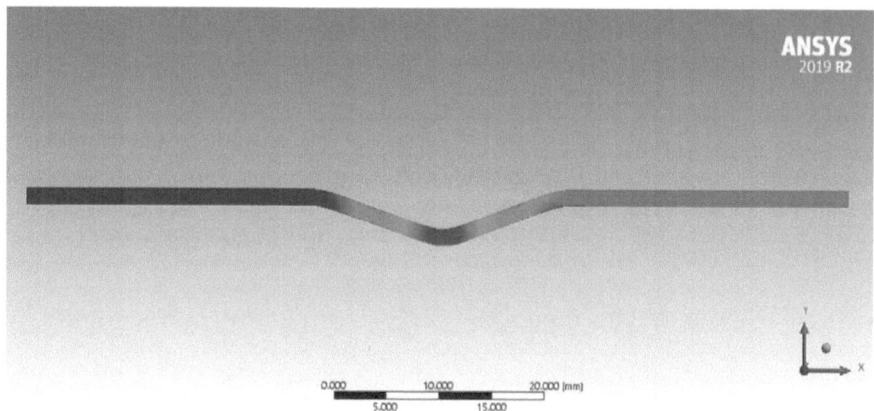

**FIGURE 10.11**    AA6061-mild-steel joint.

**TABLE 10.4**
**Springback Simulation Result for the AA6061-Mild Steel Joint**

| Material Thickness (mm) | Material 1 | Material 2 | Datum | Springback |
|---|---|---|---|---|
| 1.5 | AA6061 | Mild Steel | **ANSYS** | **2.46°** |

### 10.3.3 SPRINGBACK SIMULATION FOR THE AL-STEEL JOINT

A dissimilar material joint is of interest for the production of next-generation components with varying properties within a single body. In this section, a simulated study for the joining of the Al-steel joint is considered, although experimentally, the joint is difficult to be made with current joining methods. In this simulation, the combination of AA6061 and mild steel of 1.5 mm thickness is considered and assumed to be able to be joined. AA6061 with high magnesium content is preferred due to its high formability limit, although this property is a source of concern regarding its springback behaviour (Cinar et al., 2021). Mild steel is the most common type of steel used due to its good ductility and weldability. Figure 10.11 shows the springback modelling and simulation for AA6061-mild steel.

Using Eq. (10.1), the result for the springback simulation of AA6061-mild steel is shown in Table 10.4.

## 10.4   STUDY ON THE EFFECT OF THE PUNCHER OFFSET TO SPRINGBACK

The simulations conducted in the previous sections have considered the puncher to be located along the same axis as the joint line between the two plates. In this section, the effect of the puncher offset is studied by moving the joint line away by 2, 3, 4 and 5 mm relative to the puncher location.

### 10.4.1   EFFECT OF PUNCHER OFFSET TO SPRINGBACK FOR PLATES WITH A THICKNESS OF 1.5 MM

The combinations of AA1050-AA1050 and AA5052-AA5052 were used for plates with thicknesses of 1.5 mm. The springback modelling and simulations for AA1050-AA1050 and AA5052-AA5052 with puncher offsets are shown in Figures 10.12 and 10.13, respectively.

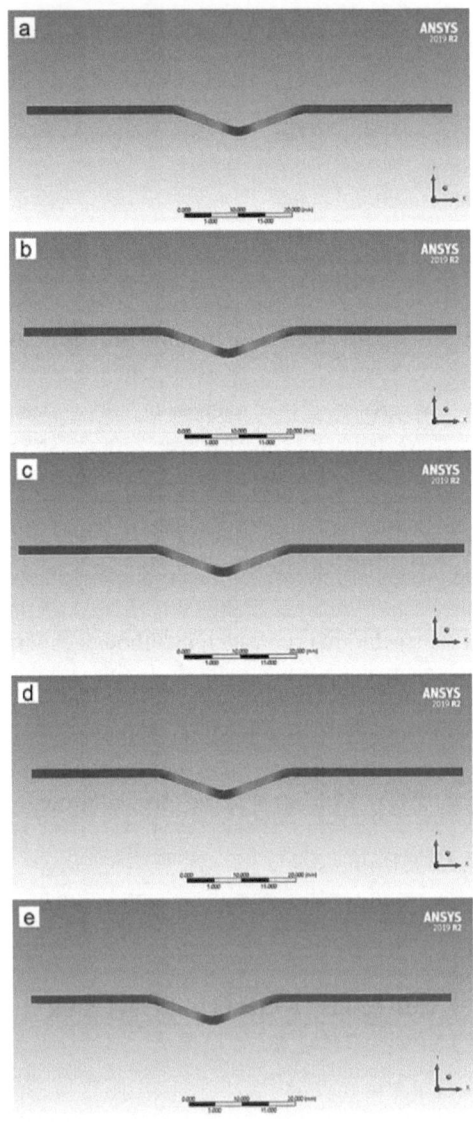

**FIGURE 10.12**   Springback simulation for AA1050-AA1050 joint with puncher offset of (a) 0 mm, (b) 2 mm, (c) 3 mm, (d) 4 mm and (e) 5 mm.

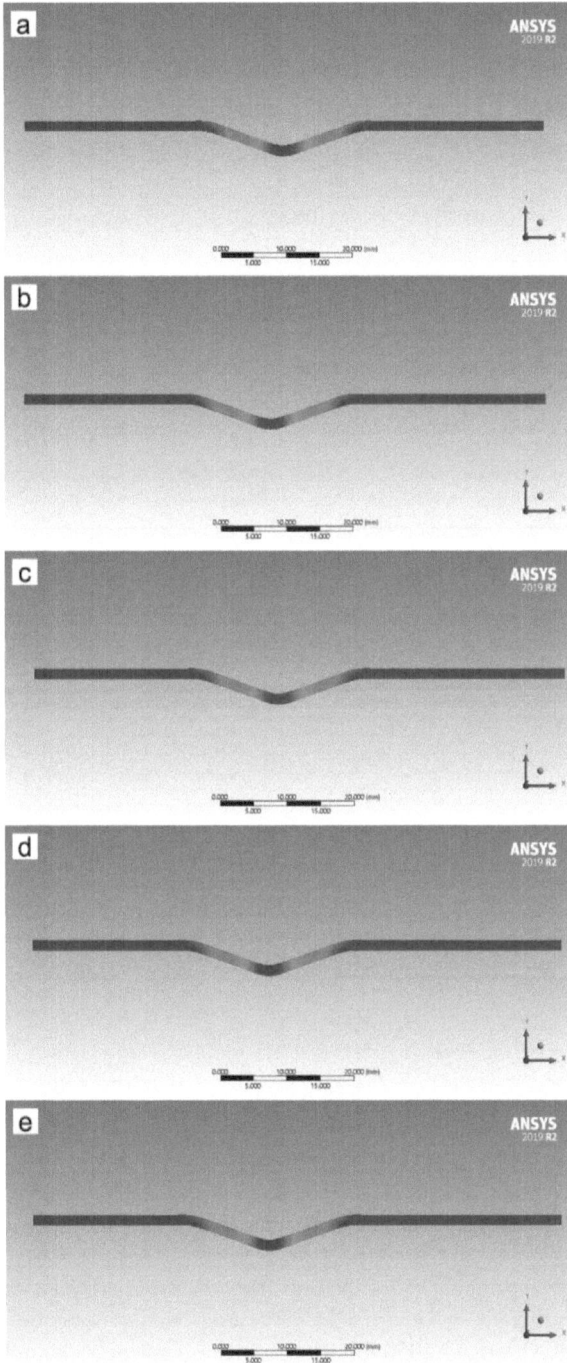

**FIGURE 10.13**  Springback simulation for AA5052-AA5052 joint with puncher offset of (a) 0 mm, (b) 2 mm, (c) 3 mm, (d) 4 mm and (e) 5 mm.

**TABLE 10.5**

**Springback Simulation Results for the Puncher Offset Study of Joints with a Plate Thickness of 1.5 mm**

| Material Thickness (mm) | Material 1 | Material 2 | Puncher Offset (mm) | Springback (°) |
|---|---|---|---|---|
| 1.5 | AA1050 | AA1050 | 0 | 0.41 |
| | | | 2 | 0.41 |
| | | | 3 | 0.41 |
| | | | 4 | 0.41 |
| | | | 5 | 0.41 |
| | AA5052 | AA5052 | 0 | 2.78 |
| | | | 2 | 2.79 |
| | | | 3 | 2.79 |
| | | | 4 | 2.79 |
| | | | 5 | 2.79 |

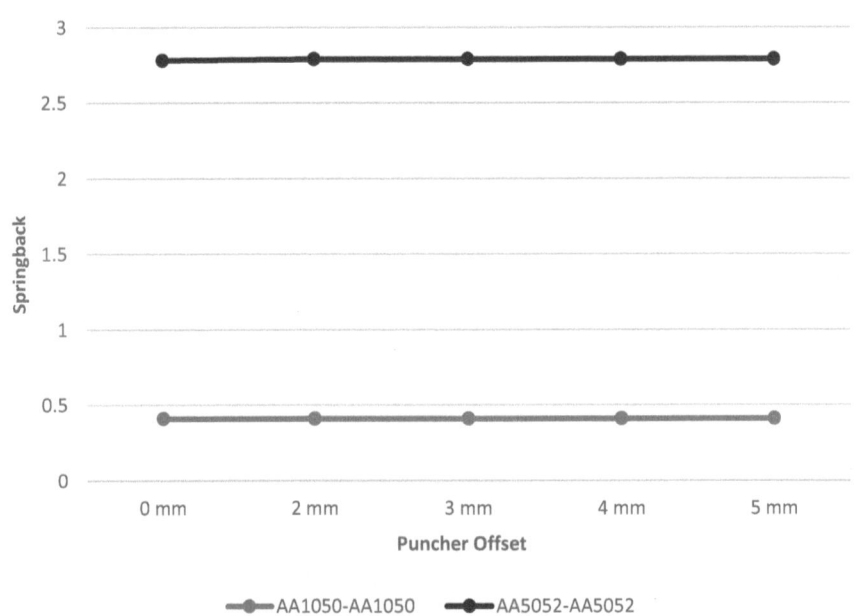

**FIGURE 10.14**    Springback vs puncher offset for joints with a thickness of 1.5 mm.

The calculated springbacks for the puncher offsets of both plate combinations are shown in Table 10.5.

In general, AA1050-AA1050 joints show a lower springback angle of 0.41° compared to 2.78° for AA5052-AA5052, which is due to the higher modulus of elasticity for AA1050. However, puncher offset does not have any effect on the springback of both material joints, as shown in Figure 10.14.

## 10.4.2 Effect of Puncher Offset on the Springback for Plates with a Thickness of 2.1 mm

For plates with a thickness of 2.1 mm, the combinations of AA6061-AA6061 and AA6061-AA5052 were used. The springback modelling and simulations for AA6061-AA6061, AA6061-AA5052 (offset towards AA6061) and AA6061-AA5052 (offset towards AA5052) are shown in Figures 10.15–10.17 respectively.

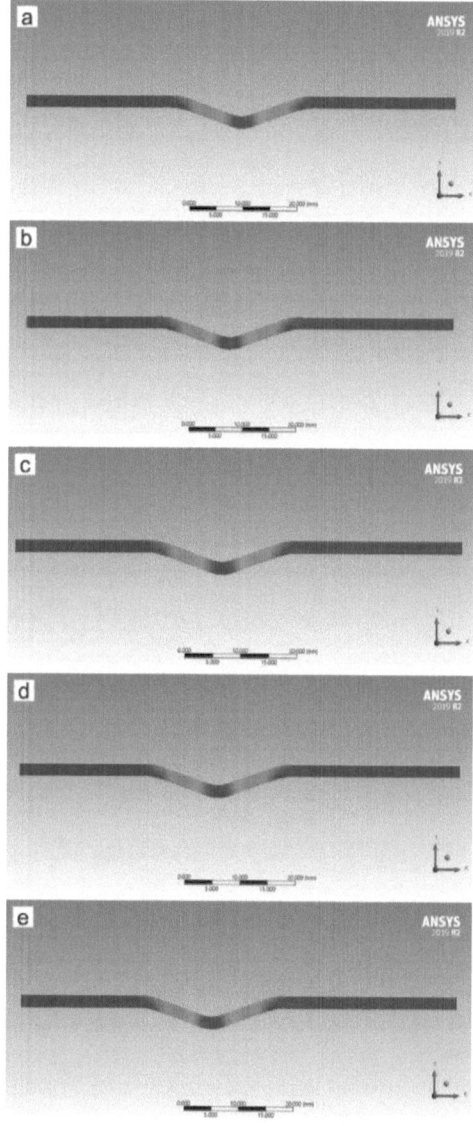

**FIGURE 10.15** Springback simulation for AA6061-AA6061 joint with puncher offset of (a) 0 mm, (b) 2 mm, (c) 3 mm, (d) 4 mm and (e) 5 mm.

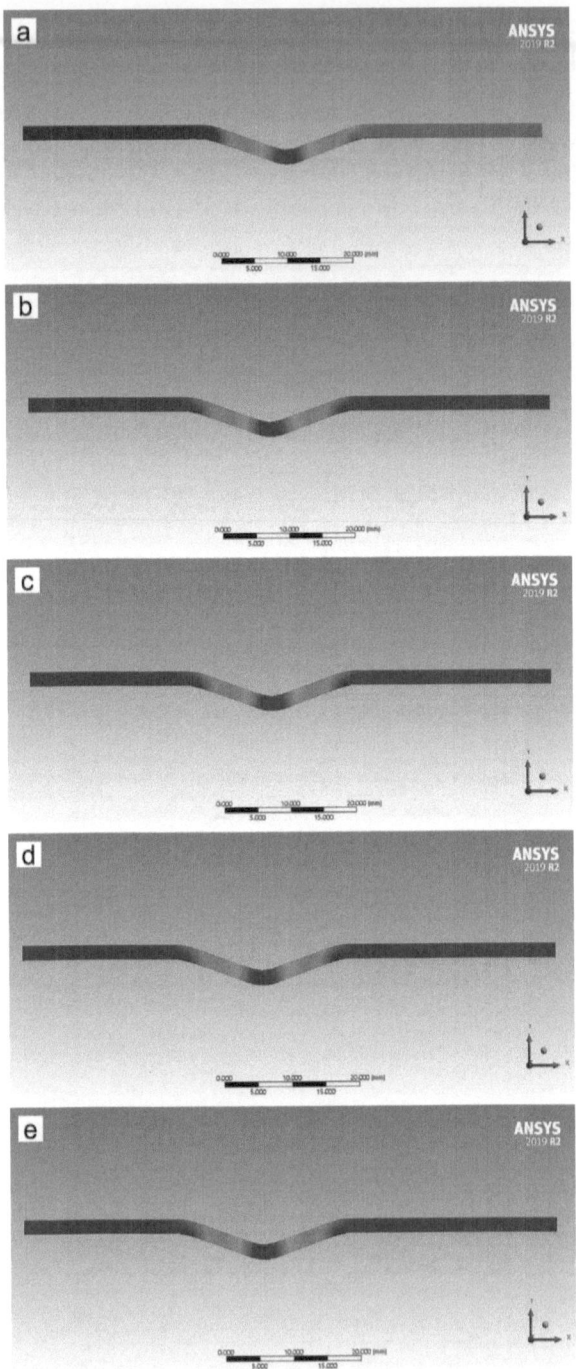

**FIGURE 10.16** Springback simulation for AA6061-AA5052 joint with puncher offset towards AA6061 of (a) 0 mm, (b) 2 mm, (c) 3 mm, (d) 4 mm and (e) 5 mm.

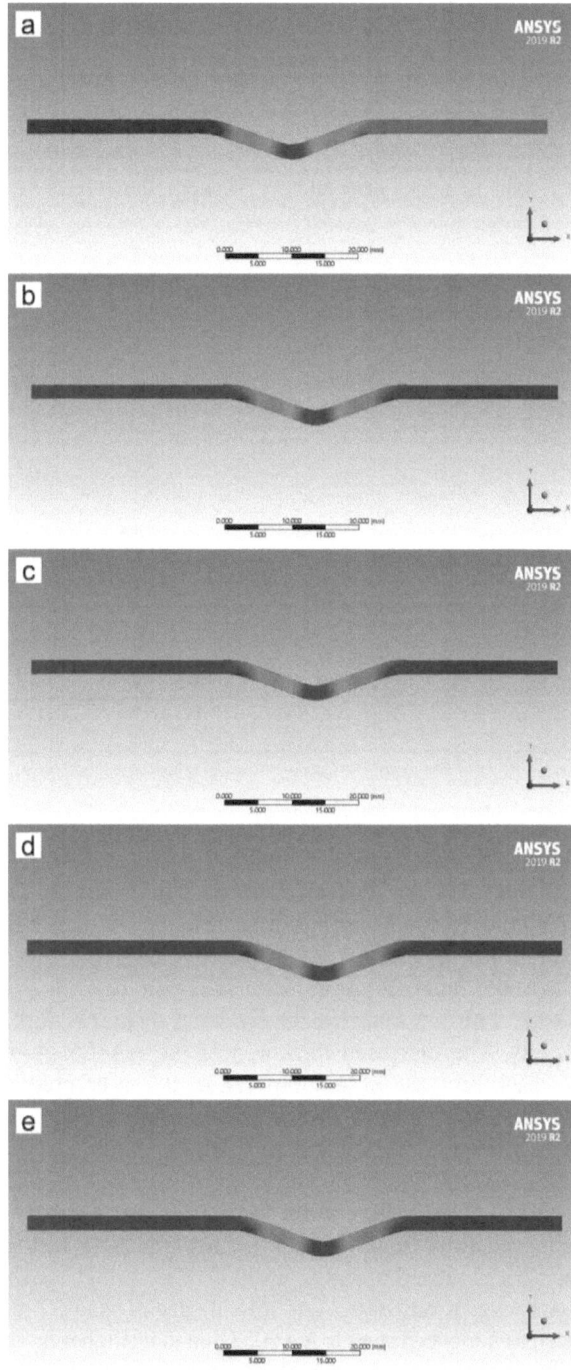

**FIGURE 10.17** Springback simulation for the AA6061-AA5052 joint with puncher offset towards AA5052 of (a) 0 mm, (b) 2 mm, (c) 3 mm, (d) 4 mm and (e) 5 mm.

**TABLE 10.6**

**Springback Simulation Results for the Puncher Offset Study of Joints with a Plate Thickness of 2.1 mm**

| Material Thickness (mm) | Material 1 | Material 2 | Puncher Offset (mm) | Springback (°) |
|---|---|---|---|---|
| 2.1 | AA6061 | AA6061 | 0 | 2.78 |
| | | | 2 | 2.78 |
| | | | 3 | 2.78 |
| | | | 4 | 2.78 |
| | | | 5 | 2.78 |
| | AA6061 | AA5052 | Offset AA6061 | |
| | | | 0 | 2.46 |
| | | | 2 | 2.65 |
| | | | 3 | 2.64 |
| | | | 4 | 2.64 |
| | | | 5 | 2.64 |
| | | | Offset AA5052 | |
| | | | 0 | 2.46 |
| | | | 2 | 2.43 |
| | | | 3 | 2.42 |
| | | | 4 | 2.42 |
| | | | 5 | 2.42 |

Table 10.6 shows the calculated springback for the AA6061-AA6061 and AA6062-AA5052 joints. For the AA6061-AA5052 joint, since the joint is made up of two different materials, the puncher was offset towards both AA6061 and AA5052 sides.

The AA6061-AA6061 with a 2.1 mm thickness shows the same springback angle as AA5052-AA5052 with a 1.5 mm thickness, even though AA6061 has a lower modulus of elasticity. The increased thickness of the AA6061-AA6061 joint has limited the springback response of AA6061. Thus, it can be predicted that if the AA6061-AA6061 joint has the same thickness of 1.5 mm as the AA5052-AA5052 joint, its springback angle will be higher than 2.78°.

For the AA6061-AA5052 joint, the calculated springback was 2.46° when the puncher was located at the joint line of the two plates, but it increased to 2.64° as the puncher was moved away from the joint line towards AA6061 and decreased to 2.42° as the puncher was moved towards AA5052. The springback increased as the puncher offset was moved towards the AA6061 side due to its lower modulus of elasticity and increased as it moved towards AA5052 due to its higher modulus of elasticity. Figure 10.18 shows the change in springback as the puncher is moved towards the AA6061 and AA5052 sides.

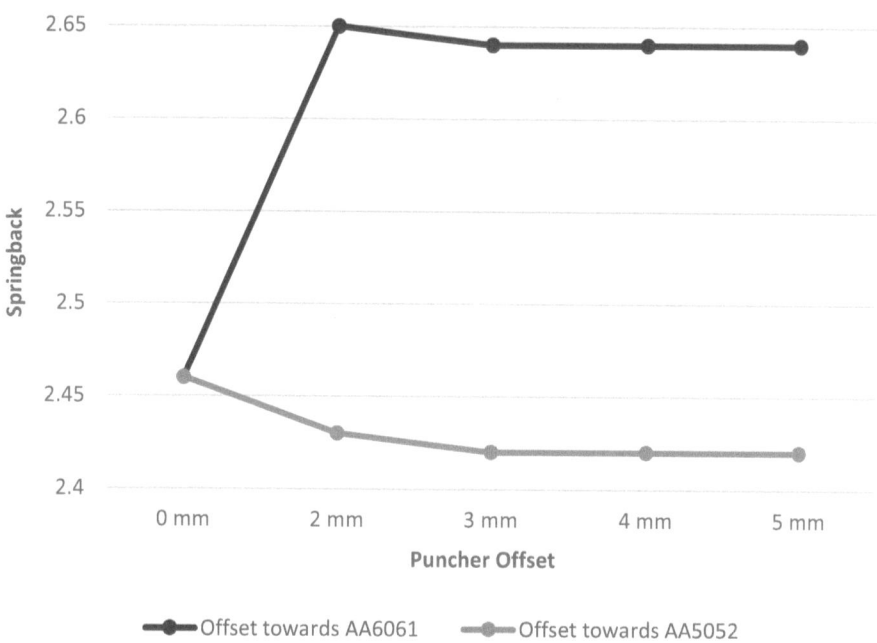

**FIGURE 10.18**    Springback vs puncher offset for the AA6061-AA5052 joint.

### 10.4.3    Effect of Puncher Offset to Springback for Al-Steel Joint

Springback simulations were conducted for AA6061-mild steel to evaluate the effect of the puncher location on the springback, as shown in Figure 10.19 (puncher offset is towards AA6061) and Figure 10.20 (puncher offset is towards mild steel).

Table 10.7 shows the calculated results of springback of AA6061-mild steel based on the offset location of the puncher relative to the joint line. The offset of the puncher was conducted towards both AA6061 and mild-steel sides.

The results show that the springback for the AA6061-mild-steel joint increased when the puncher was offset towards the AA6061 side. In contrast, the springback decreased when the puncher was offset towards the mild-steel die. This difference in springback is due to the mild steel having a significantly higher modulus of elasticity as compared to AA6061. Figure 10.21 shows the increasing difference in springback as the puncher is offset away from the joint line towards either side of AA6061 or mild-steel plates.

Springback increases with the increase of yield strength and strength factor and decreases with the increase of the strain-hardening exponent (Geng et al., 2019). An increase in strain hardening results in the reduction of the minimum bending radius (Abdullah and Samad, 2013). A less springback behaviour was observed as the elastic modulus and elastic recovery increased (Zhu et al., 2012). Springback can also be reduced by increasing the blank thickness (Chikalthankar et al., 2014). This is the reason why from the simulation results, the springback is higher for a lower sheet thickness.

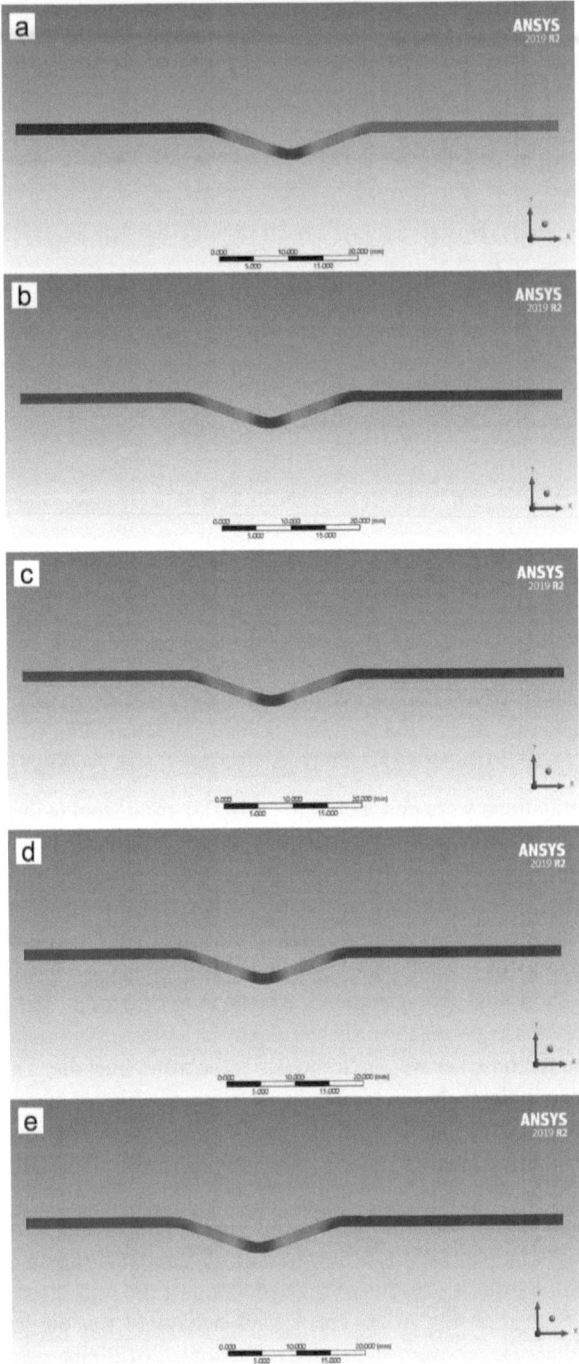

**FIGURE 10.19** Springback simulation for AA6061-mild steel joint with puncher offset towards AA6061 of (a) 0 mm, (b) 2 mm, (c) 3 mm, (d) 4 mm and (e) 5 mm.

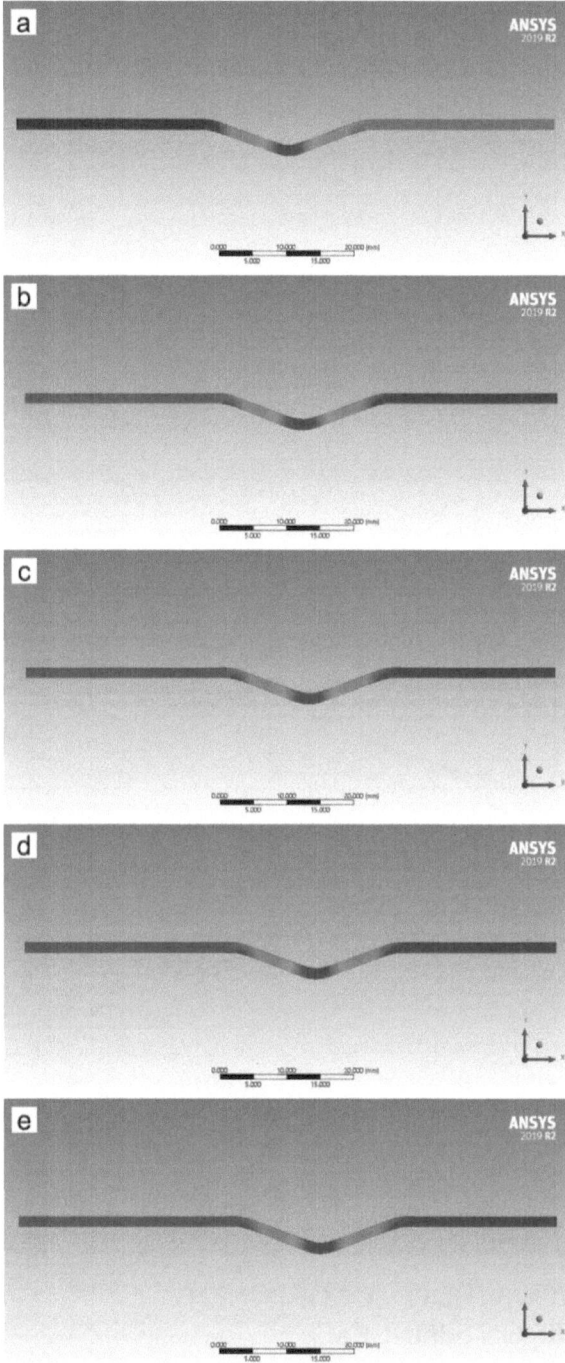

**FIGURE 10.20**  Springback simulation for the AA6061-mild-steel joint with puncher offset towards mild steel of (a) 0 mm, (b) 2 mm, (c) 3 mm, (d) 4 mm and (e) 5 mm.

**TABLE 10.7**

**Springback Simulation Results for the Puncher Offset Study of AA6061-Mild Steel**

| Material Thickness (mm) | Material 1 | Material 2 | Puncher Offset (mm) | Springback (°) |
|---|---|---|---|---|
| 1.5 | AA6061 | Mild Steel | Offset AA6061 | |
| | | | 0 | 2.46 |
| | | | 2 | 2.56 |
| | | | 3 | 2.71 |
| | | | 4 | 2.74 |
| | | | 5 | 2.76 |
| | | | Offset Mild Steel | |
| | | | 0 | 2.46 |
| | | | 2 | 2.24 |
| | | | 3 | 2.10 |
| | | | 4 | 1.99 |
| | | | 5 | 1.92 |

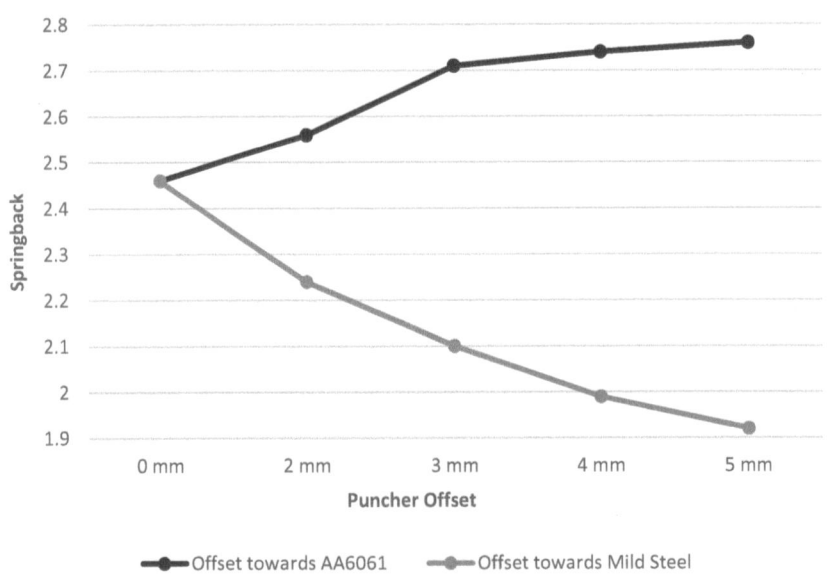

**FIGURE 10.21**   Springback vs puncher offset for the AA6061-mild-steel joint.

## 10.5   STUDY OF THE EFFECT OF PUNCH STROKE ON SPRINGBACKS

The punch stroke is the y-axis movement of the puncher onto the plate, which results in plate bending. In this section, the evaluation is made for different punch strokes of 5, 6, 7 and 8 mm.

## 10.5.1   EFFECT OF PUNCH STROKE ON SPRINGBACK FOR PLATES WITH A THICKNESS OF 1.5 MM

The effect of punch strokes was evaluated for plate thickness of 1.5 mm for the material combinations of AA1050-AA1050 and AA5052-AA5052. The modelling and springback simulation for both joints are shown in Figures 10.22 and 10.23.

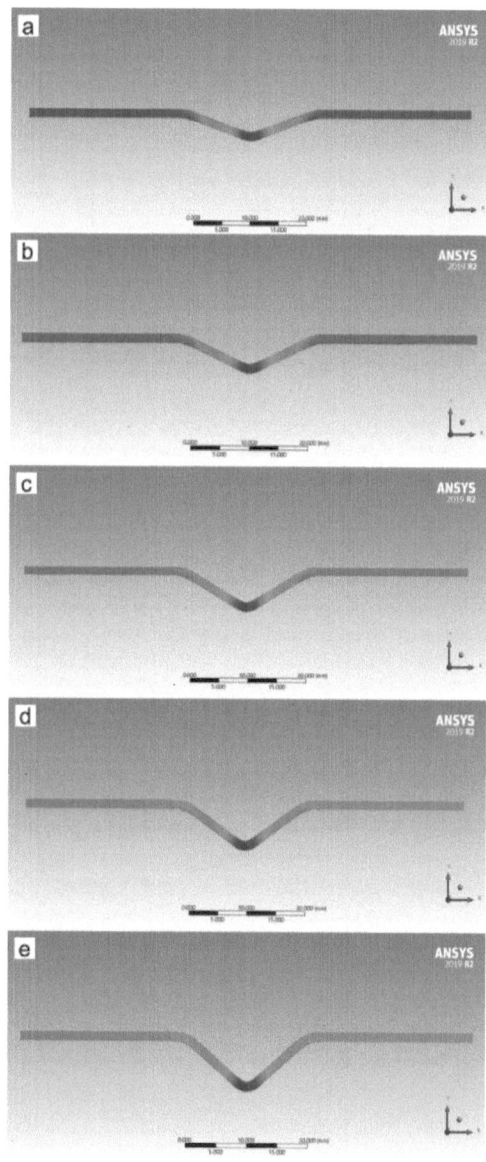

**FIGURE 10.22**   Springback simulation for the AA1050-AA1050 joint with punch strokes of (a) 4 mm, (b) 5 mm, (c) 6 mm, (d) 7 mm and (e) 8 mm.

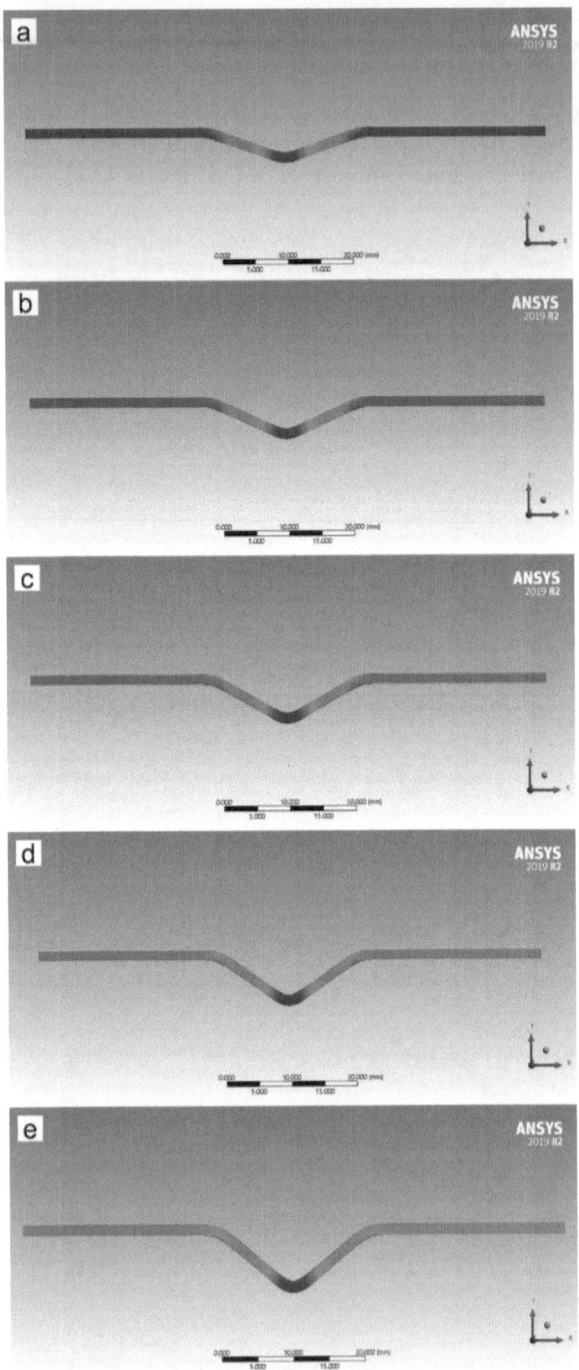

**FIGURE 10.23**   Springback simulation for the AA5052-AA5052 joint with punch strokes of (a) 4 mm, (b) 5 mm, (c) 6 mm, (d) 7 mm and (e) 8 mm.

**TABLE 10.8**

**Springback Simulation Results for the Punch Stroke Study of Joints with a Plate Thickness of 1.5 mm**

| Material Thickness (mm) | Material 1 | Material 2 | Punch Stroke (mm) | Springback (°) |
|---|---|---|---|---|
| 1.5 | AA1050 | AA1050 | 4 | 0.42 |
| | | | 5 | 0.41 |
| | | | 6 | 0.40 |
| | | | 7 | 0.37 |
| | | | 8 | 0.32 |
| | AA5052 | AA1050 | 4 | 2.78 |
| | | | 5 | 2.76 |
| | | | 6 | 2.54 |
| | | | 7 | 2.52 |
| | | | 8 | 2.41 |

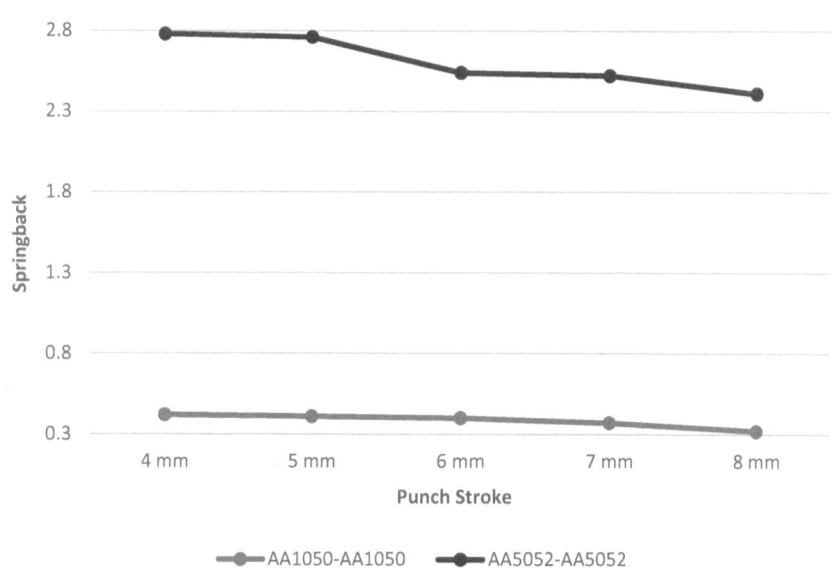

**FIGURE 10.24**   Springback vs punch stroke for joints with thickness 1.5 mm.

Table 10.8 shows the calculated springback from the simulation for a plate thickness of 1.5 mm.

In general, it was shown that the springback decreases as the punch stroke increases for both materials joints of AA1050-AA1050 and AA5052-AA5052, as shown in Figure 10.24. The AA1050-AA1050 joint exhibited lower springback values as compared to AA5052-AA5052 due to the higher modulus of elasticity of AA1050.

## 10.5.2 EFFECT OF PUNCH STROKE ON THE SPRINGBACK FOR PLATES WITH A THICKNESS OF 2.1 MM

The material combinations of AA6061-AA6061 and AA6061-AA5052 were selected for the evaluation of the punch stroke effect for 2.1 mm plate thickness. The modelling and springback simulation for both joints are shown in Figures 10.25 and 10.26.

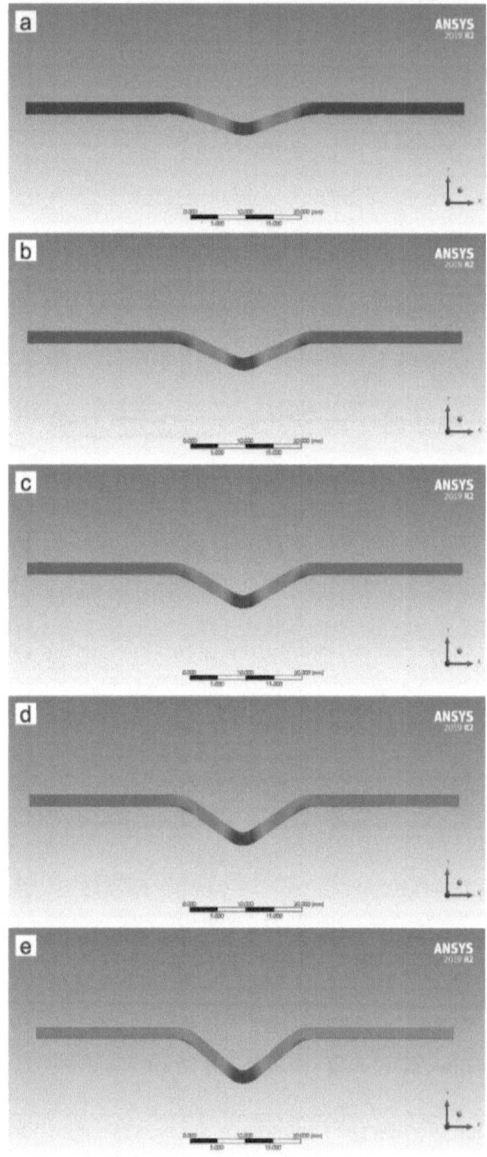

**FIGURE 10.25** Springback simulation for the AA6061-AA6061 joint with punch strokes of (a) 4 mm, (b) 5 mm, (c) 6 mm, (d) 7 mm and (e) 8 mm.

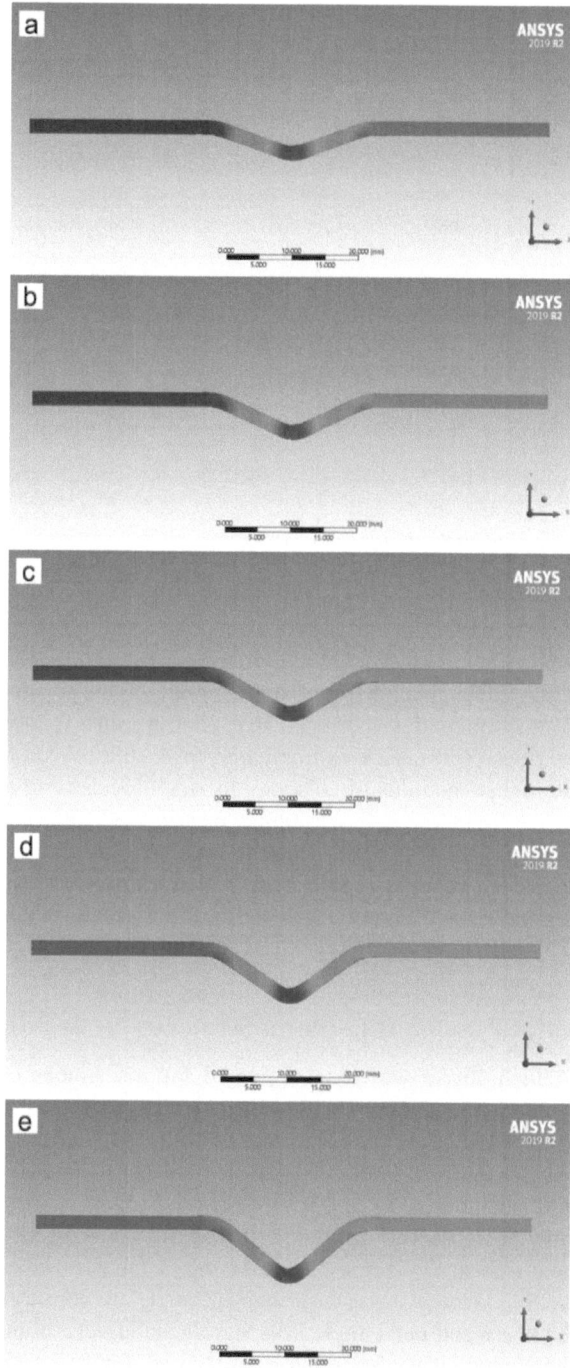

**FIGURE 10.26**    Springback simulation for the AA6061-AA5052 joint with punch strokes of (a) 4 mm, (b) 5 mm, (c) 6 mm, (d) 7 mm and (e) 8 mm.

**TABLE 10.9**

**Springback Simulation Results for the Punch Stroke Study of Joints with a Plate Thickness of 2.1 mm**

| Material Thickness (mm) | Material 1 | Material 2 | Punch Stroke (mm) | Springback (°) |
|---|---|---|---|---|
| 2.1 | AA6061 | AA6061 | 4 | 2.78 |
| | | | 5 | 2.71 |
| | | | 6 | 2.54 |
| | | | 7 | 2.52 |
| | | | 8 | 2.39 |
| | AA6061 | AA5052 | 4 | 2.46 |
| | | | 5 | 2.41 |
| | | | 6 | 2.33 |
| | | | 7 | 2.26 |
| | | | 8 | 2.15 |

The calculated springbacks for a 2.1 mm plate thickness for different punch stroke values are shown in Table 10.9.

In general, it was shown that the springback decreases as the punch stroke increases for both AA6061-AA6061 and AA6061-AA5052 2.1 mm joints. The AA6061-AA5052 shows a slightly lower springback as compared to AA6061-AA6061. This can be attributed to the higher modulus of elasticity in AA5052, causing the overall joint modulus of elasticity to increase. In addition, the springback values of 2.1 mm thickness specimens were lower than the springback values of 1.5 mm thickness specimen joint. This is due to the changes in thickness, which decrease the springback as the plate thickness is increased. Figure 10.27 shows the decrease in springback as the punch stroke is increased.

### 10.5.3   Effect of Punch Stroke on the Springback of Al-Steel Joint

The effect of punch stroke was also evaluated for the simulation of dissimilar AA6061-mild-steel joint. Figure 10.28 shows the modelling and springback of AA6061-mild-steel joints.

Table 10.10 shows the calculated springback values for different punch strokes in the AA6061-mild-steel joints.

Similar to previous observations, the springback in the AA6061-mild-steel joint decreases as the punch stroke increases. Increasing the punch stroke would increase the bend angle, thus decreasing the springback angle. According to Cinar et al. (2021), sheet thickness and bend angle have significant effects on the springback of AA6061. The bend angle was also found by Adnan et al. (2017) to be the most significant parameter that affects the springback.

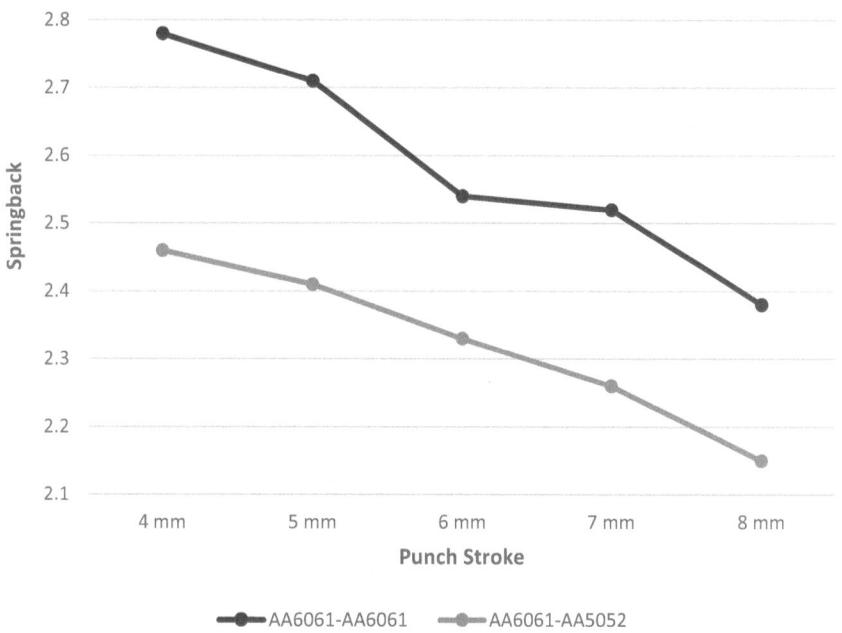

**FIGURE 10.27** Springback vs punch stroke for joints with a thickness of 2.1 mm.

**TABLE 10.10**
**Springback Simulation Results for the Punch Stroke Study of AA6061-Mild Steel**

| Material Thickness (mm) | Material 1 | Material 2 | Punch Stroke (mm) | Springback (°) |
|---|---|---|---|---|
| 1.5 | AA6061 | Mild Steel | 4 | 2.46 |
| | | | 5 | 2.40 |
| | | | 6 | 2.38 |
| | | | 7 | 2.31 |
| | | | 8 | 2.20 |

## 10.6 CONCLUSIONS

In this chapter, the effects of material properties and V-bending parameters of puncher offset and punch stroke on springback are presented. The following conclusions can be drawn:

1. For 1.5 mm thickness joints, AA1050-AA1050 joints show lower springback values than AA5052-AA5052 joints. This can be due to AA1050 having a higher modulus of elasticity than AA5052.

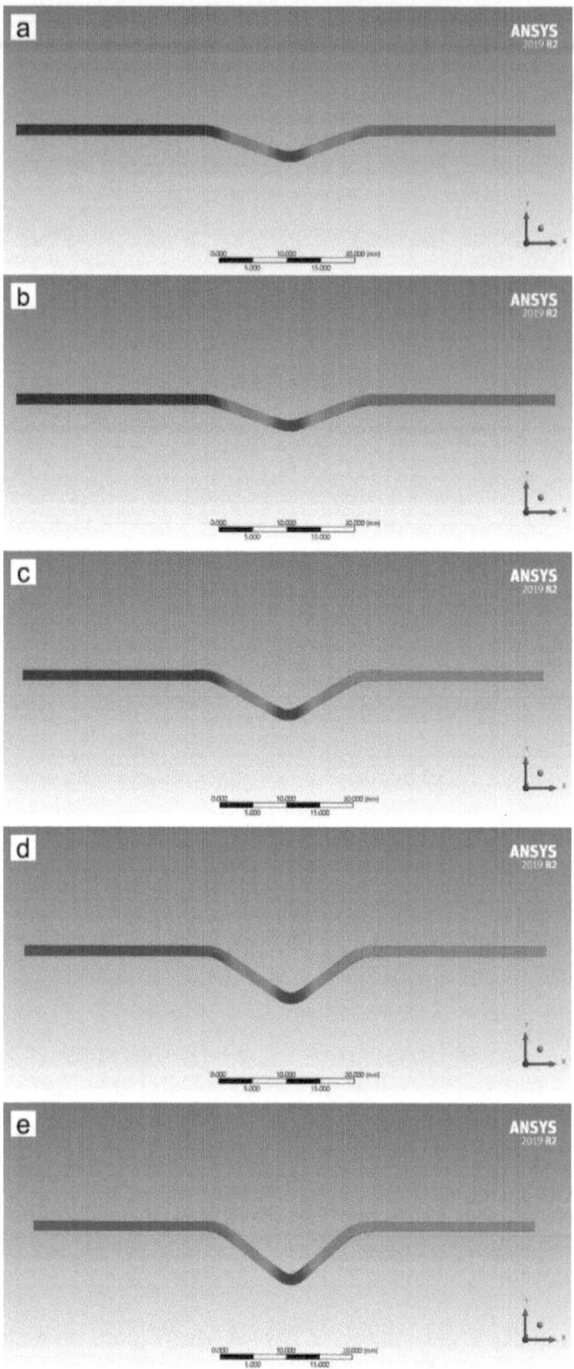

**FIGURE 10.28** Springback simulation for the AA6061-mild steel joint with punch strokes of (a) 4 mm, (b) 5 mm, (c) 6 mm, (d) 7 mm and (e) 8 mm.

2. Multi-material combinations with different elastic modulus properties, such as the joining of AA6061 with AA5052, lower the springback due to the high elastic modulus in AA5052.

3. The parameter of puncher offset did not have a noticeable effect for similar material joints such as AA1050-AA1050, AA5052-AA5052 or AA6061-AA6061. However, for dissimilar material joints, such as AA6061-AA5052, the springback increases when the puncher is offset towards the AA6061 side of lower elastic modulus. Similarly, for the AA6061-mild-steel joint, the springback is lower when the puncher is offset towards the mild-steel side having a significantly higher elastic modulus.

4. The punch stroke was seen to be the most influential parameter on springback behaviour. Increasing the punch stroke would increase the bend angle, resulting in the reduction of springback, regardless of material combinations, elastic modulus or sheet thickness values.

5. The sheet thickness is another factor that can affect the springback behaviour. In general, it was found that the joints with higher thickness show lower springback responses for both the puncher offset and punch stroke simulations.

## REFERENCES

Abdullah, A. B., & Samad, Z. (2013). An experimental investigation of springback of AA6061 aluminum alloy strip via V- Bending process. *IOP Conference Series: Materials Science and Engineering*, **50**(1). doi:10.1088/1757-899X/50/1/012069.

Adnan, A. F., Abdullah, A. B., & Samad, Z. (2017). Study of springback pattern of non-uniform thickness section based on V-bending experiment. *Journal of Mechanical Engineering and Sciences*, **11**(3), 2845–2855. doi:10.15282/jmes.11.3.2017.7.0258.

Bakhshi-Jooybari, M., Rahmani, B., Daeezadeh, V., & Gorji, A. (2009). The study of springback of CK67 steel sheet in V-die and U-die bending processes. *Materials and Design*, **30**(7), 2410–2419. doi:10.1016/j.matdes.2008.10.018.

Chikalthankar, S. B., Belurkar, G. D., & Nandedkar, V. M. (2014). Factors affecting on springback in sheet metal bending : A review. *International Journal of Engineering and Advanced Technology (IJEAT)*, **3**(4), 247–251.

Cinar, Z., Zeeshan, Q., & Safaei, B. (2021). Effect of springback on A6061 sheet metal bending: A review. *Jurnal Kejuruteraan*, **33**(1), 13–26.

Gautam, V., & Kumar, D. R. (2018). Experimental and numerical investigations on springback in V-bending of tailor-welded blanks of interstitial free steel. *Proceedings of the Institution of Mechanical Engineers, Part B: Journal of Engineering Manufacture*, **232**(12), 2178–2191. doi:10.1177/0954405416687146.

Geng, H., Wang, Y., Wang, Z., & Zhang, Y. (2019). Investigation on contact heating of aluminum alloy sheets in hot stamping process. *Metals*, **9**(12). doi:10.3390/met9121341.

Katre, S., Karidi, S., & Narayanan, G. (2014). Springback of friction stir welded sheets: Experimental and prediction. *In 5th International & 26th All India Manufacturing Technology* (pp. 318-1–318-5), Gulwahati.

Park, S., Lee, C. G., Kim, J., Han, H. N., Kim, S. J., & Chung, K. (2008). Improvement of formability and spring-back of AA5052-H32 sheets based on surface friction stir method. *Journal of Engineering Materials and Technology, Transactions of the ASME*, **130**(4), 0410071–04100710. doi:10.1115/1.2975233.

Rao, B. D., & Narayanan, R. G. (2014). Springback of friction stir welded sheets made of aluminium grades during V-bending : An experimental study. *ISRN Mechanical Engineering*, 2014, 1–15.

Simoncini, M., Panaccio, L., & Forcellese, A. (2014). Bending and stamping processes of FSWed thin sheets in AA1050 alloy. *Key Engineering Materials*, **622–623**, 459–466. doi:10.4028/www.scientific.net/KEM.622-623.459.

Wagoner, R. H., Lim, H., & Lee, M. G. (2013). Advanced issues in springback. *International Journal of Plasticity*, **45**, 3–20. doi:10.1016/j.ijplas.2012.08.006.

Zhu, Y. X., Liu, Y. L., Yang, H., & Li, H. P. (2012). Development and application of the material constitutive model in springback prediction of cold-bending. *Materials and Design*, **42**, 245–258. doi:10.1016/j.matdes.2012.05.043.

# 11 Effect of Post-Welding Heat Treatment on Springback Pattern of AA6061 TWBs

*AB Abdullah*
Universiti Sains Malaysia

*M Mohamed*
Impression Technologies Ltd
Helwan University

*AF Pauzi*
Universiti Sains Malaysia

## CONTENTS

## 11.1 INTRODUCTION

The demand for lightweight sheet metal materials (e.g., Al) in automotive industries is high. Many European automotive companies have implemented tailor welded blank (TWB) technology to achieve lighter-weight vehicular structures. Al alloys have been attracting extensive attention due to their various advantages. Each alloy is developed to meet certain criteria and characteristics. For example, AA5052 is the

DOI: 10.1201/9781003164241-11

strongest non-heat-treatable Al that is commonly used as an alternative to steel in aerospace, marine, and automotive industries because of its low weight, good formability, high strength, and high corrosion resistance [1,2]. AA6061 is suitable for applications that require good tensile strength at high temperatures and dimensional stability. In addition, this alloy possesses good corrosion resistance to seawater [3] and excellent weldability and is typically used for heavy-duty structures in shipbuilding and rail coaches [4]. Although AA1100 is rarely used in sheet metal works that require high strength because of its poor machinability, this alloy has excellent electrical conductivity and a reflective finish [5]. Automotive industries prefer to use 6000 series Al because of its high strength, excellent surface quality, and good weldability [6]. Utilizing the advantages of the above-mentioned alloys by joining them through friction stir welding (FSW) can result in the production of high-quality blanks and cost savings.

Several studies on the use of FSW to join dissimilar Al alloys have been conducted. Lin et al. [7] found that dissimilar Al alloys are difficult to join through FSW because of the differences in material properties, which result in difficulty in selecting suitable welding parameters. Some studies, however, have determined appropriate parameters. For example, Balaji et al. [8] discovered that the most suitable rotational speed for welding AA2024 and AA7075 ranges from 600 to 1200 rpm. They further conducted tensile tests, which revealed that the strength of the joint was reduced by only 10% as compared with the base materials. This result is consistent with the findings of Liu et al. [9], who reported that the maximum tensile strength that can be achieved after joining different materials is ~85% of that of the base aluminium alloy. After investigating the FSW process for AA7075-T651 and AA2024-T351, Hassan et al. [10] inferred that the weld hardness drops at the heat-affected zone of soft materials. The results of the studies of Ilangovan et al. [11] and Krishna et al. [12] involving AA6061 and AA5086 and the combination of AA2024 and AA7075, respectively, indicated that tool profile influences the performance of the weld. Similarly, Kim et al. [13] studied the profile geometries' effect on the joint. The assessment of the effect of welding parameters, such as welding and tool rotation speeds, the effect on tensile strength, and microhardness, on AA2024 and AA707, indicated that the highest tensile strength can be obtained at a feed rate of 31.5 mm/min under a rotational speed of 900 rpm. Piccini and Svoboda [14] studied the effect of tool depth on a blank composed of AA5052 and AA6063 combined using the lap joint. The findings also revealed that when AA6063 serves as the upper material, the fracture load increases.

Numerous studies regarding the effect of post-welding heat treatment (PWHT) such as annealing on the properties of welded blanks have been performed. For example, studies by Muruganandam [15], Momeni and Guillot [16] and Oztoprak et al. [17] proved that annealing improves the mechanical properties of joints after welding [18–21]. Improving properties, such as tensile strength and ductility, reduces the springback amount [20–24]. This study investigates the effect of annealing on the springback amount of Al alloy blanks fabricated via FSW at room temperature.

This chapter is organized as follows: it starts with an introduction that contains the latest update about the relevant research and analysis of gaps, followed by the methodology that contains the material and strategy to achieve the objectives, and the results are presented and discussed. It ends with conclusions and recommendation for future works.

## 11.2 METHODOLOGY

In this study, Al 6061 blanks are fabricated using FSW. The specimens are annealed and quenched under different parameter settings to analyse the effect of annealing on springback at room temperature. The specimens that underwent a V-bending test and the springback are measured. Specimens with 3 mm-thick AA6061 Al metal plates are sectioned into 200 mm×80 mm rectangular specimens. The elemental composition of AA6061 is presented in Table 11.1. The fixture and clamping components shown in Figure 11.1 are fabricated based on the design recommendation of Kamble et al. [25]. A 350 mm×250 mm clamping assembly is designed and mounted onto the worktable of a conventional milling machine. The independent left and right clamps firmly held the weld plates regardless of the thickness to prevent the lifting or bending of the specimen. The built-in side notches on the clamp prevented the lateral movements of the plates, thereby avoiding splitting in the weld region. The dimensions of the cylindrical FSW tool made from hardened tool steel are shown in Figure 11.2. The tool consists of a 2 mm pin and 10 mm shoulder profile, both of which are untreated.

**TABLE 11.1**
**The Material Composition of AA6061-T6 (% wt)**

| % | Si | Fe | Cu | Mn | Mg | Cr | Zn | Ti | Al |
|---|----|----|----|----|----|----|----|----|-----|
| Min | 0.4 | | 0.15 | | 0.8 | 0.04 | | | Remainder |
| Max | 0.8 | 0.7 | 0.4 | 0.15 | 1.2 | 0.35 | 0.25 | 0.15 | |

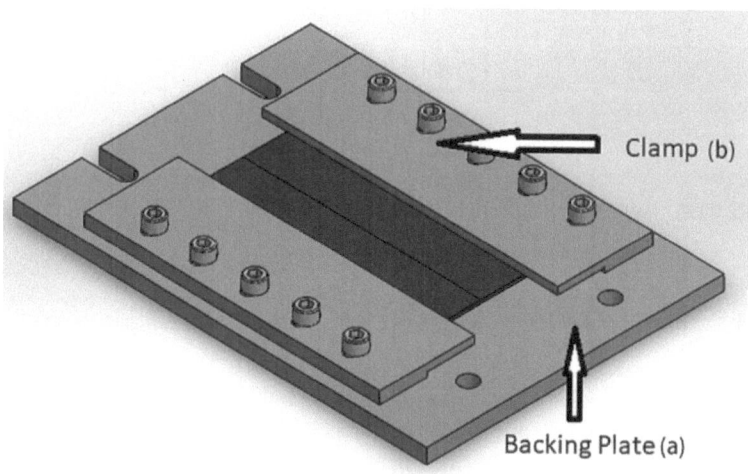

FIGURE 11.1   Design of the fixture.

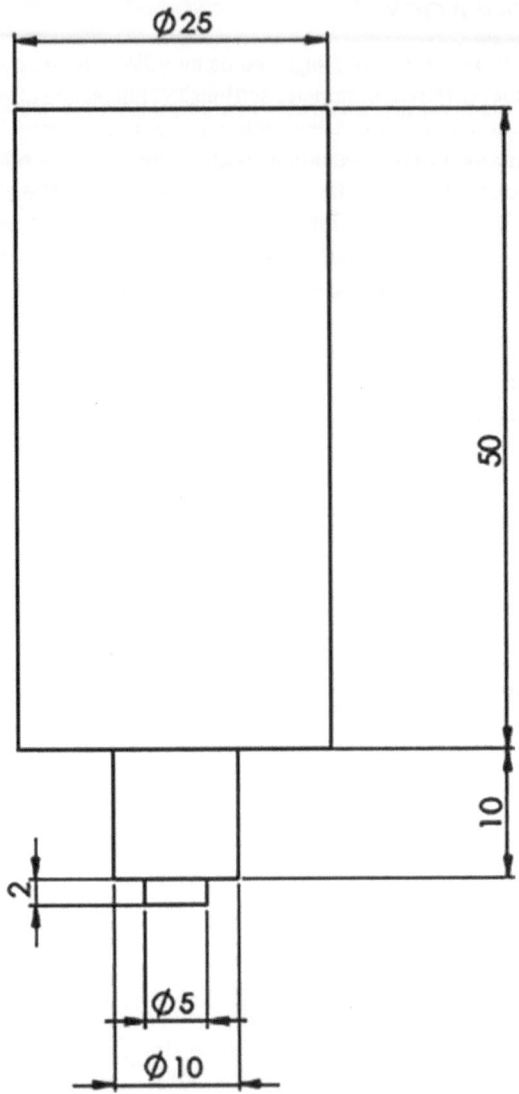

**FIGURE 11.2**   Schematic diagram of the tool (unit in mm).

### 11.2.1   SELECTION OF THE PARAMETERS AND THEIR LEVELS

The design-of-experiment tool is utilized in this study to identify the total number of experiments for the selected three parameters with three levels each. The springback parameters and the levels are shown in Table 11.2. Nine experiments identified are listed in Table 11.3. The two AA6061 plates (200 mm×80 mm×3 mm) are arranged in a butt weld configuration on the clamp, which is fixed on a conventional milling machine (Figure 11.3).

## TABLE 11.2
## Annealing Parameters Levels

| | | Level | | |
|---|---|---|---|---|
| Factor | Parameter | 1 | 2 | 3 |
| A | Annealing temperature | 200°C | 300°C | 400°C |
| B | Annealing time | 20 min | 40 min | 60 min |
| C | Quenching medium | Air at room temperature | Water | Oil |

## TABLE 11.3
## Experiment Design for the Three Parameters and Levels

| | Parameter | | | |
|---|---|---|---|---|
| Experiment | Annealing Temperature (A) | Annealing Time (B) | Quenching Medium (C) | Parameter Setting |
| 1 | 200 | 20 | Air | A1 B1 C1 |
| 2 | 200 | 40 | Water | A1 B2 C2 |
| 3 | 200 | 60 | Oil | A1 B3 C3 |
| 4 | 300 | 20 | Water | A2 B1 C2 |
| 5 | 300 | 40 | Oil | A2 B2 C3 |
| 6 | 300 | 60 | Air | A2 B3 C1 |
| 7 | 400 | 20 | Oil | A3 B1 C3 |
| 8 | 400 | 40 | Air | A3 B2 C1 |
| 9 | 400 | 60 | Water | A3 B3 C2 |

FIXTURE CLAMP

FSW TOOL

MILLING TABLE

**FIGURE 11.3**   Experimental setup on the conventional milling machine.

## 11.2.2 Materials and Equipment

The tool in the milling machine setup for FSW is tilted by 3°. Two AA6061 plates are clamped into the fixture. The tool is placed between the two plates and inserted into the workpiece at a depth of 2.1 mm. The tool traversed along the weld line at the spindle and feed rates of 865 rpm and 65 mm/min, respectively. These are the optimal welding parameters obtained from the previous study [26].

An electrical discharge machining wire-cut machine is then used to cut the welded Al blanks into 28 pieces of 50 mm × 20 mm specimens. The oven is set to specified temperatures of 200°C, 300°C and 400°C for the annealing process and the specimens are placed in the oven for different annealing times (20, 40 and 60 min). The specimens are taken out and placed directly into different quenching media, (water, oil and air) at room temperature. Steps 1–3 are repeated at different temperatures (300°C and 400°C). One specimen is left untreated for comparison. Each parameter setting contains three specimens to obtain the average value.

## 11.2.3 Hardness Measurement

The hardness values were measured along and perpendicular to the weld line, as shown in Figure 11.4. A Rockwell hardness tester was used to determine the hardness using a steel ball with a diameter of 1.588 mm and 100 kg of loading.

## 11.2.4 Experimental Setup

For the V-bending test, the die opening (32 mm) and the stroke are set at 3 mm. The centre of each specimen is marked with a straight line and the V-bending test is performed on the marked area for each specimen. Figure 11.5 shows the setup of the die set. The specimen is cut into 50 mm × 20 mm. The experiment is analysed under three strokes (3, 6, and 9 mm). To measure the springback amount, a coordinate-measuring machine (CMM) with a resolution of 0.0005 mm.

### 11.2.4.1 Springback Measurement

The specimen is fixed on the CMM table using fixtures. In the measurement, the probe is moved along the left side of the specimen to obtain two reference points to create line 1. Similarly, the probe is moved along the right side of the specimen to obtain two reference points to create line 2. The difference between these two lines is defined as springback. Steps are repeated for the 27 remaining specimens. Figure 11.6a illustrates the obtained springback amount, which is measured on the

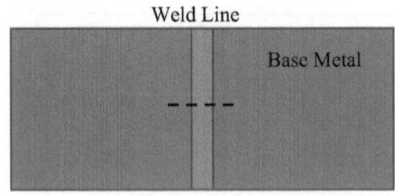

**FIGURE 11.4**   Hardness test perpendicular to the weld line.

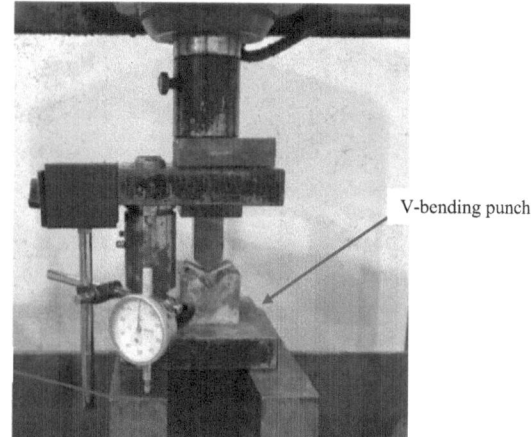

**FIGURE 11.5**    V-bending setup on the hydraulic press machine.

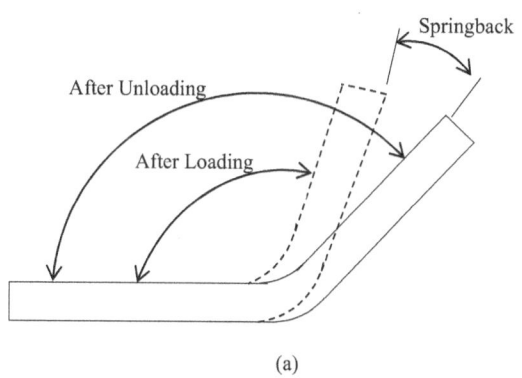

(a)

(b)

**FIGURE 11.6**    Schematic diagram of (a) springback measurement and (b) bent specimens.

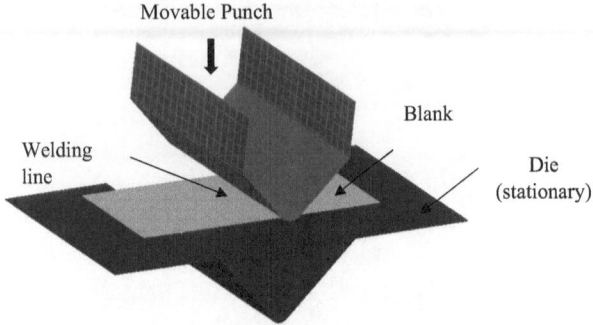

**FIGURE 11.7**  Tool setup with boundary conditions in the FE model.

basis of the difference between the angles of the bent strip (Figure 11.6b) after loading and unloading as described above.

## 11.2.5 FE SIMULATION MODEL

FE simulation of the bending process of TWBs under different heat treatment conditions was conducted for the AA6061 aluminium alloy in the PAM-STAMP FE package. n is the simulation model, the blank is meshed by shell elements having 3 mm thickness and five integration points through-thickness using Belytschko–Tsay element formulation (Type 2). The mesh size for the blank was set to 1 mm. The boundary and loading conditions are applied as shown in Figure 11.7. Studies have been carried out to validate the isotropic springback model.

## 11.2.6 ISOTROPIC HARDENING SPRINGBACK MODEL

In the isotropic hardening model, the hardening part of the material behaviour obeys the Krupkowsky law, Eq. (11.1), while the plasticity law obeys Hill's plasticity [27].

$$\sigma = k(\varepsilon_o + \varepsilon)^n \tag{11.1}$$

where $k$ is the strength coefficient, $n$ is the work hardening, $\varepsilon$ is the effective plastic strain, and $\sigma$ is the yield stress [28]. The model parameters for different temperatures are given in Table 11.6.

To accurately model and describe the sheet metal behaviour during the simulation process, the material constants of the Krupkowsky strength model should be determined precisely. The calibration procedures to determine the numerical values of material constant ($k$, $n$ and $\varepsilon_0$) at different temperatures as presented in Table 11.4 are as follows:

a. Solve the constitutive equation of the model numerically using Excel or MATLAB® programme.

**TABLE 11.4**

**Material Constants of the Krupkowsky Model for AA6061**

| Annealing Temperature | 200 | 300 | 400 |
|---|---|---|---|
| $K$ | 0.28 | 0.22 | 0.192368 |
| $\varepsilon_o$ | 0.012 | 0.002 | 0.008836 |
| $n$ | 0.2 | 0.12 | 0.126996 |

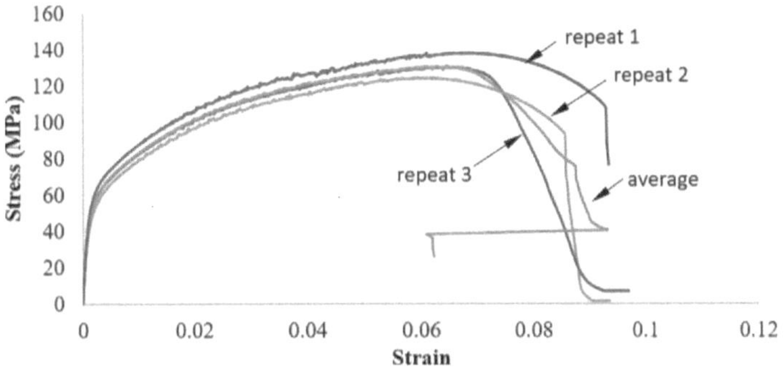

**FIGURE 11.8**  Experimental tensile data of AA6061-T6 Al metal plates at an annealing temperature of 400°C.

   b. Determine the material constants by minimizing the sum square error between the numerical solution and the experimental data of the sheet metal at the different temperatures (200°C, 300°C and 400°C).

   c. Build up the material card for the model.

   d. Test the implementation process of the material card in the PAM-STAMP FE package by comparing the FE results of a baseline one-element tension test with the corresponding model's numerical solution.

   e. Apply the material model in the FE simulation of the bending process of TWB.

For instance, Figure 11.8 shows the experimental tensile data of AA6061 plates at an annealing temperature of 400°C. while Figure 11.9 illustrates the calibration of the Krupkowsky model by fitting the model numerical solution with the experimental stress–strain curve of the AA6061-T6 material at an annealing temperature of 400°C.

## 11.3  RESULTS AND DISCUSSION

For the discussion, this section will be divided into two: experimental and simulation results.

### 11.3.1  EXPERIMENTAL

In our previous study, the combination of two AA6061 alloys with a rolling direction of 0° gave a maximum tensile strength of 172 MPa [26]. Even though the joint

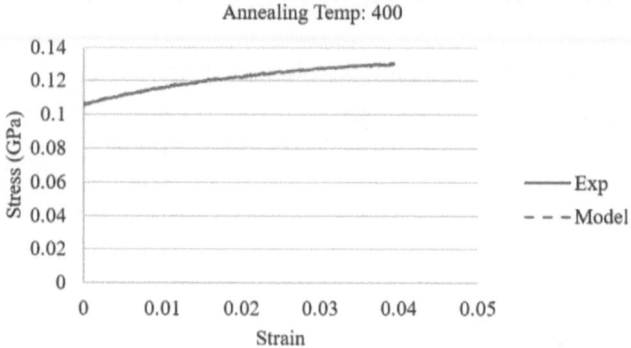

**FIGURE 11.9**  Calibration of the models by fitting the data with the stress–strain curve of the material at an annealing temperature of 400°C.

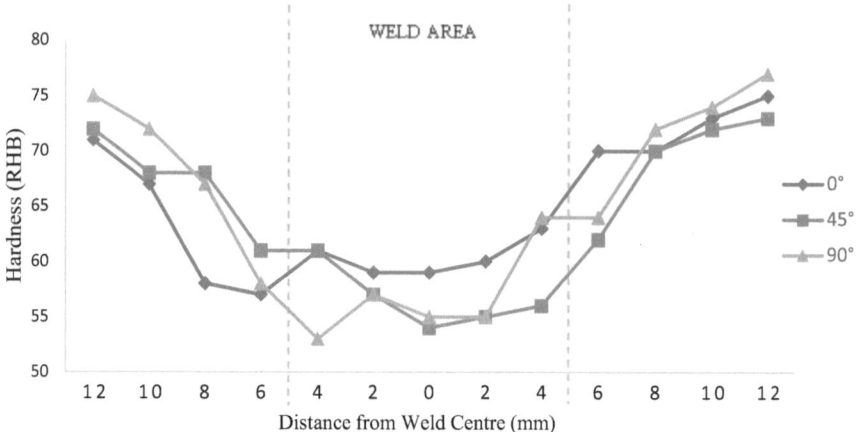

**FIGURE 11.10**  Hardness pattern cross-sectional to the welded zone at different rolling directions. Note: hardness for the AA6061-T6 base metal is 72HRB [22].

efficiency is relatively low, i.e., ~55.4%, a similar pattern can be observed for FSW of AA6061-T6 using a cylindrical flat tool as reported by Chandu et al. [29] and Maneiah et al. [30] at 52.9%–60.0% and 61.6%, respectively. The hardness decreases at the weld area and then increases at 5 mm from the centre. The hardness of the weld zone is lower than that of the base materials (Figure 11.10). Moreover, cracks form along the welded line after certain strokes. The cracks in the metal part are more visible at a high temperature (400°C) than at a low one (200°C). Post-welding heat treatments result in crack formation. The difference in the temperature gradient due to the uneven heating rate may be detrimental to the component. When the metal is heated rapidly or when the annealing temperature in the furnace is not constant, the metal is not heated at a proper rate and therefore produces stress cracks and residual stress when cooled. Furthermore, rapid cooling can affect the hardness and make the metal brittle.

**TABLE 11.5**
**Results of the V-Bending Experiment**

| Experiment | Parameter Setting | Springback (°) | Springback (%) | S/N Ratio |
|---|---|---|---|---|
| 1 | A1B1C1 | 1.293 | 6.089 | −2.6018 |
| 2 | A1B2C2 | 1.230 | 5.791 | −4.2143 |
| 3 | A1B3C3 | 0.727 | 3.421 | 1.6144 |
| 4 | A2B1C1 | 1.710 | 8.051 | −5.4366 |
| 5 | A2B2C2 | 1.223 | 5.760 | −2.4685 |
| 6 | A2B3C3 | 1.817 | 8.553 | −5.3068 |
| 7 | A3B1C1 | 2.257 | 10.625 | −7.7853 |
| 8 | A3B2C2 | 3.203 | 15.082 | −10.1736 |
| 9 | A3B3C3 | 3.027 | 14.250 | −9.6603 |

The displacement of the molecules and the stress and strain of the material causes the springback. When the aluminium metal piece is bent, the compression of the inner region and the expansion of the outer one occur. The internal surface has a higher molecular density than the external surface. When the load is removed, the compressive force becomes lower than the tensile force, which allows the metal piece to go back to its initial state.

Three parameters, namely, annealing time, annealing temperature, and cooling medium, affect the robustness of resistance of the process against the variation caused by the noise factor. The process is designed to determine which parameter induces the lowest springback on the aluminium workpiece. Orthogonal arrays are used to balance the different parameter settings so that the factor levels are equally weighed to access each parameter independently. This step reduces the time and cost associated with the experiment. The effect of noise factors on the characteristic of the welded blank is evaluated using the signal-to-noise ratio (SNR). The springback value and the initial bend angle are determined from the experiment to calculate the springback percentage and SNR (Table 11.5). Table 11.4 indicates that the A1B3C3 method yields the lowest springback, with a positive value of 1.6144. A high SNR signifies a low noise factor.

The three different parameters that affect the springback are expressed in an integrated form (Table 11.6). Delta is calculated by subtracting the high value from the levels with low values. The ranking will determine which parameter exerts the greatest effect on the springback value. Table 11.5 shows that the delta for annealing temperature is the highest (7.472); hence, this parameter is the main influencing factor of the springback in this experiment. The quenching medium ranks second, and the annealing time ranks the last among the three. The optimal performance of the springback ratio under parameters A1, B3 and C3 is obtained when oil is used as the quenching medium at an annealing temperature and time of 200°C and 60 min, respectively (Figure 11.8). A low SNR is desired because the objective is to reduce the response (i.e., springback). The high values in the SNR graph represent the control factor settings that minimize the effects of the noise factors.

**TABLE 11.6**
**SNR for Springback Percentage Observations**

| Level | Annealing Temperature | Annealing Time | Quenching Medium |
|-------|----------------------|----------------|------------------|
| 1     | −1.734               | −5.275         | −6.027           |
| 2     | −4.404               | −5.619         | −6.437           |
| 3     | −9.206               | −4.451         | −2.880           |
| Delta | 7.472                | 1.168          | 3.557            |
| Rank  | 1                    | 3              | 2                |

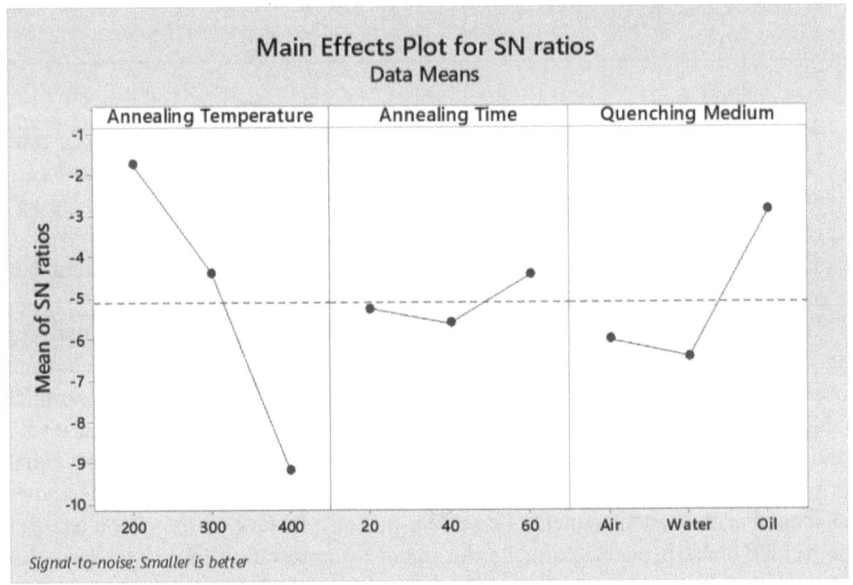

**FIGURE 11.11**   Effect of process parameters on springback.

As previously mentioned, the annealing temperature is the main factor that affects the springback effect. Figure 11.11 shows that this parameter exhibits the highest SNR at 200°C; the SNR, however, constantly drops as the temperature increases. This finding suggests that the higher the annealing temperature, the lower the springback effect. A small fluctuation is observed in the annealing time readings. The result shows that 60 min is the optimal annealing time for minimizing the springback effect. The SNR increases from −5.275 at 20 min to −4.451 at 60 min, with a slight decrement to −5.619 at the 40 min mark. The lowest springback angle is obtained when oil is used as the cooling medium, with a ratio of −2.88. The highest springback angle is obtained when water is used as the medium to cool the annealed metal piece, followed by air. Moreover, the lowest springback effect is achieved when the aluminium metal is annealed at 200°C. The springback angle and percentage of the untreated Al metal are 1.62° and 7.627%, respectively (Table 11.6). Experiments 1, 2, 3, and 5 display lower springback angles in comparison with the untreated Al metal piece. The parameters that demonstrated high occurrences among the four experiments include A1 (annealing temperature at 200°C), B2 (annealing time of 40 min), and C3 (oil medium). Experiment 3 exhibited the lowest springback effect, with a springback angle of 0.727° and a springback percentage of 3.421%. In summary, the best setting for reducing the springback effect is when oil is used as the medium to prevent rapid cooling at an annealing temperature of 200°C and an annealing time of 60 min. Moreover, this finding is in agreement with the statement that lowering the annealing temperature, will lower the springback effect as indicated by Mehta and Badheka [10]. For untreated blank, the measured springback is 7.627% or ~1.62 degrees in angle. The result indicated that the springback percentage decreases from 6.4% at 200°C to 4.2% at 400°C, whereas the present experiment suggests that the springback percentage increases from 1.743% at 200°C to 9.206% at 400°C. The finding is in agreement with the result obtained by Zhang et al. [31].

The temperature consistently increases throughout the experiment, which indirectly causes data changes. In this experiment, the surrounding temperature that may affect the data is disregarded. The assumptions for the experiments are as follows: (1) the temperature is stable and consistent throughout the entire annealing process, (2) the aluminium strip is heated evenly, (3) no contamination is present in the cooling media, and (4) the aluminium strip pieces are consistently quenched 20 s after leaving the furnace.

## 11.3.2   FE Simulation

Two different parameters (flow curve for different annealing temperatures, 200°C, 300°C, and 400°C and different strokes, 3, 6, and 9 mm) were considered in the simulation to investigate their effect on springback. An example of the simulation results is shown in Figure 11.12 which shows the comparison of the part profile before and after springback at different strokes, 3, 6, and 9 mm at the annealing temperature condition of 400°C. It is noticed that upon increasing the stroke, the springback at the edges of the bank increases. However, the springback at the centre area of the blank is nearly similar.

To explore this further, Figure 11.13 shows the calculated distance between the part before and after springback at an annealing temperature of 400°C and different strokes, (a) 3 mm, (b) 6 mm and (c) 9 mm. In the figure, it is shown that the black and dark grey colour distribution represents a negative springback (part edges) and the light grey-coloured distribution represents a positive springback (the Centre of the part).

Figure 11.14 represents the relation between the distance between the parts before and after springback at an annealing temperature of 400°C at different strokes. As noticed, the central area has a positive springback and is constant at a different stroke. However, the edge of the part has a negative springback and is highly dependent on the amount of deformation (stroke). Under high deformation conditions, the negative springback is high and decreases with low deformation.

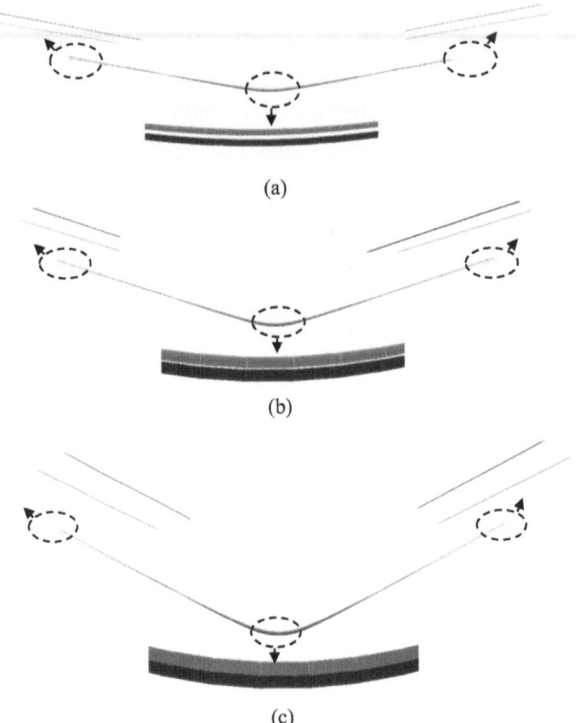

FIGURE 11.12   Simulation results showing the comparison of the part before and after springback at an annealing temperature of 400°C and different strokes: (a) 3 mm, (b) 6 mm and (c) 9 mm.

The same effect occurred at annealing temperatures of 300°C and 200°C, as shown in Figures 11.15 and 11.16. However, there is another remark recognized from the results in Figures 11.14 and 11.15. At low-temperature conditions, the amount of negative springback increases appreciably and particularly at a large stroke. One of the main factors that affect the springback is the ductility and hardness of the material. As the temperature increases, a significant increase in the ductility and decrease in the hardness of welded joint was observed as reported by Rao et al. [32]; therefore as the result, the springback will increase.

### 11.3.3   COMPARISON BETWEEN EXPERIMENTAL AND SIMULATION RESULTS

Experimental measurements (springback angles) are used to validate the FE model. As mentioned previously, the difference between the angles of the bent strip was measured using the CMM. In the experimental measurement, the 3 mm stroke at different temperatures is the only data measured and used to validate the model. Therefore, the simulated parts at a stroke of 3 mm and different annealing temperatures are exported from PAM-STAMP and used in the comparison. Figure 11.17 shows the comparison of measured angles of the FE simulations and experiments for a punch

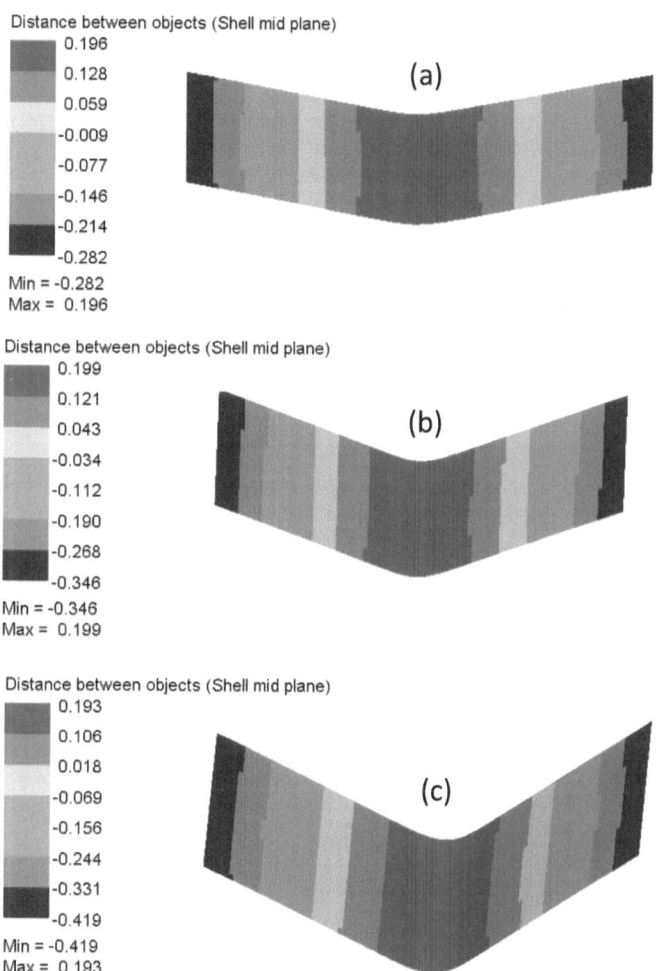

**FIGURE 11.13**  FE simulation results showing the distance between the parts before and after springback at an annealing temperature of 400°C and different strokes: (a) 3 mm, (b) 6 mm and (c) 9 mm.

stroke of 3 mm and annealing temperatures of 200°C, 300°C and 400°C. From the results shown in Figure 11.16, the error (%) between the experimental and simulation is calculated, and it is within 8%. This is can be considered a good agreement between the model and experimental data. The springback model used in this study is an isotonic hardening model which cannot be used to describe the cyclic behaviour of the material. It cannot describe the early re-yielding and Bauschinger effect. It usually gives an overestimation of stress level while cyclic loading. Therefore, kinematic hardening rules were introduced to overcome the limitation of the isotropic hardening rules.

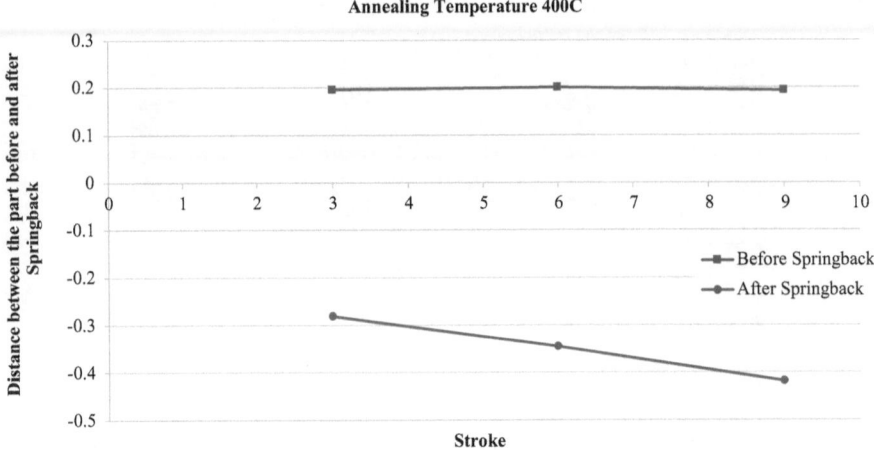

**FIGURE 11.14** The relation between the distance between the parts before and after springback at an annealing temperature of 400°C at different strokes.

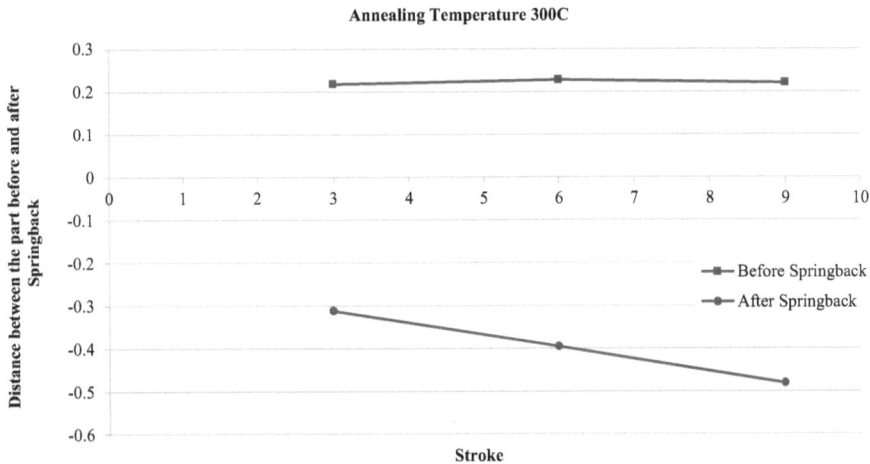

**FIGURE 11.15** The relation between the distance between the parts before and after springback at an annealing temperature of 300°C at different strokes.

## 11.4 CONCLUSIONS AND FUTURE WORKS

This study investigated the springback pattern and the effect of PWHT parameters (i.e., annealing parameters) to reduce the springback on friction stir welded blank. The joint efficiency is expected to be relatively low for FSW of AA6061 using cylindrical flat, i.e., ~55.4%. However, the hardness, on average, after welding was reduced by up to 30%. The results showed that the optimal annealing temperature

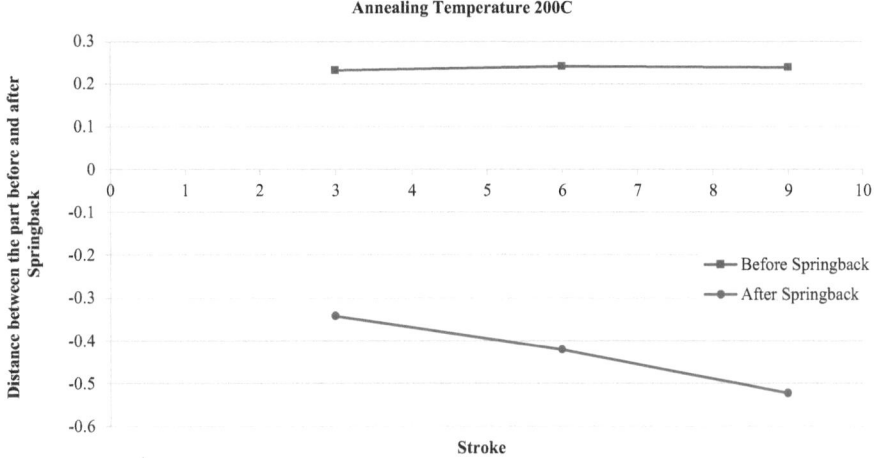

**FIGURE 11.16**   The relation between the distance between the parts before and after springback at an annealing temperature of 200°C at different strokes.

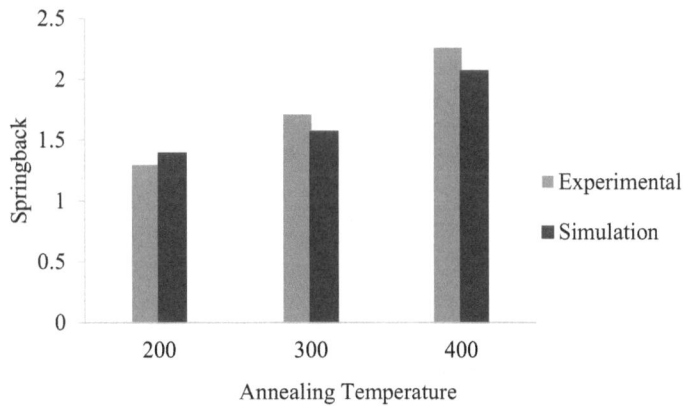

**FIGURE 11.17**   Shows the comparison of measured angles of the FE simulations and experiments for a punch stroke of 3 mm and annealing temperatures of 200°C, 300°C and 400°C.

and time for producing the lowest springback angle are 200°C and 60 min, respectively, and the most suitable quenching medium is oil. The annealing temperature is directly proportional to the springback percentage, whereas the annealing time is inversely proportional to the latter. For the cooling medium, a slow cooling rate yields a low springback percentage. For future works, we will investigate the effect of post-welding treatment on the springback pattern of dissimilar materials, as well as on the thickness variation of the welded blanks. These findings were supported by simulation results with a <10% difference.

## REFERENCES

1. S. Shanavas and J.E. Raja Dhas, Weldability of AA 5052 H32 aluminium alloy by TIG welding and FSW process: A comparative study, *IOP Conf. Ser. Mater. Sci. Eng.*, vol. 247, no. 1, p. 012016, 2017.
2. S. Shanavas, J.E. Raja Dhas and N. Murugan, Weldability of marine grade AA 5052 aluminium alloy by underwater friction stir welding, *Int. J. Adv. Manuf. Technol.*, vol. 95, no. 9–12, pp. 4535–4546, 2018.
3. T.V. Christy, N. Murugan and S. Kumar, A comparative study on the microstructures and mechanical properties of Al 6061 alloy and the MMC Al 6061/TiB2/12P, *J. Miner. Mater. Charact. Eng.*, vol. 09, no. 01, pp. 57–65, 2010.
4. K.S. Kumar, Tensile strengthand hardness teston friction stir welded aluminum 6061-T6 and 5083-H111-O alloys, *Int. J. Sci. Dev. Res.*, vol. 2, no. 1, pp. 88–93, 2017.
5. R. Kumar, S. Varghese and M. Sivapragash, A comparative study of the mechanical properties of single and double sided friction stir welded aluminium joints, *Procedia Eng.*, vol. 38, pp. 3951–3961, 2012.
6. J. Hirsch, Aluminium in innovative light-weight car design, *Mater. Trans.*, vol. 52, no. 5, pp. 818–824, 2011.
7. H. Lin, Y. Wu, S. Liu and X. Zhou, Effect of cooling conditions on microstructure and mechanical properties of friction stir welded 7055 aluminium alloy joints, *Mater. Charact.*, vol. 141, pp. 2116–2120, 2018.
8. N. Balaji, S. Kannan and S. Arun, Performance analysis of friction stir welding on aluminium Aa7075 and Aa2024 alloy material, *Int. J. Eng. Res. Adv. Technol.*, vol. 3, no. 4, pp. 10–16, 2017.
9. X. Liu, S. Lan and J. Ni, Analysis of process parameters effects on friction stir welding of dissimilar aluminum alloy to advanced high strength steel, *Mater. Des.*, vol. 59, pp. 50–62, 2014.Hassan, R., Hassan, G., and Sudarsanam, B. (2016). Effects of Heat Affected Zone Softening Extent on Strength of Advanced High Strength Steels Resistance Spot Weld. 10th International Conference on Trends in Welding Research, Tokyo, Japan
11. M. Ilangovan, S. Rajendra Boopathy and V. Balasubramanian, Effect of tool pin profile on microstructure and tensile properties of friction stir welded dissimilar AA 6061–AA 5086 aluminium alloy joints, *Def. Technol.*, vol. 11, no. 2, pp. 174–184, 2015.
12. K.G. Krishna, A. Devaraju, and B. Manichandra, Study on mechanical propreties of friction stir welded dissimilar AA2024 and AA7075 aluminum alloy joints, *Mater. Sci. Eng. B*, vol. 11, no. 3, pp. 285–291, 2017.
13. J.R. Kim, E.Y. Ahn, H. Das, Y.H. Jeong, S.T. Hong, M. Miles and K.J. Lee, Effect of tool geometry and process parameters on mechanical properties of friction stir spot welded dissimilar aluminum alloys, *Int. J. Precis. Eng. Manuf.* vol. 18, pp. 445–452, 2017.
14. J.M. Piccini and H.G. Svoboda, Effect of the tool penetration depth in friction stir spot welding (FSSW) of dissimilar aluminum alloys, *Procedia Mater. Sci.*, vol. 8, pp. 868–877, 2015.
15. D. Muruganandam, Influence of post weld heat treatment in friction stir welding of AA6061 and AZ61 alloy, *Russ. J. Nondestr. Test*, vol. 54, pp. 294–301, 2018.
16. M. Momeni and M. Guillot, Post-weld heat treatment effects on mechanical properties and microstructure of AA6061-T6 butt joints made by friction stir welding at right angle (RAFSW), *J. Manuf. Mater. Process.*, vol. 3, no. 2, p. 42, 2019.
17. N. Oztoprak, C.E. Yeni and B.G. Kiral, Effects of post-weld heat treatment on the microstructural evolution and mechanical properties of dissimilar friction stir welded AA6061+SiCp/AA6061-O joint, *Lat. Am. J. Solids Struct.*, vol.15, no.8, pp. 1–17, 2018.

18. M. Cabibbo, A. Forcellese, M. Simoncini, M. Pieralisi and D. Ciccarelli, Effect of welding motion and pre-/post-annealing of friction stir welded AA5754 joints, *Mater. Des.*, vol. 93, pp. 146–159, 2016.

19. A. Suri, R. Setia and K.H. Raj. Pre and post annealing effect on dissimilar friction stir welding of Al 5083/6063 using new pedal shape pin tool, *IOP Conf. Ser. J. Phys. Conf. Ser.*, vol. 1240, p. 012167, 2019.

20. M. Cabibbo, A. Forcellese, E. Santecchia, C. Paoletti, S. Spigarelli and M. Simoncini, New approaches to friction stir welding of aluminum light-alloys, *Metals*, vol. 10, no. 2, p. 233, 2020.

21. P. Prasanna, C.H. Penchalayya and D.A. Rao, Effect of tool pin profiles and heat treatment process in the friction stir welding of AA 6061 aluminum alloy, *Am. J. Eng. Res.*, vol. 2, no. 1, pp. 7–15, 2013.

22. Z. Zhang, H. Zhang, Y. Shi, N. Moser, H. Ren, K.F. Ehmann and J. Cao, Springback reduction by annealing for incremental sheet forming, *Procedia Manuf.*, vol. 5, pp. 696–706, 2016.

23. M.F. Adnan, A.B. Abdullah and Z. Samad, Effect of annealing, thickness ratio and bend angle on springback of AA6061-T6 with non-uniform thickness section, *MATEC Web Conf.*, vol. 90, no. 01002, 2017.

24. C. Wang, S. Wang, S. Wang, G. Chen and P. Zhang, Investigation on springback behavior of Cu/Ni clad foils during flexible die micro V-bending process, *Metals*, vol. 9, no. 8, p. 892, 2019.

25. L.V. Kamble, S.N. Soman and P.K. Brahmankar, Understanding the fixture design for friction stir welding research experiment, *Mater. Today Proc.*, vol. 4, no. 2, pp. 1277–1284, 2017.

26. A.F. Pauzi, A.B. Abdullah, M.F. Jamaluddin, Pre-forming evaluation of dissimilar aluminium alloys blank fabricated using friction stir welding technique, *IOP Conf. Series Mater. Sci. Eng.*, vol. 670, no. 1, p. 012077, 2019.

27. A. Elsayed, M. Mohamed, M. Shazly and A. Hegazy, An investigation and prediction of springback of sheet metals under cold forming condition, *IOP Conf. Series Mater. Sci. Eng.*, vol. 280, no. 1, p. 021021, 2017.

28. M. Amer, M. Shazly and M. Mohamed, et al. Ductile damage prediction of AA 5754 sheet during cold forming condition, *J. Mech. Sci. Technol.*, vol. 34, pp. 4219–4228, 2020.

29. K.V.P.P. Chandu, , E. Venkateswara Rao, A. Srinivasa Rao and B.V. Subrahmanyam, The strength of friction stir welded aluminium alloy 6061, *IJRMET*, vol. 4, no. 1, pp. 119–122. 2014.

30. D. Maneiah, D. Mishra, K. Prahlada Rao and K. Brahma Raju, Process parameters optimization of friction stir welding for optimum tensile strength in Al 6061-T6 alloy butt welded joints, *Mater. Today Proc.*, vol. 27, part 2, pp. 904–908, 2020.

31. Z. Zhang, H. Zhang, Y. Shi, N. Moser, H. Ren, K.F. Ehmann and J. Cao, Springback reduction by annealing for incremental sheet forming, *Procedia Manuf.*, vol. 5, pp. 696–706, 2016.

32. P.N. Rao, D. Singh and R. Jayaganthan, Effect of annealing on microstructure and mechanical properties of Al 6061 alloy processed by cryorolling, *Mater. Sci. Technol.*, vol. 29, no. 1, pp. 76–82, 2013.

# 12 Springback Analysis of Low-Power Laser Welded Aluminum Tailor Welded Blanks

*MF Jamaludin*
Universiti Malaya

*AB Abdullah*
Universiti Sains Malaysia

*F Yusof*
Universiti Malaya

## CONTENTS

## 12.1 INTRODUCTION

Tailor welded blanks (TWBs) are semi-finished stock materials made from welding two or more metal sheets together to form a single blank, which is then formed into the desired shape. A TWB can be made up of combinations of different materials, treatments, thicknesses, and/or coating in a single blank. The technology was introduced in the 1980s by automotive manufacturers as a lightweight construction method that optimizes the locations of materials for weight savings and localized strengthening of the component. Figure 12.1 shows an example of the application of TWBs to produce a vehicle door.

DOI: 10.1201/9781003164241-12

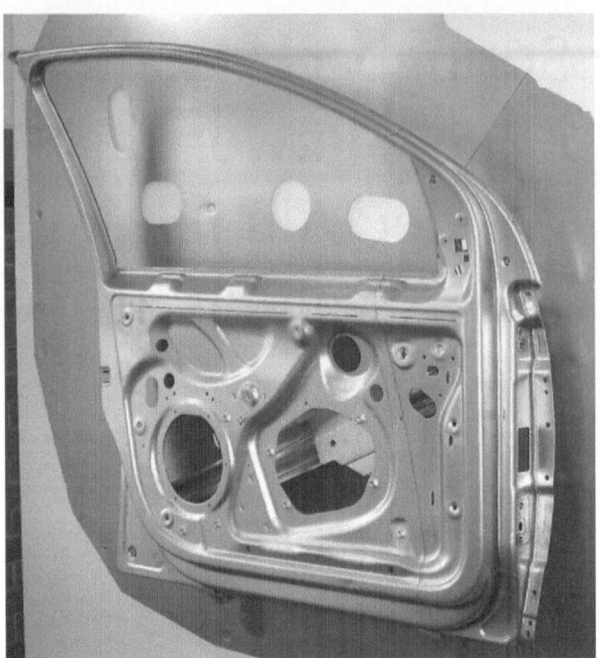

**FIGURE 12.1**   TWB behind a lightweight door inner panel, Tailored Blank Hoesch Museum. (Photograph by Stahlkocher, distributed under a CC BY-SA 3.0 licence.)

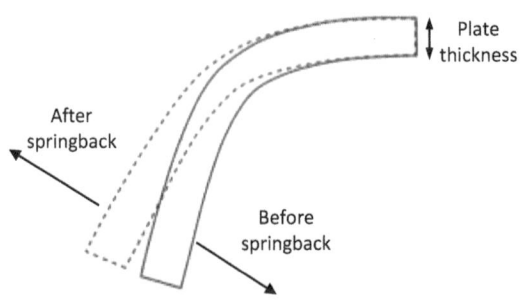

**FIGURE 12.2**   Illustration of the springback effect for a simple bending.

As with all metal forming processes, forming of TWBs is subjected to springback effects, due to elastic recovery of the material just after completion of the bending. Springback is generally referred to as the change of part shape that occurs upon removal of constraints after forming (Xia and Cao, 2014). This is illustrated in Figure 12.2 for a simple bending process.

Springback directly affects the geometrical precision of the workpiece, which results in a deviation from the shape and size of the working die. The final part's dimensional shape does not match the desired nominal shape, falling outside of required tolerances and making subsequent assembly difficult.

Factors that affect springback include material properties, material thickness, part shape, bending location and forming techniques. A review on the effect of springback on AA6061 sheet metal bending conducted by Cinar et al. (2021) has described some of the parameters considered by researchers such as the bent angle, die shoulder radius, blank holder force, sheet thickness, punch thickness, Young's modulus, Poisson's ratio, annealing temperature, applied load and bending operations. It was noted that sheet thickness and bend angle parameters have substantial effects on the springback. It has been noted that the level of springback was closely related to the tangential stress distribution in the top and bottom surfaces of the blank after forming. A smaller difference between the stresses in the top and bottom layers of the blank would result in lower levels of springback (Wang et al., 2016).

Springback is a semi-predictable factor which can be compensated by controlling the bending parameters such as the bending angle, pressures, and tool configurations. In addition, finite element (FE) software can also be used to compute and compensate for springback in complex 3D geometries.

### 12.1.1 METHODS FOR DETERMINING SPRINGBACK

There are several methods used for the evaluation of springback. These test methods are usually based on the bend testing of the material. As listed by Kędzierski and Popławski (2019), different bending devices are based on the ISO 7348, ASTM E290, VDA 238-100, JIS Z 2248, and AWS D1.1, among others. The ASTM E290 describes procedures for four conditions of constraints on the bend portion of the specimens, namely, (1) a guided bend test, (2) a semi-guided bend test, (3) a free-bend test, and (4) a bend and flatten test (E28 Committee, 2014). Different test set-ups have been proposed and standardized for the evaluation of bending performance. Among the most used method is the guided bending into a V-shaped die using a similarly shaped stamp. However, as most standards do not restrict the device geometries of the bending set-up, such as the die widths and opening angles, some researchers have fabricated their own V-bending punch and die designs. Some recent studies on deformation behaviour and springback for aluminium alloys are shown in Table 12.1.

### 12.1.2 RECENT STUDIES ON SPRINGBACK OF ALUMINIUM TWBs

Springback evaluations of aluminium TWBs are expanding, especially with the introduction of solid-state joining methods, such as friction stir welding. For example, Kim et al. (2011) have compared the springback of three aluminium automotive alloys, 6111-T4, 5083-H18, and 5083-O produced by FSW, each having one or two different thicknesses. Post-forming evaluation, inclusive of springback for dissimilarly joined aluminium to steel TWB by FSW, has also been investigated by a number of researchers, as described by Rizlan et al. (2021) (Table 12.2).

In this study, the springback of aluminium alloys AA5052-H32 and AA6061-T6 TWBs would be analysed. The evaluation would be for AA5052-AA5052 joints of dissimilar thicknesses and AA5052-AA6061 joints of similar thicknesses. The springback of the TWBs would be compared with those of the base materials.

**TABLE 12.1**

**Recent Studies on the Deformation and Springback Behaviour of Aluminium Alloys**

| Method | Materials | Factors Evaluated | References |
|---|---|---|---|
| V-bending | AA7075 | Deformation temperature, punch radius, blank holder force | Zhou et al. (2022) |
| | | Deformation temperature and deformation speed | Kilic (2019) |
| | AA2024 | Effect of electric current pulses | Mohammadtabar et al. (2021) |
| | AA5754 | Effect of pre-straining on springback | Toros et al. (2011) |
| | AA6016 | Temperature, blank holding force, die corner radius | Ma et al. (2019) |
| | AA6061 | Effect of arc pretreatment | Ren et al. (2017) |
| | | Length and thickness of specimen, bending angle | Abdullah and Samad (2013) |
| | AA6082-T6 | Bending radius, temperature | Suckow et al. (2019) |
| | | Laser-assisted bending | Gisario et al. (2011) |
| 3-point bending test | AA6061 | Ageing time | Guan et al. (2020) |
| U-shapes bending, L-shaped bending | AA5754 | Springback under hot stamping conditions, high forming speeds, effect of die gap, corner radius | Wang et al. (2016) |
| | AA5052-O | Effect of forming a gap, discharge voltage of electromagnetic-assisted bending on springback | Xiao et al. (2019) |
| Customized stretch bending device | AL5086-H111 | Effect of punch velocity, rolling direction, punch-die clearance, holding time, heat treatment conditions | Hakimi and Soualem (2021) |

## 12.2 METHODOLOGY

### 12.2.1 SAMPLE PREPARATIONS

TWB specimens were prepared by double-sided autogenous butt weld using a fibre laser weld (Rofin Starfiber 300). The welding configuration is shown in Figure 12.3.

The selected aluminium alloys were AA5052-H32 and AA6061-T6, with thickness variations of 1.0 and 1.5 mm. The laser processing parameters were set for variations in welding power from 270 to 300 W and the welding speeds of 15 and 20 mm/s. In this study, two cases of TWBs would be evaluated, namely, (1) similar alloy, dissimilar thicknesses and (2) dissimilar alloy, similar thicknesses (Table 12.3).

Test specimens for springbacks were prepared according to the dimensions specified in Figure 12.4 by a Wire Electrical Discharge Machining (WEDM) process . Base material samples were also prepared for comparison.

**TABLE 12.2**

**Selected Studies on Springback for Non-Uniform Aluminium TWBs**

| Method | Specimen | Factors Evaluated and Findings | References |
|---|---|---|---|
| V-bending and Taguchi method | Dissimilar thickness AA6061 prepared by forging | • Thickness ratio, alignment, bend radius.<br>• Contribution of parameters to springback were 92.51% (bend radius), 4.54% (thickness ratio), and 0.93% (alignment) | Adnan et al. (2017) |
| V-bending | AA6061-T6 to 5052-H32 prepared by FSW | • Dissimilar grade combinations<br>• Rolling direction<br>• Shoulder diameter | Durga Rao and Ganesh Narayanan (2014) |
| Cylindrical bending test, draw bending test | AA6111-T4, 5083-H18, 5083-O prepared by FSW | • Material property dependence on springback was more apparent for unconstrained cylindrical bending test<br>• Bending was localized at the weld zone for 6111-T4 and 5083-O | Kim et al. (2011) |
| FE analysis | AA2024-T352 properties as input data for FSW blank geometry | Three different models were evaluated:<br>(1) no weld details<br>(2) with mechanical properties of weld nugget<br>(3) with mechanical properties of the weld nugget and HAZ | Zadpoor et al. (2007) |

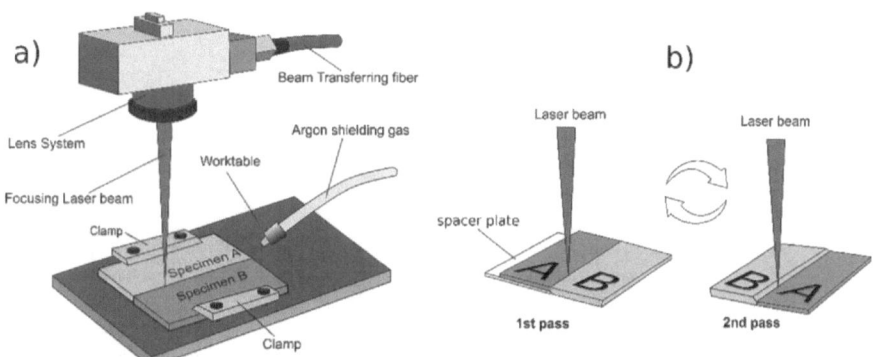

**FIGURE 12.3** Preparation of the dissimilar sample by a low-power fibre laser showing (a) the laser welding configuration and (b) successive double-sided laser welding pass.

A V-bending test was conducted using a V-bend die and punch jig mounted on a universal testing machine (Instron UTM, model 3369), as shown in Figure 12.5. To account for the thickness variation in the dissimilar thickness specimens, a 0.5 mm steel metal shim was placed into the die to account for the gap difference, accordingly.

Angular measurements of the bend were evaluated using a profile projector (Mitutoyo PJ-R3000), as shown in Figure 12.6 and compared with the angle of the

## TABLE 12.3
## TWB Specimen Welding Parameters

| Laser Welding Parameters | Group A (Similar Alloy, Dissimilar Thickness) | Group B (Dissimilar Alloy, Similar Thickness) |
|---|---|---|
| Material pair | AA5052-H32 (1.5 mm) to AA5052-H32 (1.0 mm) | AA5052 (1.0 mm) to AA 6061-T6 (1.0 mm) |
| Laser power (W) | 270, 290, 310 | 270, 290, 310 |
| Welding speed (mm/s) | 15, 20 | 15, 20 |

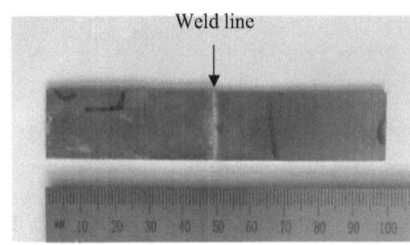

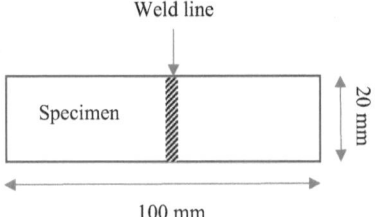

**FIGURE 12.4**   Test specimen for V-bending.

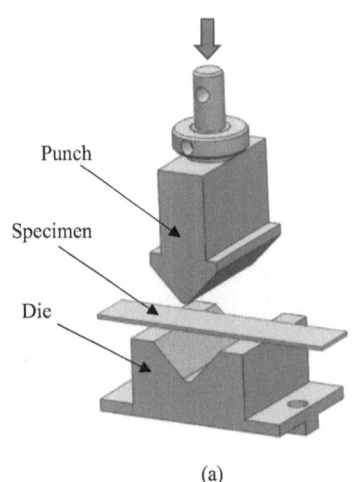

(a)                                               (b)

**FIGURE 12.5**   V-bending test showing (a) schematic of the test and (b) testing jig mounted on an Instron 3369 UTM.

punch to determine the springback angle. The average springback is determined by comparing the internal angle of the bend during loading and after unloading, schematically shown in Figure 12.7. The evaluations were made for the TWB samples and compared with the results for the base materials.

(a)                                   (b)

**FIGURE 12.6**  Measurement of the springback angle using a profile projector, showing the (a) apparatus used and (b) measurement of angular deviation from 90° (measured internal radius).

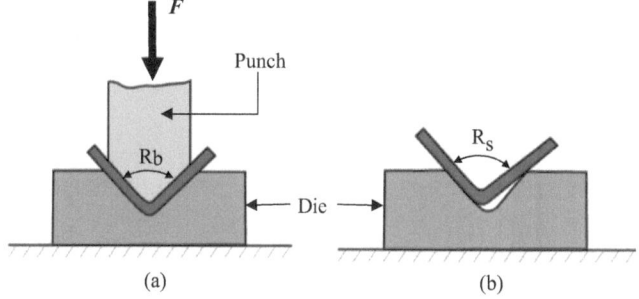

(a)                                   (b)

**FIGURE 12.7**  Determination of the springback angle by comparison of the bending angle (a) during loading, $R_b$ and (b) after unloading, $R_s$.

## 12.3  RESULTS AND DISCUSSION

### 12.3.1  SPRINGBACK OF BASE MATERIALS

Springback evaluation was conducted for base material specimens with thicknesses of 1.0, 1.5, and 2.0 mm with rolling orientations of 0°, 45° and 90°. Each springback value was calculated from the average of three specimen results. The springback values of the base materials are shown in Figure 12.8.

In general, the springback of AA6061-T6 is larger than AA5052-H32, which is due to the larger yield strength of the material. It can also be seen that the springback increases with increasing thickness. The influence of rolling direction was minimal, except for 1.0 mm thickness of AA5052-H32 specimens, where the springback angle increases with increasing rolling direction angle, from 0 to 90.

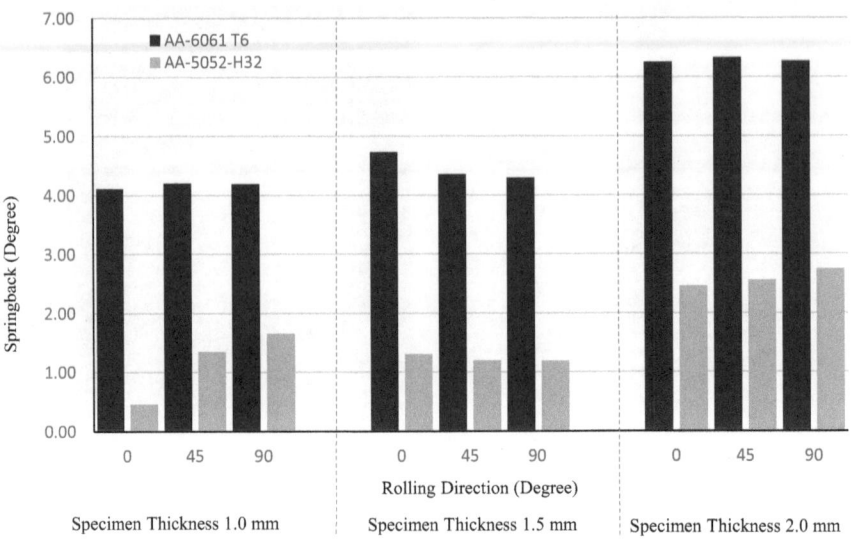

**FIGURE 12.8** Springback of base materials (AA5052-H32 and AA6061-T6).

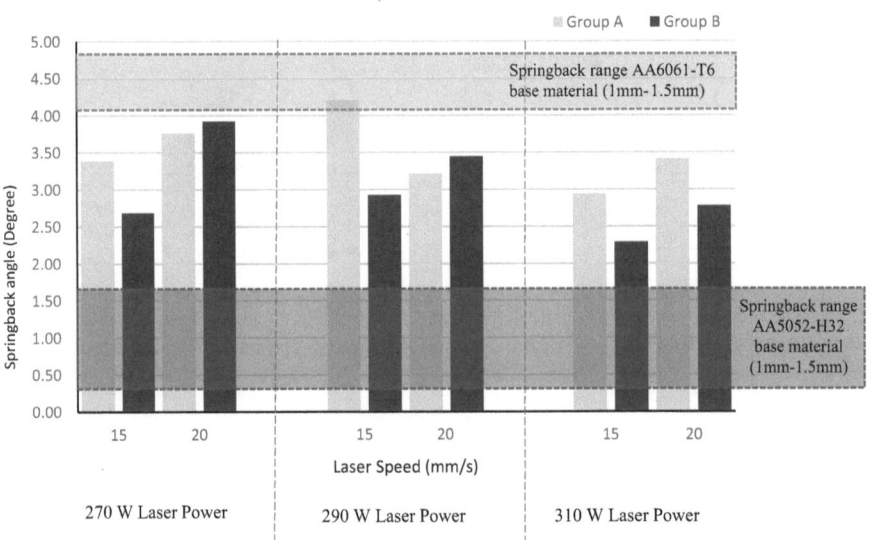

**FIGURE 12.9** Springback of aluminium TWBs for group A and group B specimens.

### 12.3.2 SPRINGBACK OF ALUMINIUM TWB SPECIMENS

Figure 12.9 shows that the springback values of the aluminium TWBs lie in between the springback range of their constituent base materials.

The TWB specimens show better springback performance as compared to the AA6061-T6 base material but are inferior to AA5052-H32. This is similar to the finding by Durga Rao et al. (2014) in their assessment of the 6061-T6–5052-H32 FSW joint. In the case of dissimilar thickness, increasing the laser power tends to reduce the springback of AA5052-H32 joints. While for the case of different material pairs of AA5052-H32 and AA6061-T6, for a particular laser power, increasing the laser speed would increase the springback angle.

## 12.4  CONCLUSION

From the analysis of the springback of laser-welded AA5052-H32 and AA6061-T6, it can be concluded that

- The TWB specimens show better springback performance as compared to 6061-T6 base material but are inferior to AA5052-H32.
- For a particular laser power, increasing the laser welding speed would increase the springback angle for the AA5052-AA6061 joint.

## REFERENCES

Abdullah, A. B., & Samad, Z. (2013). An experimental investigation of springback of AA6061 aluminum alloy strip via V-bending process. *IOP Conference Series: Materials Science and Engineering, 50*(1), 012069.

Adnan, M. F., Abdullah, A. B., & Samad, Z. (2017). Springback behavior of AA6061 with non-uniform thickness section using Taguchi method. *The International Journal of Advanced Manufacturing Technology, 89*(5–8), 2041–2052. doi: 10.1007/s00170-016-9221-0.

Cinar, Z., Asmael, M., Zeeshan, Q., & Safaei, B. (2021). Effect of springback on A6061 sheet metal bending: a review. *Journal of Kejuruteraan, 33*(1), 13–26.

Durga Rao, B., & Ganesh Narayanan, R. (2014). Springback of friction stir welded sheets made of aluminium grades during V-bending: An experimental study. *ISRN Mechanical Engineering, 2014*, 1–15. doi: 10.1155/2014/681910.

E28 Committee. (2014). ASTM E290-14 standard test methods for bend testing of material for ductility. ASTM International.

Gisario, A., Barletta, M., Conti, C., & Guarino, S. (2011). Springback control in sheet metal bending by laser-assisted bending: Experimental analysis, empirical and neural network modelling. *Optics and Lasers in Engineering, 49*(12), 1372–1383. doi: 10.1016/j.optlaseng.2011.07.010.

Guan, W., Ting, L., Linyuan, K., Guangxu, Z., Xuejun, Z., Xin, S., & Zhiwen, L. (2020). Influence of artificial aging time on bending characteristics of 6061 aluminum sheet. *Materialwissenschaft Und Werkstofftechnik, 51*(11), 1533–1542. doi: 10.1002/mawe.202000100.

Hakimi, S., & Soualem, A. (2021). Evaluation of the sensitivity of springback to various process parameters of aluminum alloy sheet with different heat treatment conditions. *Engineering Solid Mechanics, 9*(3), 323–334. doi: 10.5267/j.esm.2021.1.005.

Kędzierski, P., & Popławski, A. (2019). Development of test device for springback study in sheet metal bending operation. *AIP Conference Proceedings, 2078*, 020016. doi: 10.1063/1.5092019.

Kilic, S. (2019). Experimental and Numerical investigation of the effect of different temperature and deformation speeds on mechanical properties and springback behaviour in Al-Zn-Mg-Cu alloy. *Mechanics, 25*(5), 406–412. doi: 10.5755/j01.mech.25.5.22689.

Kim, J., Lee, W., Chung, K.-H., Kim, D., Kim, C., Okamoto, K., Wagoner, R., & Chung, K. (2011). Springback evaluation of friction stir welded TWB automotive sheets. *Metals and Materials International, 17*(1), 83–98.

Ma, W., Wang, B., Xiao, W., Yang, X., & Kang, Y. (2019). Springback analysis of 6016 aluminum alloy sheet in hot V-shape stamping. *Journal of Central South University, 26*(3), 524–535. doi: 10.1007/s11771-019-4024-8.

Mohammadtabar, N., Bakhshi-Jooybari, M., Gorji, H., Jamaati, R., & Szpunar, J. A. (2021). Effect of electric current pulse type on springback, microstructure, texture, and mechanical properties during V-bending of AA2024 aluminum alloy. *Journal of Manufacturing Science and Engineering, 143*(1), 011004.

Ren, D., Zhao, D., Fan, R., Zhao, K., Gang, S., & Chang, Y. (2017). Bending of sheet aluminum alloy assisted by arc pretreatment. *The International Journal of Advanced Manufacturing Technology, 92*(1–4), 1291–1298. doi: 10.1007/s00170-017-0190-8.

Rizlan, M. Z., Abdullah, A. B., & Hussain, Z. (2021). A comprehensive review on pre-and post-forming evaluation of aluminum to steel blanks via friction stir welding. *The International Journal of Advanced Manufacturing Technology, 114*(7), 1871–1892.

Suckow, T., Günzel, J., Schell, L., Sellner, E., Dagnew, J., & Groche, P. (2019). Temperature influence in aluminum sheet metal forming: Springback behavior and process limits during V-bending of EN AW-6082 and EN AW-7075. *WT Werkstattstechnik, 109*(10), 733–739.

Toros, S., Alkan, M., Ece, R. E., & Ozturk, F. (2011). Effect of pre-straining on the springback behavior of the AA5754-0 alloy. *Materiali in Tehnologije, 45*(6), 613–618.

Wang, A., Zhong, K., Fakir, O. E., Liu, J., Sun, C., Wang, L.-L., Lin, J., & Dean, T. A. (2016). Springback analysis of AA5754 after hot stamping: Experiments and FE modelling. *The International Journal of Advanced Manufacturing Technology, 89*(5–8), 1339–1352. doi: 10.1007/s00170-016-9166-3.

Xia, Z. C., & Cao, J. (2014). Springback. In L. Laperrière & G. Reinhart (Eds.), *CIRP Encyclopedia of Production Engineering* (pp. 1133–1138). Springer: Berlin Heidelberg. doi: 10.1007/978-3-642-20617-7_6500.

Xiao, W., Huang, L., Li, J., Su, H., Feng, F., & Ma, F. (2019). Investigation of springback during electromagnetic-assisted bending of aluminium alloy sheet. *The International Journal of Advanced Manufacturing Technology, 105*(1–4), 375–394. doi: 10.1007/s00170-019-04161-8.

Zadpoor, A. A., Sinke, J., & Benedictus, R. (2007). Springback behavior of friction stir welded tailor-made blanks. *Proceedings of IDDRG07*, Győr, Hungary.

Zhou, J., Yang, X., Wang, B., & Xiao, W. (2022). Springback prediction of 7075 aluminum alloy V-shaped parts in cold and hot stamping. *International Journal of Advanced Manufacturing Technology, 119*(1–2), 203–216. doi: 10.1007/s00170-021-08204-x.

# 13 Recent Studies and Related Issues in Tailor Blanks – A Short Review

*AB Abdullah*
Universiti Sains Malaysia

*MF Jamaludin*
Universiti Malaya

## CONTENTS

## 13.1 INTRODUCTION

The tailored blanks have evolved drastically over the years since their first introduction in the 1980s by a German company, ThyssenKrupp. Since then, the tailored blanks have expanded into many forms for numerous applications. The blank is either joined by forming or welding. Buffa et al. (2022) performed a comprehensive study on the blanks joint by forming. Benefits from the utilization of tailored blanks include weight savings (Merklein et al., 2014), component strengthening and reduction in material usage. These advantages can translate to improved environmental impact, both in the production process and utilization of the products, such as in reduced emissions due to fuel savings in lightweight vehicles. This chapter will review the recent updates in tailored blanks fabrication and applications, focusing on the work within the last 5 years.

## 13.2 RECENT FABRICATION METHODS

Established technologies for tailored blanks are based on tailor welded blanks (TWBs) and tailor-rolled blanks (TRBs). Evolution of technology has brought new and hybrid forming methods to produce these blank geometries. For example, Hetzel et al. (2020)

DOI: 10.1201/9781003164241-13

**TABLE 13.1**
**Recent Technologies on Tailored Blank Fabrication**

| References | Name of the Technology | Advantages and Limitation |
|---|---|---|
| Rosochowskia and Olejnik (2017) | Tailor-sheared blank (TSB) | Save more material and reduce the weight of the material as compared to typical welding-based tailored blanks |
| Singar and Banabic (2021) | Tailored hybrid blank (THB) | Offer more material with different properties and behaviour to be joint |
| Graser et al. (2019) | Tailored heat-treated blank (THTB) | Enhancing the formability of high-strength aluminium alloys |

have introduced orbital forming methods to produce tailored blanks with different thickness distributions. Table 13.1 summarizes recent technologies in the fabrication of tailored blanks. There are specific advantages offered by the technologies; for example, material saving via tailor-sheared blanks (TSBs) allows for more material to be joined as offered by tailor hybrid blanks (THBs) and better in terms of formability for certain high-strength alloys that can be achieved by tailor heat-treated blanks (THTBs).

## 13.3   RECENT APPLICATION OF TAILORED BLANKS

Until recently, it was thought that full forward extrusion for a geometry of complicated metal sheet is difficult to be achieved. However, in this case, Reck and Merklein (2021) have proposed the use of sheet-bulk metal forming, combining the operations of bulk forming operations on sheet metals. Testing and evaluation of the method have proved that it can produce better blanks in terms of mechanical properties, die filling and surface integrity, as well as formability. In other applications, tailored blanks were tested on a single-point incremental forming (SPIF) process and have shown promising performance in terms of formability (Panahi-liavoli et al., 2020). Similarly, Tucci et al. (2021) have also explored the plastic behaviour of the joints using SPIF and considered the microhardness of the cross section as a basis for the evaluation. Hetzel et al. (2020) manipulated the advantages offered by tailored blanks in orbital forming by highlighting the material usage efficiency. They introduced the heat treatment process to enhance the formability of the AA6016 to meet the geometrical complexity. Pollock et al. (2020) observed in terms of application in additive manufacturing. While, Saito et al. (2019) studied the bending behaviour of the laser- and plasma-welded blanks, mainly to observe the weld line quality. While Tajul et al. (2017) focused on the forging process capability of tailored blanks with different thicknesses, in another work, Zhang et al. (2021) utilized the spinning process on scrap 2195 Al-Li alloy TWBs produced by friction stir welding under different process parameter designs. Liang et al. (2018) proceeded further to study the performance of the blank during applications, focusing on the aspect of crashworthiness. Klinke et al. (2022) introduced a systematic selection method or parts for TRBs, while Badgujar and Bobade (2021) discussed about stamping of car door inner panel using TWB.

## 13.4 FORMABILITY

A tailored blank sheet consists of at least two different sheet metal components, which can differ in terms of material properties and thicknesses. Thus, each section of the blank will have distinctive mechanical properties and forming behaviours. Thus, the formability of tailored blanks remains a challenge as new materials are being introduced into the technology, such as advanced high-strength steel and aluminium such as AA5182/AA6061 (Parente et al., 2016). In the review by Deepika et al. (2022), it was stated that the method or joining technique will affect the formability of tailored blanks. Several improvements to the technique were recently suggested, such as the proposed friction stir vibration welding (FSVW) by Abbasi et al. (2021) to improve the formability. In general, the two most common test methods to evaluate formability are based on the Nakajima or dome test, and the cupping test methods. Table 13.2 summarizes recent formability studies conducted on tailored blanks and the material types tested.

Simulation techniques can also be utilized to evaluate the formability of tailored blanks. With the advancement of computational power and software, simulation studies are becoming the preferred approach in the last few years. Recent studies by Wu et al. (2021) and Kumar (2020) have focused on the formability of the tailored blanks, although looking at different aspects related to the process, they were similar in terms of determination of limiting dome height (LDH). Other works by Kumar et al. (2019) have utilized finite element analysis to predict the residual stress in biaxial stretching of tailored blanks, while Battina et al. (2022) have studied the effect of the pin profile to the formability of aluminium blanks using both experimental and simulation analyses.

### TABLE 13.2
### Recent Formability Studies on Tailored Blanks

| References | Combined Materials (Thickness) | Joint Technique | Observation |
|---|---|---|---|
| Singar and Banabic (2021) | HC340LA (0.8 mm) -AA6014-T4 (1.2 mm) | CMT | Specimens tend to fail in the more ductile material, in this case, aluminium |
| Singhal (2020) | GNDQAK steel (1.2 mm) | Laser welding | Welding does not significantly affect the formability |
| Panahi-liavoli et al. (2020) | St12 (1 mm) – St14 (1.5 mm) | Nd:YAG pulsed-laser welding | The forming limit angle of TWBs is much lower, in comparison with the base metals |
| Abbasi et al. (2021) | AA6061 (same thickness) | FSW and FSVW | Introduction to vibration-assisted welding improves mechanical properties and formability of the blanks |
| Lalvani and Mandal (2021) | AA5251 (H22) – AA6082 (T6) | LBM and EBM | PWHT improved the formability of Al-6082 (T6) and the level of localised thinning observed after stamping is much improved |

## 13.5   SUSTAINABILITY ISSUE IN TAILORED BLANKS

The initial aim of tailored blank utilization was to reduce the weight of the vehicle for improved fuel efficiency. Issues on sustainability and energy savings are becoming more relevant in the production and forming processes of the tailored blank itself. Several studies have proposed the use of controlled welding and heat-assisted forming to reduce the costs of fabrication. For example, Fadzil et al. (2021) have proposed the use of a low-power fibre laser for joining dissimilar aluminium-tailored blanks. By introducing a double-sided joining, the limited weld depth can be compensated to produce a good joint. In another work, Dwibedi et al. (2018) introduced the micro-plasma-transferred arc ($\mu$-PTA) process in the fabrication of stainless-steel joints of different thicknesses. This process minimizes the area of the heat-affected zone using less energy compared to Gas Tungsten Arc Welding (GTAW), better cost-effectiveness as compared to GTAW and better cost-effectiveness as compared to laser beam welding (LBM) and electron beam welding (EBM) (Chaudhary et al., 2019). In terms of formability, warm forming methods can be utilized to reduce the forces needed for the forming process. It was found by Satya-Suresh et al. (2020) that tailor welding for warm forming resulted in a 33% reduction in material usage and reduced energy requirement in the blank pressing stage.

## 13.6   CONCLUSIONS

Tailored blank technology offers potential weight savings without compromising the mechanical performance for the fabrication of metal-formed components, which is a major process in the automotive industry. This reduces the utilization of materials in production, increases fuel efficiency and reduces emissions during the service life of the vehicle. Future production of automobiles should consider increasing the use of tailored blanks in the fabrication and expanding its use in non-steel materials. As shown in this chapter, there are several issues that can be highlighted for future improvement of tailored blank technology.

1. Apart from the automotive industry, the use of tailored blank technology can be expanded to other industries, such as consumer, machinery, electronic, rail, and aerospace sectors. However, the usage of technology for these industries should be evaluated as limiting issues in the forming of dissimilar materials remain challenging. Lightweight and material savings can be important factors to consider, for example, in the production of high-speed train locomotive structures, although performance aspects of the component, such as the strength, should not be neglected.
2. Environmental impacts of tailored blank production can be reduced by exploring energy-efficient methods for production. For example, low-energy techniques such as highly efficient low-power laser and $\mu$-PTA technologies can be further developed for tailored blank production.
3. New fabrication approaches, especially in the joining of dissimilar and high-performance alloys such as titanium and magnesium, can be proposed.

Solid-state welding, such as friction stir welding (FSW), are potential techniques that have recently entered the automotive production ecosystem. Hybrid techniques, such as the integration of additive manufacturing and FSW, as suggested by Schulte et al. (2020), can open new opportunities for tailored blank productions in the future.

## REFERENCES

Abbasi, M., Bagheri, B., Abdollahzadeh, A., & Moghaddam, A. O. (2021). A different attempt to improve the formability of aluminum tailor welded blanks (TWB) produced by the FSW. *International Journal of Material Forming*, 14(5), 1189–1208. doi: 10.1007/s12289-021-01632-w.

Badgujar, T. Y. & S. A. Bobade (2021). Advances in sheet metal stamping technology: A case of design and manufacturing of a car door inner panel using a tailor welded blank. In: *Manufacturing and Industrial Engineering: Theoretical and Advanced Technologies* (ed. P. Agarwal, et al.) 1st ed. CRC Press: Hoboken, NJ. doi: 10.1201/9781003088073.

Battina, N. M., Hari Krishna, C., & Vanthala, V. S. P. (2022). Influence of pin profile on formability of friction stir welded aluminium tailor welded blanks: An experimental and finite element simulation analysis. *Transactions of the Canadian Society for Mechanical Engineering*. 46(3), 602–613. doi: 10.1139/tcsme-2022-0031.

Bobade, S. A. & Badgujar, T. Y. (2017). Tailor Welded Blanks (TWBs) for a sheet metal industry: An overview. *International Journal of Advanced Research and Innovative Ideas in Education*, 3(3), 3771–3779.

Buffa, G., Fratini, L., La Commare, U., Römisch, D., Wiesenmayer, S., Wituschek, S., & Merklein, M. (2022). Joining by forming technologies: Current solutions and future trends. *International Journal of Material Forming*, 15, 27. doi: 10.1007/s12289-022-01674-8.

Chaudhary, J., Jain, N. K., Pathak, S., & Koria, S. C. (2019). Investigations on thin SS sheets joining by pulsed micro-plasma transferred arc process. *Journal of Micromanufacturing*, 2(1), 15–24.

Deepika, D., Lakshmi, A. A., Buddi, T., & Rao, C. S. (2022). Effect of formability parameters on tailor-welded blanks of light weight materials. In: *Light Weight Materials: Processing and Characterization.* (eds. K. Kumar, B. S. Babu, & J. P. Davim). John Wiley and Sons, River Street, NJ. doi: 10.1002/9781119887669.ch7.

Dwibedi, S., Jain, N. K., & Pathak, S. (2018). Investigations on joining of stainless-steel tailored blanks by μ-PTA process. *Materials and Manufacturing Processes*, 33(16), 1851–1863.

Fadzil, M., Abdullah, A. B., Samad, Z., Yusof, F., & Manurung, Y. H. P. (2021). Application of lightweight materials toward design for sustainability in automotive component development. In: *Design for Sustainability* (pp. 435–463). (eds. S. M. Sapuan & M. R. Mansor). Elsevier: Amsterdam, Netherlands.

Graser, M., Wiesenmayer, S., Müller, M. D., & Merklein, M. (2019). Application of tailor heat treated blanks technology in a joining by forming process. *Journal of Materials Processing Technology*, 264, 259–272.

Hetzel, A., Merklein, M., & Lechner, M. (2020). Enhancement of the forming limits for orbital formed tailored blanks by local short-term heat treatment. *Procedia Manufacturing*, 47, 1197–1202. doi: 10.1016/j.promfg.2020.04.177.

Klinke, N., Kobelev, V., & Schumacher, A. (2022). Rule and optimization-based selection of car body parts for the application of tailor rolled blank technology. *Structural and Multidisciplinary Optimization* 65, 60. doi: 10.1007/s00158-021-03111-x.

Kumar, A. (2020). Formability of tailor welded blanks of interstitial free and draw quality steels in hydraulic bulging, PhD Thesis, Indian Institute of Technology Delhi.

Kumar, A., Kumar, D. R., & Gautam, V. (2019). Prediction of residual stresses in biaxial stretching of tailor welded blanks by finite element analysis. *IOP Conference Series: Materials Science and Engineering*, 651(1), 012039.Lalvani, H., & Mandal, P. (2021). Cold forming of Al-5251 and Al-6082 tailored welded blanks manufactured by laser and electron beam welding, *Journal of Manufacturing Processes*, 68(A), 1615–1636.

Liang, J., Powers, J., & Stevens, S. (2018). A tailor welded blanks design of automotive front rails by ESL optimization for crash safety and lightweighting (No. 2018-01-0120). SAE Technical Paper. doi: 10.4271/2018-01-0120.

Merklein, M., Johannes, M., Lechner, M., & Kuppert, A. (2014). A review on tailored blanks: Production, applications and evaluation. *Journal of Materials Processing Technology*, 214(2), 151–164.

Panahi-liavoli, R., Gorji, H., Bakhshi-Jooybari, M., & Mirnia, M. J. (2020). Investigation on formability of tailor-welded blanks in incremental forming. *IJE Transactions B: Applications*, 33(5), 906–915.

Parente, M., Safdarian, R., Santos, A. D., Loureiro, A., Vilaca, P., & Jorge, R. M. (2016). A study on the formability of aluminum tailor welded blanks produced by friction stir welding. *The International Journal of Advanced Manufacturing Technology*, 83(9), 2129–2141.

Pollock, T. M., Clarke, A. J., & Babu, S. S. (2020). Design and tailoring of alloys for additive manufacturing. *Metallurgical and Materials Transactions A*, 51A, 6000–6019.

Reck, M. & Merklein, M. (2021). Investigation on tailored blanks in a full forward extrusion process of sheet-bulk metal forming. *ESAFORM 2021, MS15 (Incremental Forming)*. doi: 10.25518/esaform21.3673.

Rosochowskia, A. & Olejnik, L. (2017). New method of producing tailored blanks with constant thickness. *Procedia Engineering*, 207, 1433–1438.

Saito, M., Nakazawa, Y., Otsuka, K., Yasuyama, M., Tokunaga, M., & Yoshida, T. (2019). Bendability of weld metal: Development of application technology of tailor-welded blanks 1st report. *Materials Transactions*, 60(10), 2137–2142.

Satya-Suresh, V. V. N., Suresh, A., Regalla, S. P., Ramana, P. V., & Vamshikrishna, O. (2020). Sustainability aspects in the warm forming of tailor welded blanks. *E3S Web of Conferences*, 184, 01042. doi: 10.1051/e3sconf/202018401042.

Schulte, R., Papke, T., Lechner, M., & Merklein, M. (2020). Additive manufacturing of tailored blank for sheet-bulk metal forming processes. *IOP Conference Series: Materials Science and Engineering*, 967(1), 012034.Singar, O. & Banabic, D. (2021). Formability of Tailored hybrid blanks. *The Romanian Journal of Technical Sciences. Applied Mechanics*, 66(2), 93–101.

Singhal, H. (2020). Formability Evaluation of Tailor Welded Blanks (TWBs). Master Thesis, The Ohio State University.

Suresh, V. S., Regalla, S. P., & Gupta, A. K. (2019). Minimization of weld line movement in heat-assisted forming of tailor welded blanks. *Proceedings of the Institution of Mechanical Engineers, Part C: Journal of Mechanical Engineering Science*, 233(11), 3760–3768. doi: 10.1177/0954406218810308.

Tajul, L., Maeno, T., Kinoshita, T., & Mori, K. I. (2017). Successive forging of tailored blank having thickness distribution for hot stamping. *The International Journal of Advanced Manufacturing Technology*, 89(9), 3731–3739. doi: 10.1007/s00170-016-9356-z.

Tucci, F., Andrade-Campos, A., Thuillier, S., & Carlone, P. (2021). On the elastoplastic behavior of friction stir welded tailored blanks for single point incremental forming. *Paper Presented at ESAFORM 2021: 24th International Conference on Material Forming*, Liège, Belgique. doi: 10.25518/esaform21.437.

Wu, J., Hovanski, Y., & Miles, M. (2021). Investigation of the thickness differential on the formability of aluminum tailor welded blanks. *Metals*, 11(6), 875. doi: 10.3390/met11060875.

Zhang, H., Zhan, M., Zheng, Z., Li, R., & Lei, Y. (2021). A systematic study on effects of process parameters on spinning of thin-walled curved surface parts with 2195 Al-Li alloy tailor welded blanks produced by FSW. *Frontiers in Materials*, 529. doi: 10.3389/fmats.2021.809018.

Zou, Y., Zuo, K., Liu, H., & Zhou, D. (2020). Laser-based precise measurement of tailor welded blanks: A case study. *International Journal of Advanced Manufacturing Technology*, 107, 3795–3805. doi: 10.1007/s00170-020-05090-7.

# Index

Note: **Bold** page numbers refer to tables and *italic* page numbers refer to figures.